ODOUR IN TEXTILES

Textile Institute Professional Publications

Series Editor: Helen D. Rowe for The Textile Institute

Care and Maintenance of Textile Products Including Apparel and Protective Clothing
Rajkishore Nayak and Saminathan Ratnapandian

Radio Frequency Identification (RFID) Technology and Application in Fashion and Textile Supply Chain
Rajkishore Nayak

The Grammar of Pattern
Michael Hann

Standard Methods for Thermal Comfort Assessment of Clothing
Ivana Špelić, Alka Mihelić Bogdanić, and Anica Hursa Sajatovic

Fibres to Smart Textiles: Advances in Manufacturing, Technologies, and Applications
Asis Patnaik and Sweta Patnaik

Flame Retardants for Textile Materials
Asim Kumar and Roy Choudhury

Textile Design: Products and Processes
Michael Hann

Science in Design: Solidifying Design with Science and Technology
Tarun Grover and Mugdha Thareja

Textiles and Their Use in Microbial Protection: Focus on COVID-19 and Other Viruses
Jiri Militky, Aravin Prince Periyasamy, and Mohanapriya Venkataraman

Dressings for Advanced Wound Care
Sharon Lam Po Tang

Medical Textiles
Holly Morris and Richard Murray

Odour in Textiles: Generation and Control
G. Thilagavathi and R. Rathinamoorthy

For more information about this series, please visit: www.routledge.com/Textile-Institute-Professional-Publications/book-series/TIPP

ODOUR IN TEXTILES

Generation and Control

Edited by
G. Thilagavathi and R. Rathinamoorthy

CRC Press is an imprint of the
Taylor & Francis Group, an **informa** business

First edition published 2022
by CRC Press
6000 Broken Sound Parkway NW, Suite 300, Boca Raton, FL 33487-2742

and by CRC Press
2 Park Square, Milton Park, Abingdon, Oxon, OX14 4RN

CRC Press is an imprint of Taylor & Francis Group, LLC

Library of Congress Cataloguing-in-Publication Data
A catalog record for this title has been requested

ISBN: 978-0-367-69336-7 (hbk)
ISBN: 978-0-367-69333-6 (pbk)
ISBN: 978-1-003-14142-6 (ebk)

DOI: 10.1201/9781003141426

Typeset in Times
by MPS Limited, Dehradun

Contents

Textile Institute Professional Publications

The aim of the ***Textile Institute Professional Publications*** is to provide support to textile professionals in their work and to help emerging professionals, such as final year or Master's students, by providing the information needed to gain a sound understanding of key and emerging topics relating to textile, clothing and footwear technology, textile chemistry, materials science, and engineering. The books are written by experienced authors with expertise in the topic and all texts are independently reviewed by textile professionals or textile academics.

The textile industry has a history of being both an innovator and an early adopter of a wide variety of technologies. There are textile businesses of some kind operating in all counties across the world. At any one time, there is an enormous breadth of sophistication in how such companies might function. In some places where the industry serves only its own local market, design, development, and production may continue to be based on traditional techniques, but companies that aspire to operate globally find themselves in an intensely competitive environment, some driven by the need to appeal to followers of fast-moving fashion, others by demands for high performance and unprecedented levels of reliability. Textile professionals working within such organisations are subjected to a continued pressing need to introduce new materials and technologies, not only to improve production efficiency and reduce costs, but also to enhance the attractiveness and performance of their existing products and to bring new products into being. As a consequence, textile academics and professionals find themselves having to continuously improve their understanding of a wide range of new materials and emerging technologies to keep pace with competitors.

The Textile Institute was formed in 1910 to provide professional support to textile practitioners and academics undertaking research and teaching in the field of textiles. The Institute quickly established itself as the professional body for textiles worldwide and now has individual and corporate members in over 80 countries. The Institute works to provide sources of reliable and up-to-date information to support textile professionals through its research journals, the ***Journal of the Textile Institute***[1] and ***Textile Progress***[2], definitive descriptions of textiles and their components through its online publication ***Textile Terms and Definitions***[3], and contextual treatments of important topics within the field of textiles in the form of self-contained books such as the ***Textile Institute Professional Publications***.

REFERENCES

1. http://www.tandfonline.com/action/journalInformation?show=aimsScope&journalCode=tjti20
2. http://www.tandfonline.com/action/journalInformation?show=aimsScope&journalCode=ttpr20
3. http://www.ttandd.org

Preface

Odour in Textiles: Generation and Control specifically covers the odour generation and control mechanism in textile materials. The editors of this book have a decade of experience in odour research. Though malodour formation in textile material is very common in day-to-day life, the research in this area is minimal. Even though odour-related research has been performed in the cosmetics and food industries, odour-related studies in textiles are still in their infancy. Textile fabric has always played a major role in odour propagation due to its retention ability of sweat and the ability to act as a substrate for bacterial growth. However, awareness among customers on the influence of textile fibers, structures, and finishes on odour formation in the application is null. The existing research data available are also spread across different domains like healthcare textiles, microbiology, forensic studies, and the biomedical industry.

Hence, it's been our primary objective to amass such research findings in one place for the next generation of students, academicians, and technocrats. This book provides insight to readers about the influence of textiles and their properties on the formation of sweat in the human body, along with the role of textiles and their fiber content type and structure on odour formation characteristics in apparel. The book also covers various finishing methods, odour control chemicals, and the fragrance used in this sector to control textile malodour. The book is made up of contributions from international experts in this field, covering cutting-edge research and developments. This book is a first of its kind in the textile and fashion discipline that entirely covers the fundamentals of the odour formation mechanism and its control technologies.

Chapter 1 details consumer perception of body odour and textile malodour problems along with the societal viewpoint and importance of the issue. The chapter is contributed by Prof. Ingun Grimstad Klepp and Dr. Kirsi Laitala, from Oslo Metropolitan University, Norway, and Dr. R. Rathinamoorthy from PSG College of Technology, India.

Chapter 2 summarises the 'human scent', the current knowledge on the odour formation pathways from the skin and recent methods adapted by Dr. Paola Prada-Tiedemann, from Texas Tech University, USA. Detailed information on the odour-forming compounds in the skin, along with its characterization techniques, are reviewed in Chapter 3, by Dr. Mourad Krifa, from Kent State University, Ohio, USA.

Chapter 4 is devoted to the various textile fibers and fabric structures and their interaction with odourscausing components on the skin. The mechanism of odour formation, along with the role of textile materials, is detailed by Dr. R. Rathinamoorthy and Prof (Dr.). G. Thilagavathi, PSG College of Technology, India. Chapter 5 provides insight into the odour adsorption techniques to control odour formation with various chemicals. Ayşegül Çetmeli Bakadur from Üniteks Textile R&D Center and Dr. Ahmet Çay from Ege University, Turkey, have provided a detailed mechanism of odour sorption materials. Chapter 6 describes odour control using synthetic antibacterial agents, polymers, and bacteria application in textile. The interaction mechanisms to control odour molecules and textiles are detailed by Rosie Broadhead, Laure Craeye, and Dr. Chris Callewaert from Ghent University, Belgium.

Chapter 7 mainly concentrates on the application of herbal extracts as a sustainable antibacterial agent on different textile materials. The chapter, which details the effectiveness of these herbal extracts against various odour-causing bacterial strains, is authored by Dr. R. Rathinamoorthy and Prof (Dr.). G. Thilagavathi, PSG College of Technology, India. The role of fragrance finishes in the odour controlling process is detailed in Chapter 8, by T. Hemamalini and Dr. V.R. Giri Dev from Anna University, India. Further, they detailed the various methods used for finishing textiles with fragrance material. Chapter 9 outlines the impact of various other odours like cigarette smoke odour, laundry odour, pet odour, etc. It was written by Dr. R. Rathinamoorthy and Prof (Dr.). G. Thilagavathi, PSG College of Technology, India.

Chapter 10 reviews the effectiveness of laundry on the odour removal process. It was consolidated by Dr. Kirsi Laitala, Prof. Ingun Grimstad Klepp, and Vilde Haugrønning, from Oslo Metropolitan University, Norway. Chapter 11, which was written by Ms. Sima Shakoorjavan and Dr. Dawid Stawski from Lodz University of Technology, Lodz, Poland, and Dr. Somaye Akbari from Amirkabir University of Technology (Tehran Polytechnic), Tehran, Iran, explains the role of E-nose in textile odour measurement. They also outlined the various analytical tools used in odour measurements.

Chapter 12, contributed by Anton P. van Harreveld, from Sensenet Holding BV, The Netherlands, Rajal Shinkre and Saisha Naik, from Odournet Holding India Pvt. Ltd., India, and Carmen Villatoro, from Odournet S.L., Spain, discusses the various standards and testing methods used in the field of odour measurement of textiles specifically. It also reports real-time cases of odour measurements.

This book provides a window through which the latest advancements in odour research, along with detailed information about basics tenets of odour formation with respect to textile material, can be seen. It is the hope and confidence of the editors and contributors that this book will serve as a ready reckoner for researchers and common people for understanding the role of clothing on odour formation. This book sows the seeds for advanced researches.

Prof (Dr.) G. Thilagavathi
Dr. R. Rathinamoorthy

Editors

Prof (Dr.) G. Thilagavathi has been involved in teaching and research in the field of textiles for more than 28 years. She is working as Professor and Head, in the Department of Textile Technology, PSG College of Technology, Coimbatore, India. She completed her Ph.D in 2003, under Bharathiar University, Coimbatore. As Principal Investigator, she established a Centre of Excellence in Industrial Textiles, project is sponsored by the Ministry of Textiles, the Government of India with a budget of Rs. 28 Crores. In addition, she handles several sponsored projects with a combined budget of INR 30 Crores. She developed new products and technology through research projects funded by DST, DRDO, AICTE, UGC, and MOT. Based on her research publications, Prof (Dr.). G. Thilagavathi holds a h-index of 24 and an i-10 index of 52 with 1932 citations. She has guided 10 Ph.D Scholars, and currently 5 scholars are pursuing doctoral research under her guidance. She was the recipient of the prestigious Dr. R. Sundaramoorthy Award for the Outstanding Academician of PSG College of Technology, successively from 2012 to 2016. Under her leadership, the Department of Textile Technology, PSG CT, received the AICTE-CII Award for the best industry-linked engineering institute 2015 and NBA Accreditation (both UG and PG) for 5 years. She is a Chairman and Council Member of the Textile Divison Board, Institution of Engineers (I), India, Kolkata. She also holds membership in various professional/academic bodies and technical/advisory committees. She has visited foreign countries like Germany, the United Kingdom, Egypt, and Spain for academic and research purpose.

Dr. R. Rathinamoorthy has worked as an Associate Professor in the Department of Fashion Technology, PSG College of Technology, Coimbatore, India, since 2009. He completed his Ph.D. on medical textiles in 2016. He received the Young Achiever Award for the year 2019 by the Institute of Engineers India (IEI), Coimbatore chapter. In 2017, he also received a national-level Young Engineer Award for the year 2016–17 in the Textile Engineering domain, by the Institute of Engineers India (IEI), Kolkata, West Bengal, India. He has a Google H index of 14 and a Scopus H index of 10, with more than 650 citations. He has published 19 national and 65 international research articles in various refereed and non-refereed journals. He authored 7 technical books in the area of apparel and fashion technology and 23 book chapters with various international publishers like Woodhead Publisher, Springer Verlag, Springer Nature, Springer Singapore, Taylor & Francis, and Elsevier publishers. In 2011, he sanctioned a research project from University Grants Commission for the tune of Rs. 10.15 Lakhs and completed it in 2013. He has 12 years of teaching experience in the area of Textile and Fashion discipline. His research interest is on sustainable material for the textile and fashion industries. He is currently working on biomaterials and environmental pollution due to textiles and researching sustainable alternative materials and methods to reduce their environmental impact.

Contributors

Somaye Akbari
Textile Engineering Department
School of Materials and Advanced Processes Engineering
Amirkabir University of Technology (Tehran Polytechnic)
Tehran, Iran

Ayşegül Çetmeli Bakadur
Üniteks Textile R&D Center
İzmir, Turkey

Rosie Broadhead
Ghent University
Center for Microbial Ecology and Technology
Coupure Links
Ghent, Belgium

Chris Callewaert
Ghent University
Center for Microbial Ecology and Technology
Coupure Links
Ghent, Belgium

Ahmet Çay
Department of Textile Engineering Faculty of Engineering
Ege University
Bornova İzmir, Turkey

Laure Craeye
Ghent University
Center for Microbial Ecology and Technology
Coupure Links
Ghent, Belgium

V.R. Giri Dev
Department of Textile Technology
Anna University
Chennai, India

Anton P. van Harreveld
Sensenet Holding BV
Amsterdam, The Netherlands

Vilde Haugrønning
Consumption Research Norway (SIFO)
Oslo Metropolitan University
Oslo, Norway

T. Hemamalini
Department of Textile Technology
Anna University
Chennai, India

Laitala Kirsi
Consumption Research Norway (SIFO)
Oslo Metropolitan University
Oslo, Norway

Ingun Grimstad Klepp
Consumption Research Norway (SIFO)
Oslo Metropolitan University
Oslo, Norway

Mourad Krifa
School of Fashion & Advanced Materials and Liquid Crystal Institute
Kent State University
Kent, Ohio, USA

Saisha Naik
Odournet Holding India Pvt. Ltd.
Goa, India

Paola A. Prada-Tiedemann
Department of Environmental Toxicology
Texas Tech University
Texas, United States

R. Rathinamoorthy
Department of Fashion Technology
PSG College of Technology
Coimbatore, India

Mathilda Savocchia
School of Fashion
Kent State University
Kent, Ohio, USA

Sima Shakoorjavan
Lodz University of Technology
Lodz, Poland, and Amirkabir University of Technology (Tehran Polytechnic)
Tehran, Iran

Rajal Shinkre
Odournet Holding India Pvt. Ltd.
Goa, India

Dawid Stawski
Division of Chemistry and Physical Chemistry of Polymers
Institute of Materials Science of Textiles and Polymers Composites
Lodz University of Technology
Lodz, Poland

G. Thilagavathi
Department of Textile Technology
PSG College of Technology
Coimbatore, India

Carmen Villatoro
Odournet S.L.
Barcelona, Spain

1 The Consumer Perception of Odour

Ingun Grimstad Klepp and Kirsi Laitala
Consumption Research Norway (SIFO), Oslo Metropolitan University, Oslo, Norway

R. Rathinamoorthy
Department of Fashion Technology, PSG College of Technology, Coimbatore, India

CONTENTS

1.1 INTRODUCTION

Textiles have a central role in the lives of consumers in all cultures. We use them for numerous purposes, such as keeping us warm, dry, clean and decent, decorating our homes, and even structuring roads and for fishing. Clothing is worn not only to protect us from the physical environment, but also to fit into the social setting and to express ourselves. Odour, cleanliness, and clothing will be the recurring theme in this book and will be illuminated based on technical-scientific knowledge about textiles. They are at the same time biological and cultural phenomena, which means that how they are experienced, described, emphasized, and understood varies between cultures and people (Schilling et al. 2010). Odour is caused by volatilized chemical compounds that we can perceive through our sense of smell. The way we perceive smells in textiles today differs from past times and in other societies, and it has therefore been a topic of cultural and historical studies. At the same time, olfaction, the sense of smell, is a trait that humans share, and which can be studied on the basis of biological and psychological knowledge traditions. This makes odour in textiles a rich but complex field to study. It is not possible to study it solely based on natural science methods, isolated from cultural aspects. The very goal of removing odours, or producing textiles that smell less, is to meet culturally defined

DOI: 10.1201/9781003141426-1

needs. There are cultural differences in what are considered to be strong, weak, pleasant, or appalling smells. We are all consumers, and consumption can be defined as acquiring, using – and thus also caring for – and disposing of goods. For manufacturers of clothing and other textiles, consumers are the first and foremost customers. Textiles and clothes are an integral part of consumers' day-to-day life. Apart from wearing clothes, we wash, clean, air, brush, and evaluate the textiles around us, including their potential odours. The clothes we wear thus become not only a product of the textile industry, but also of consumers' labor in laundry rooms and wardrobes. This book will discuss both odour-related knowledge about production as well as knowledge about how odour can be measured and removed. Odour is associated with cleanliness, as the general belief is that pleasant odour, or lack of odour, is considered to be clean. The history of odour and its understanding is deeply connected with the understanding of purity and cleanliness. We will therefore begin with some considerations on these related topics that all chapters in this book go narrower and deeper into: odour, clothes, and cleanliness. We will do this by gathering knowledge about odour from a wide range of scientific traditions, from anthropology, sociology, philosophy, and history to biology, physiology, marketing, and psychology. As odour can be found everywhere, it can be studied through various scientific methods and traditions, and it is relevant in different applied fields such as health and the environment.

1.2 OLFACTION: THE SENSE OF SMELL

The sense of smell or olfaction, the ability to perceive odours, is one of the most complex human senses as it depends on unique interactions between the olfactory system and diverse chemical substances. The olfactory system works as a defense mechanism for the human body against malodours and irritants. Much of its function and functionality is still unexplored (Reinarz 2014). The sense of smell is affected by chemical compounds in gaseous form. The olfactory epithelium, located at the top of the nasal cavity on both sides of the nasal septum, is a mucous membrane with olfactory cells, a type of nerve cell that is associated with the olfactory region of the brain. Similar to taste, the sense of smell is a chemical sense. The olfactory cells are stimulated only by volatile substances, i.e., by substances that evaporate and release molecules into the air. The odorants bind to odour receptors on the olfactory cells. This will trigger a biochemical reaction chain that leads to a nerve impulse, which in turn is sent to the brain (Winther 2018).

Humans can distinguish up to 10,000 different odours. It is unclear whether we have one receptor for each odorant, or whether the smell feels specific because several receptors are activated simultaneously (Winther 2018). However, several hundred different odour receptors have been found, where each is likely to be activated by a definite type of odorant. The sense of smell is extremely sensitive to certain odorants. The strongest smelling substance known is methylmercaptan, which is found in garlic and rotting meat. It can be detected in a concentration of less than one nanogram (one billionth of a gram) per litre of air (Winther 2018) and has a distinct odour awareness level of 0.002 ppm (Inaba and Inaba 1992). However, not everything we perceive as smell is conveyed by the sense of smell.

The pungent odour sensations that are triggered by substances such as menthol, chlorine, ammonia, and nitric acid are due to irritation and pain transmission of other nerves in the nose (Winther 2018).

One important aspect for understanding how the sense of smell functions is called adaptation. We do not sense the same odour over time. After a while, even the strongest and most unpleasant odours can become imperceptible. This habituation (adaptation) may be due to some form of fatigue of both the sensory receptor cells and the olfactory cells in the brain. The adaptation is specific; that is, we only stop noticing the one smell we are used to; the sensitivity to other odours is preserved unimpaired (Winther 2018). This property we talk about in everyday speech when we, for example, say that we do not detect our own smell, or when we 'get used to' a perfume. The property means also that we may not notice the smell of a textile in one environment, but it may have a strong, specific smell if it is moved to a new environment, for example, clothing used by a bonfire. The olfactory cells are located in the top of the nasal cavity, and with normal breathing, much of the air will pass through further down. This is why you sniff when you want to detect a faint odour. With this reflex, the nostrils are opened, and you breathe in more strongly so that the airflow has greater speed and is directed up toward the olfactory cells at the top of the nasal cavity (Winther 2018).

It is often said that animals have a stronger sense of smell than humans, and also that animals use theirs for vital functions, such as orienting themselves, avoiding danger, attracting partners, scaring away enemies, seeking out food, and more. These allegations are disputed. It is also claimed that the human olfactory organ is far more advanced than previously thought and the aspects that are crucial for the ability to smell do not distinguish humans from other mammals (McGann 2017). There is also increasing knowledge about how not only animals but also humans use the sense of smell to orient themselves and to communicate, choose a partner, and create security and bonds between people (McBurney, Streeter, and Euler 2012). By using human social chemosignals, we form and maintain social bonds and even optimize the health status of the immune system in the population (Pause 2012).

The sense of smell is included together with other senses in our social signals. The special thing about this chemosignal system is that it takes place without our awareness of it. It does not require our attentional resources; thus, it is also something we do not notice (Pause 2012). It has been shown that smell is capable of triggering an automatic response to stimuli without consumer awareness of it (Hirsch 1995). In addition to the fact that the sense of smell works below the threshold of consciousness, there are also other historical and cultural reasons why the human sense of smell is underestimated. Compared to sight and hearing, it has had a very low status in Western culture and philosophy since ancient times. To present the sense of smell as something that particularly characterizes animals thus fits into the image of humans in the Western philosophical tradition.

It is difficult to identify even common things solely based on their odour, and it is difficult to describe odours. As already noted by the ancient Greek philosophers, it is easier to report the perception of pleasantness, edibility, emotionality, and familiarity of odours. These features help us to recognize what to eat, and who or what to approach or avoid (Zucco, Herz, and Schaal 2012). Losing the sense of smell can cause problems in daily life, as it impairs the ability to detect spoiled food, gas leaks, or smoke as well as the activities of eating and cooking. The disability reduces quality of life, especially in the areas of safety and eating (Miwa 2001). Newborn babies have a highly developed and acute sense of smell that is important for their learning (Zucco, Herz, and Schaal 2012). Our sense of smell only changes little as a function of time (Büttner 2017), although children below the age of 9 and people aged 70+ score significantly lower than the age groups in between (Sorokowska et al. 2015). Above all, smell is a sense that we cannot turn off completely; it is constantly on, no matter what we may be doing otherwise (Low 2008).

1.3 HISTORY OF THE SENSE OF SMELL

"Smell is a cultural phenomenon" is a recurring theme of the book *Past Scents: Historical Perspectives on Smell* (Reinarz 2014). Reinarz writes about the history of smell from ancient to modern times. He shows how attention to the history of smell arose in the 1980s, together with a general shift towards greater attention to the cultural history of the senses. Much of the inspiration for this turn came from anthropology. Many studies of smell in various Global South societies start by noting how the importance of the olfactory experience is underestimated in the Global North and its intellectual traditions (Reinarz 2014). Conventionally, the senses of taste and smell were regarded as inferior to sight and hearing. This hierarchy of the senses is traced back to ancient Greek philosophers like Plato and Aristotle. Plato declared sight to be the most worthy and important sense (Dugan 2011; Reinarz 2014). According to him, smell lacked any definite pattern, most smells being half-formed entities, which made them good or bad but otherwise cryptic or difficult to classify (Classen, Howes, and Synnott 1994). Similarly, Aristotle pointed to difficulty in identifying odours, turning descriptions often into emotional classifications, such as those that were pleasurable or those that generated feelings of disgust (Guérer 1994). However, Aristotle tried to classify smells into five olfactory categories: sweet, harsh, astringent, pungent, and rich (Classen, Howes, and Synnott 1994; Reinarz 2014). These ideas survived for many generations and continued to be taken as definitive (Reinarz 2014).

In Europe, for centuries up to the beginning of the 1900s, people with a good sense of smell were considered to be closer to animals and savages than those who were not able to detect subtle variations in odours. At the same time, individuals with superior senses were considered less civilized and dirtier, often connected to racial differentiation (Reinarz 2014). Smell was treated as a property of animals and only considered superficially by philosophers like Immanuel Kant and psychologists like Sigmund Freud. Fortunately, it seems that this negativity has reversed, and especially since the 1980s, knowledge about people's sense of smell and the way we use it has increased. This does not mean that we know everything now. It just means that odour as well as studies of odour were not taken seriously some decades ago, but this is no longer the case (Classen, Howes, and Synnott 1994). This can also be seen in literature reviews related to the effect of scent on consumer behaviour, where the number of studies has increased significantly since the year 2000 (Rimkute, Moraes, and Ferreira 2016).

Despite the low status that smell has had, and lack of knowledge about it, smell is important in our daily lives. This is a field that has been studied by anthropologists and sociologists. Within social sciences, a central point of reference in research on cleanliness is Mary Douglas's theory on dirt as a matter out of place. "There is no such thing as absolute dirt" (Douglas 1984, 66), she wrote, and she argued that dirt is the result of a system of classification. She made "clean" and "dirty" cultural constructs that are suitable objects for cultural and social studies. In the research on cleanliness, Leach's taboo theory, Elias's civilization theory, and Foucault's perspective on power have been sources of inspiration for studies on hygiene historical topics like culture and class struggle (Elias 1969; Foucault and Gordon 1980; Leach 1989). The fight against dirt, disorder, bad manners, and bacteria is not only understood as a way to promote people's health, but also as a way of advancing the regulated, modern, bourgeois society. Several studies on the history of mentality give insight into changes that occur over time (Hutton 1981).

1.4 PERCEPTION OF ODOUR

Scents and odours influence how we think, act, and behave. Bodily odours are, therefore, fundamental elements that denote personality and self-identity (Low 2008). Even though understanding of the importance of odour is increasing, knowledge is hindered by lack of words and difficulties in categorising them. Odours are frequently described as smelling of something else, like freshly baked bread, a rose, a forest, or rotten egg, for example. In contrast, humans have a large vocabulary for different categories and nuances of colours beyond the basic ones. Research indicates that odours could be classified similarly to basic odours from which mixtures of other nuances can be created, but these basic odours are still not discovered (Berglund and Höglund 2012). We can perceive a large number of odorous molecules, and most of us seem to have very similar odour impressions that originate from a specific stimulus (Büttner 2017). The individual differences are likely connected to the physiological reaction to odorants as well as the learned response. Our expectations and beliefs about the odour's impact, how it is perceived, and the expectations can even override truthful perception (Dalton 2012). It seems that learned and innate behavioural responses to odorants are treated by different cortical areas, which enables studying emotional components of odour perception in more detail (Büttner 2017).

Odours not only evoke memories but also create them. Psychological investigations have shown that odour-evoked memory is based on associative learning (Zucco, Herz, and Schaal 2012). The field of odour-memory comprises both memories for odours and the ability for odours to evoke autobiographical recollections. The sense of smell has a fundamental biological significance that can provoke emotional meaning and therefore motivate behaviours (Zucco, Herz, and Schaal 2012). Malodour can also be of serious concern. For example, odour emissions from factories can affect quality of life and can lead to psychological stress and symptoms such as problems with sleep, loss of appetite, and irrational behaviour (Belgiorno, Naddeo, and Zarr 2012). Body odour is often referred to as a problem and something negative.

Human body odour contains numerous volatile compounds related to gender and individual-specific hormones, and it also represents genetic details (Penn et al. 2007). As with olfaction, human body odour is often not understood clearly, and our cultures have an anxious connection with it. Attributes like lower social status and out-group bias developed widespread negative impressions on body odour in the society (McBurney, Levine, and Cavanaugh 1976; Schleidt, Hold, and Attili 1981). A pleasant body odour is believed to be a sign of a healthy and fit body (Thornhill and Gangestad 1999), whereas an unpleasant body odour is associated with negative interpersonal relationships and undesirable social traits (McBurney, Levine, and Cavanaugh 1976). At the same time, a bad body odour was also reported as the perception of vulnerability and concern for others. Prehn-Kristensen et al. (2009) supported these findings in their report and mentioned that the odour created by an anxious person creates empathy in others. This was again confirmed through experimentation by Camps et al. (2014). A new person's bad body odour positively affects the third person's helping behaviour. Through experimental study, they recorded that bad body odour creates pitifulness more than a neutral smelling person, which supports Prehn-Kristensen et al. (2009). The study also accounted for other additional smells like alcohol along with sweat odour, demonstrating that bad odour that the person was not in control of caused more sympathy than unpleasant odours that the person could have avoided (Camps et al. 2014).

The psychological effect of masking body odour using fragrance showed an improved relaxation to the wearer (Abriat et al. 2007). Body odour masking using active ingredients like antimicrobial products with scent also developed self-confidence and self-perceived attractiveness in the user. These changes were reported by other observers through images and videos of the user in the absence of verbal and olfactory cues (Craig Roberts et al. 2009). Other researchers have reported increased self-confidence, increased confidence-related attributes, and increased attraction to the opposite gender while using fragrance (Higuchi et al. 2005). The majority of the researchers showcased the simplest methods to evaluate the body odour in terms of generic terminologies in the area of pleasantness and intensity (Allen et al. 2016). Hence, underlying issues like specific characteristics and impact of the odour were not addressed in detail. Allen et al. (2018) analysed body odour with perfume experts and derived a lexicon for body odour perception by humans, as provided in Table 1.1. The developed lexicon was found useful in discriminating the male and female odour by differentiating the sex. Spicy, animalic, and metallic are the three descriptors that strongly differentiated male and female body odours. This confirms the perceptual and chemical difference between male and female body odour (Allen et al. 2018).

Several traditional approaches mainly used armpit or axillary sweats from worn clothing, specifically T-shirts and cotton pads, to evaluate olfaction behaviours. However, the method mainly assesses odour in the closest range, but in real-life situations, odour of the entire body is measured or perceived at a socially acceptable distance. Similarly, most of the studies focused only on the natural odour of the individuals after the conditioning period to avoid other influencing parameters like food habits, smoking, regular living habits, etc. Hence, to assess real-life situations, studies evaluated the diplomatic odour of an individual (represents the body odour modified

TABLE 1.1
Definitions of the odour descriptors developed by Allen et al. (2018) in body odour assessment (Reprinted with permission)

Descriptor	Standard definitions developed for assessment
Musty	Stale air, old furniture
Mouldy	Household mould, mould found on clothes, bread mould (not cheese mould)
Earthy	Soil, wet forest floor, mud, wet tree bark
Onion	The smell of raw onion, red, white, spring, and leeks
Spicy	Refers only to culinary spices such as clove, nutmeg, cumin, anise, pepper, etc.
Fatty	Cold fats and oils used for cooking, including butter and lard, margarine, olive oil, vegetable oil, and rendered beef fat
Oily	Oil paint, violet leaf absolute, car engine oil, WD40, non-edible oils
Greasy	Dirty human scalp and/or hair
Chipfat	Fat from a deep fat fryer used to cook potato
Animalic	Odours from an animal source, including goat, horse, sweat, skin, fur, leather, etc.
Vegetable	Savoury vegetable aroma, vegetable stock or soup, cooked vegetables, raw vegetables, including potato, carrot, celery
Heavy	Nonvolatile odours, similar olfactive feel to larger musk molecules
Milky	Lactonic, milk from all animal sources
Sweet	Vanilla, chocolate, sugar
Metallic	Smells like metal, hot metal, tin, iron

by fragranced products, deodorants, dietary choices, and personal habits) instead of the natural odour (Gaby and Zayas 2017). The assessors used blindfolds, including keeping ears closed, and the subjects' whole body odour was assessed at a social distance by sitting nearby. The consumer's perception of body odour was highly influenced by the person's diplomatic odour and helped in judging the characteristics of the person. This judgmental decision was mainly influenced by the preference of the fragrances, genetic and dietary habits, and health information, but a poor relationship between the assessment performed based on natural and diplomatic olfactory cues of the same individual was reported. These findings suggested the proposed method is better at addressing the interactions between humans in their normal physical environment than the previous research methods about body odour perception (Gaby and Zayas 2017). In addition, people preferred scent formulations that complement and enhance their body odour. In this aspect, the resulting body odour is unique and possesses characteristics of both body odour and perfume used. However, the resulting odour has considerable differences from both the constituents (Milinski and Wedekind 2001). The body odour interacts with fragrances and creates specific odour mixtures for individuals. More pleasantness was observed when individual body odour is mixed with their selection of perfume compared to randomly allocated perfume. The fragrances also interact with body odour and develop individualistic odour apart from masking the body odour (Lenochova et al. 2012).

Experiments by Mitro et al. (2012) showed that armpit body odour from old-age individuals was less intense and less unpleasant compared to middle-aged and young people. The reduction in the lipid concentration on the skin surface was reported as the main cause for the reduced body odour of old-age people. The participants were able to identify the old people by their body odour to a larger degree than they were able to discriminate middle-aged and younger people (Mitro et al. 2012). The pleasantness of body odour is also capable of creating trustworthiness on the person compared to the person with unpleasant body odour (Janek et al. 2020). Hence, the smell is also associated with positive emotions, memories, and unity between people. For example, it is quite common to smell clothing used by a loved one when separated for comfort (McBurney, Streeter, and Euler 2012). These positive aspects of odour are less frequently discussed.

1.5 SCENTS ENHANCING MARKETING

Using odour is relevant within consumer studies and marketing (Rimkute, Moraes, and Ferreira 2016). Multi-sensory or experiential marketing can target touch, taste, and smell, in addition to the more traditional vision- and hearing-based strategies (Hultén, Broweus, and Van Dijk 2009). The effectiveness of ambient scent (also referred to as atmospheric odour or olfactory cue) in influencing consumer behaviour has been studied in the fields of marketing and psychology (Rimkute, Moraes, and Ferreira 2016).

Sensory cues influence consumer decision making in various ways. They can create mental associations that strengthen their recall of brands and products (Morrin and Ratneshwar 2003, 2000). Recorded responses include more positive attitudes towards the store or products and their quality, while behavioural changes include increased likelihood of visiting the store, staying longer in there, paying higher prices for the product, and even increased intention to buy a product, thus resulting in an increased likelihood of spending more money (Fiore, Yah, and Yoh 2000; Morrin and Ratneshwar 2003; Spangenberg, Crowley, and Henderson 1996). Under different odour situations, higher spending of money was reported for simple one-odour-only situations. In the case of complex odour and completely odourless situations, the study did not find a significant difference in spending patterns (Haberland 2010). The ambient scent combined with music highly persuaded the consumers towards impulsive buying and money spent (Mattila and Wirtz 2001; Morrison et al. 2011).

Reports also show a higher level of memory enrichment in association with odour. When the customers are exposed to pleasant scents, they retrieve their happy memories associated with the same scent (Ehrlichman and Halpern 1988). When the participants of the study were exposed to the particular scent of the museum, they were able to recollect such scent from their previous visit to the museum approximately six years before. However, the participants without scent were not able to recollect such detailed information (Aggleton and Waskett 1999). Such evidence of memory recollection due to the exposure of odour was reported by several other researchers and confirms the importance of the odour on consumer perception (Ball, Shoker, and Miles 2010; Morrin and Ratneshwar 2003, 2000). The pleasant ambient odour was also aided in the bias decision of the product quality, brand, and positive

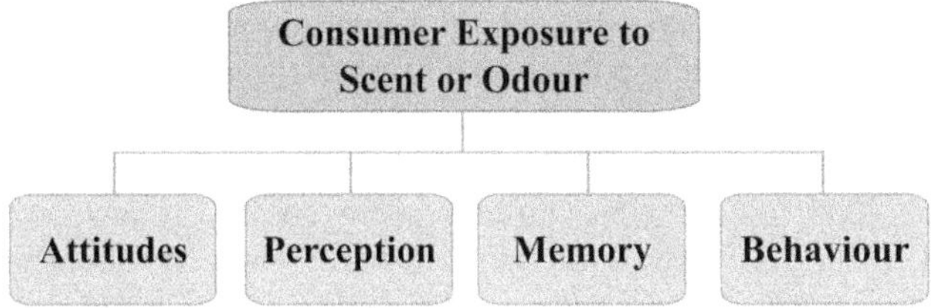

FIGURE 1.1 Consumer behaviour affected by odour/scent.

retail environment (Bone and Jantrania 1992; Chebat and Michon 2003; Spangenberg et al. 1996).

In addition to the pleasantness of the scent, the intensity is important. Spangenberg, Crowley, and Henderson (1996) found that the relationship between perceived pleasantness of scent and its intensity followed an inverted U-shaped curve, indicating that too strong an odour reduced the pleasantness. If consumers are aware of the scent, and further, recognise it as persuasive, they can use defensive mechanisms to correct its influence (Baron 1983; Bosmans 2006). It has also been noted the influence of olfactory cues is not direct but goes through various mediating and moderating variables (Baron and Kenny 1986). Mediating variables include impacting the affective state or cognitive processes of the consumer, which in turn impact the behaviour, while moderating variables include consumer factors (gender, scent preferences), scent properties, and whether the consumers are aware of the scent. Cognitive and affective responses were reported as the mediating variable by Rimkute et al. (2016). Figure 1.1 represents the various responses reported by the consumer on exposure to odour as reported by other researchers (Rimkute et al. 2016).

1.6 THE MEANING OF CLEAN CLOTHES

An understanding of the connection between good odour, purity, and absence of disease is very old. This was the core of what is often referred to as Miasma theory (Karamanou et al. 2012). The notion that foul smelling, humid, and stagnant air caused disease originated in ancient Greek medicine, from the time of Hippocrates, and was associated with the so-called humoral pathology. Some variants of Miasma theory were central to medicine until bacteriology became the dominant way of understanding infection in the late 1800s. With the discovery of bacteria, the work of cleansing the body, clothes, and everything else with the intention of eradicating tuberculosis and other infectious diseases gained importance and resulted in increased awareness of hygiene. Hygiene became the overall objective and included much more than health, as we usually perceive it today. The fight was not only against bad smells, dirt, and bacteria, but also against undesirable people and trends in society. These allegations of malodour often affect those who are also otherwise marginalized, such as homeless people. The claim or experience that something smells bad is complex, and it includes also a way of perceiving the unknown or disliked aspects as smelly. Poor hygiene and malodours are seen as a reflection of the practices, appearance, and actions of people.

Clothes have many different purposes in addition to protecting us from the physical environment. They are important for self-esteem, make our bodies socially acceptable, and influence the opinions of others (Kaiser 1997; Lamb and Kallal 1992). In addition to appearance, it is important that clothes are not regarded as dirty or smelly (Klepp 2007, 2003, 2005). Different bodily secretions, such as sweat, urine, dandruff, and other discharges, are often considered culturally problematic (Mauss 1979). The contamination may cause visible stains and malodour. Today, bodily malodours are especially considered socially stigmatising (Classen, Howes, and Synnott 1994; Frykman and Löfgren 1994; Klepp 2005). When odour is considered to be a problem, it is often associated with filth. Unpleasant odour, especially body odour, is often understood to be closely related to lack of cleanliness and the ability or willingness to integrate into society. In addition to the appearance and the tone of voice, smell has fundamental sociological importance. There is no doubt that the smell of other people is crucial to the extent to which we like them or find them attractive (Low 2008). Detailed information of different textile materials and their role in odour generation are discussed in Chapter 4.

1.7 SUMMARY

The perception of odours is culturally and situationally dependent, and it has changed throughout history. Human olfaction sense is one of the highly underestimated senses, as it influences how we think, act, and behave. At the same time, knowledge of odour seems to be emerging among the scientific community and the public at large. This review demonstrates the importance of odour or scent in improving mind relaxation, self-confidence, and attractiveness in individuals. Smell is an important attribute that guides consumer behaviour, purchase decisions, and expenditures. Similarly, body odour can be used to deem an individual's personality, character, and personal hygiene. A pleasant body odour is generally perceived as an attractive and healthy sign of an individual, whereas an unpleasant body odour is associated with lack of personal hygiene and social skills. At the same time, lack of verbal descriptions complicates scientific odour assessment methods to evaluate and understand the body odour. Textile materials are commonly used, as they are in contact with the body throughout the day, and clothes are also reported as an efficient medium for body odour formation or transfer. Hence, most body odour analysis used textile samples to analyse the odour of individuals. Though body odour is usually connected with the cleanliness of textiles and individuals, the significance on consumer perceptions cannot be ignored.

REFERENCES

Abriat, A., Barkat, S., Bensafi, M., Rouby, C., and Fanchon, C. 2007. Psychological and physiological evaluation of emotional effects of a perfume in menopausal women. *International Journal of Cosmetic Science*, 29: 399–408.

Aggleton, J.P., and Waskett, L. 1999. The ability of odours to serve as state dependent cues for real-world memories: Can Viking smells aid the recall of Viking experiences? *British Journal of Psychology*, 90: 1–7.

Allen, C., Cobiey, K.D., Havlíček, J., and Roberts, S.C. 2016. The impact of artificial fragrances on the assessment of mate quality cues in body odor. *Evolution and Human Behavior*, 37: 481–489.

Allen, C., Havlíček, J., Williams, K., and Roberts, S.C. 2018. Perfume experts' perceptions of body odors: Toward a new lexicon for body odor description. *Journal of Sensory Studies*, e12314. 10.1111/joss.12314

Ball, L.J., Shoker, J., and Miles, J.N.V. 2010. Odour-based context reinstatement effects with indirect measures of memory: The curious case of rosemary. *British Journal of Psychology*, 101: 655–678.

Baron, R.A. 1983. "Sweet smell of success"? The impact of pleasant artificial scents on evaluations of job applicants. *Journal of Applied Psychology*, 68(4): 709.

Baron, R.M., and Kenny, D.A. 1986. The moderator–mediator variable distinction in social psychological research: Conceptual, strategic, and statistical considerations. *Journal of Personality and Social Psychology*, 51(6): 1173.

Belgiorno, V., Naddeo, V., and Zarr, T. 2012. *Odour Impact Assessment Handbook*. John Wiley & Sons, Ltd.

Berglund, B., and Höglund A. 2012. Is there a measurement system for odour quality? In G.M. Zucco, R.S. Herz, and B. Schaal (eds.), *Olfactory Cognition: From Perception and Memory to Environmental Odours and Neuroscience* (pp. 3–22). John Benjamins Publishing.

Bone, P.F., and Jantrania, S. 1992. Olfaction as a cue for product quality. *Marketing Letters* 3: 289–296.

Bosmans, A. 2006. Scents and sensibility: When do (in) congruent ambient scents influence product evaluations? *Journal of Marketing*, 70(3): 32–43.

Büttner, A. 2017. *Springer Handbook of Odor*. Springer International Publishing: Imprint: Springer.

Camps, J., Stouten, J., Tuteleers, C., and Son, K.V. 2014. Smells like cooperation? Unpleasant body odor and people's perceptions and helping behaviors. *Journal of Applied Social Psychology*, 44: 87–93.

Chebat, J.C., and Michon, R. 2003. Impact of ambient odors on mall shoppers' emotions, cognition, and spending. *Journal of Business Research*, 56: 529–539.

Classen, C., Howes, D., and Synnott, A. 1994. *Aroma: The Cultural History of Smell*. Routledge.

Craig Roberts, S., Little, A.C., Lyndon, A., Roberts, J., Havlicek, J., and Wright, R.L. 2009. Manipulation of body odour alters men's self-confidence and judgements of their visual attractiveness by women. *International Journal of Cosmetic Science*, 31: 47–54.

Dalton, P. 2012. There's something in the air: Effects of beliefs and expectations on response to environmental odors. In G.M. Zucco, R.S. Herz, and B. Schaal (eds.), *Olfactory Cognition: From Perception and Memory to Environmental Odours and Neuroscience* (pp. 23–38). John Benjamins Publishing.

Douglas, M. 1984. *Purity and Danger: An Analysis of the Concepts of Pollution and Taboo*. Ark Paperbacks.

Dugan, H. 2011. *The Ephemeral History of Perfume: Scent and Sense in Early Modern England*. Johns Hopkins University Press.

Ehrlichman, H., and Halpern, J.N. 1988. Affect and memory: Effects of pleasant and unpleasant odors on retrieval of happy and unhappy memories. *Journal of Personality and Social Psychology*, 55: 769–779.

Elias, N. 1969. *The Civilizing Process, Vol. I. The History of Manners*. Blackwell.

Fiore, A.M., Yah, X., and Yoh, E. 2000. Effects of a product display and environmental fragrancing on approach responses and pleasurable experiences. *Psychology & Marketing*, 17(1): 27–54.

Foucault, M., and Gordon, C. 1980. *Power/Knowledge: Selected Interviews and Other Writings 1972-1977*. Pantheon Books.

Frykman, J., and Löfgren, O. 1994. *Det Kultiverte Mennesket*. Pax.

Gaby, J.M., and Zayas, V. 2017. Smelling is telling: Human olfactory cues influence social judgments in semi-realistic interactions. *Chemical Senses*, 42(5): 405–418. doi:10.1093/chemse/bjx012

Guérer, A.L. 1994. *Scent: The Mysterious and Essential Powers of Smell*. Kodansha International.

Haberland, M.F. 2010. The power of odor: Empirical field studies of olfactory cues on purchase behaviour. Master thesis, University of St. Gallen, St. Gallen, 233–243.

Higuchi, T., Shoji, K., Taguchi, S., and Hatayama, T. 2005. Improvement of nonverbal behaviour in Japanese female perfume-wearers. *International Journal of Psychology*, 40: 90–99.

Hirsch, A.R. 1995. Effects of ambient odors on slot-machine usage in a Las Vegas casino. *Psychology and Marketing*, 12: 585–594.

Hultén, B., Broweus, N., and Van Dijk, M. 2009. What is sensory marketing? In *Sensory Marketing* (pp. 1–23). Springer.

Hutton, P.H. 1981. The history of mentalities: The new map of cultural history. *History and Theory*, 20(3): 237–259. doi: 10.2307/2504556.

Inaba, M., and Inaba, Y. 1992. Relationship between bromidrosis and sense of smell. In *Human Body Odor: Etiology, Treatment, and Related Factors* (pp. 67–74). Springer Japan.

Janek, S.L., Probst, F., Fischbacher, U., Wirthmüller, U., and Knoch, D. 2020. Pleasant body odours, but not genetic similarity, influence trustworthiness in a modified trust game. *Scientific Reports*, 10: 3388. 10.1038/s41598-020-60407-6 1

Kaiser, S.B. 1997. *The Social Psychology of Clothing. Symbolic Appearances in Context*. 2nd ed. Fairchild Publications.

Karamanou, M., Panayiotakopoulos, G., Tsoucalas, G., Kousoulis, A.A. , and Androutsos, G. 2012. From miasmas to germs: A historical approach to theories of infectious disease transmission. *Infezioni in Medicina*, 20(1): 58–62.

Klepp, I.G.. 2003. *Fra rent til nyvasket: skittent og rent tøy (Clean to new-washed: Dirty clothes - clean habits)*. National Institute for Consumer Research.

Klepp, I.G.. 2005. The meaning of cleanliness: Modern demonstrations of female purity. In G. Hagemann, and H. Roll-Hansen (eds.), *Twentieth-century housewives: Meanings and Implications of Unpaid Work* (pp. 191–216). Unipub.

Klepp, I.G.. 2007. Patched, louse-ridden, tattered: Clean and dirty clothes. *Textile: Journal of Cloth and Culture*, 5(3): 254–275. doi: 10.2752/175183507X249459.

Lamb, J.M., and Kallal, M.J. 1992. A conceptual framework for apparel design. *Clothing and Textiles Research Journal,* 10(2): 42–47. doi: 10.1177/0887302X9201000207.

Leach, E. 1989. Anthropological aspects of language: Animal categories and verbal abuse. *Anthrozoös,* 2(3): 151–165.

Lenochova, P., Vohnoutova, P., Craig Roberts, S., Oberzaucher, E., Grammer, K., and Havlıcek, J. 2012. Psychology of fragrance use: Perception of individual odor and perfume blends reveals a mechanism for idiosyncratic effects on fragrance choice. *PLoS ONE*, 7(3): e33810. doi:10.1371/journal.pone.0033810

Low, K.E. 2008. *Scent and Scent-Sibilities: Smell and Everyday Life Experiences*. Cambridge Scholars Publishing.

Mattila, A.S., and Wirtz, J. 2001. Congruency of scent and music as a driver of in-store evaluations and behavior. *Journal of Retailing*, 77: 273–289.

Mauss, M. 1979. *Sociology and Psychology: Essays*. Routledge & Kegan Paul.

McBurney, D.H., Levine, J.M., and Cavanaugh, P.H. 1976. Psychophysical and social ratings of human body odor. *Personality and Social Psychology Bulletin*, 3: 135–138.

McBurney, D.H., Streeter, S.A., and Euler, H.A. 2012. Olfactory comfort in close relationships You aren't the only one who does it. In G.M. Zucco, R.S. Herz, and B. Schaal (eds.), *Olfactory Cognition: From Perception and Memory to Environmental Odours and Neuroscience* (pp. 59–72). John Benjamins Publishing Company.

McGann, J.P. 2017. Poor human olfaction is a 19th-century myth. *Science*, 356(6338): eaam 7263.

Milinski, M., and Wedekind, C. 2001. Evidence for MHC-correlated perfume preferences in humans. *Behavioral Ecology*, 12: 140–149.

Mitro, S., Gordon, A.R., Olsson, M.J., and Lundstro¨m, J.N. 2012. The smell of age: Perception and discrimination of body odors of different ages. *PLoS ONE*, 7(5): e38110. doi:10.1371/journal.pone.0038110

Miwa, T. 2001. Impact of olfactory impairment on quality of life and disability. *JAMA: the journal of the American Medical Association*, 286(7): 772.

Morrin, M., and Ratneshwar, S. 2000. The impact of ambient scent on evaluation, attention, and memory for familiar and unfamiliar brands. *Journal of Business Research*, 49(2): 157–165.

Morrin, M., and Ratneshwar, S. 2003. Does it make sense to use scents to enhance brand memory? *Journal of Marketing Research*, 40(1): 10–25.

Morrison, M., Gan, S., Dubelaar, C., and Oppewal, H. 2011. In-store music and aroma influences on shopper behavior and satisfaction. *Journal of Business Research*, 64: 558–564.

Pause, B.M. 2012. Processing of body odor signals by the human brain. *Chemosensory Perception*, 5(1): 55–63. doi: 10.1007/s12078-011-9108-2.

Penn, D.J. Oberzaucher, E., Grammer, K., Fischer, G., Soini, H.A., Wiesler, D., and Brereton, R.G. 2007. Individual and gender fingerprints in human body odour. *Journal of the Royal Society Interface*, 4: 331–340.

Prehn-Kristensen, A., Wiesner, C., Bergmann, T.O., Wolff, S., Jansen, O., and Mehdorn, H.M., et al. 2009. Induction of empathy by the smell of anxiety. *PLoS ONE*, 4: e5987.

Reinarz, J. 2014. *Past Scents: Historical Perspectives on Smell*. University of Illinois Press.

Rimkute, J., Moraes, C., and Ferreira, C. 2016. The effects of scent on consumer behaviour. *International Journal of Consumer Studies*, 40: 24–34.

Rimkute, J., Moraes, C., and Ferreira, C.2016. The effects of scent on consumer behaviour. *International Journal of Consumer Studies*, 40(1): 24–34. doi: 10.1111/ijcs.12206.

Schilling, B., Kaiser, R., Natsch, A., and Gautschi, M. 2010. Investigation of odors in the fragrance industry. *Chemoecology*, 20(2): 135–147.

Schleidt, M., Hold, B., and Attili, G. 1981. A cross-cultural study on the attitude towards personal odors. *Journal of Chemical Ecology*, 7: 19–31.

Sorokowska, A., Schriever, V.A., Gudziol, V., Hummel, C., Hähner, A., Iannilli, E., Sinding, C., Aziz, M., Seo, H.S., Negoias, S., and Hummel, T. 2015. Changes of olfactory abilities in relation to age: Odor identification in more than 1400 people aged 4 to 80 years. *European Archives of Oto-Rhino-Laryngology,*, 272(8): 1937–1944. doi: 10. 1007/s00405-014-3263-4.

Spangenberg, E.R., Crowley, A.E., and Henderson, P. W. 1996. Improving the store environment: do olfactory cues affect evaluations and behaviors? *Journal of Marketing*, 60(2): 67–80.

Thornhill, R., and Gangestad, S.W. 1999. The scent of symmetry: A human sex pheromone that signals fitness? *Evolution and Human Behavior*, 20(3): 175–201.

Winther, F.Ø. 2018. *Luktesansen Store Medisinske Leksikon*. snl.no. Accessed 22nd April 2021. https://sml.snl.no/luktesansen

Zucco, G.M., Herz, R.S., and Schaal, B. 2012. *Olfactory Cognition: From Perception and Memory to Environmental Odours and Neuroscience. Vol. 85*: John Benjamins Publishing.

2 Human Odor: An Overview of Current Knowledge and Experimental Designs

Paola A. Prada-Tiedemann
Department of Environmental Toxicology, Forensic Analytical Chemistry and Odor Profiling Laboratory, Texas Tech University, Texas, United States

CONTENTS

2.1 INTRODUCTION

Numerous sources contribute to the generation, composition, and ultimate definition of an individual's "odor signature". From a chemical perspective, human odor is a complicated odorant mixture of several thousand compounds that significantly differ. Body odor generation pathways can be attributed to a diverse number of variables that include physiological processes, genetic makeup, and microbial interactions. Furthermore, several influencing factors that affect the generated odor include age, gender, and even environmental or lifestyle habits (see Figure 2.1). Due to the diversity of odor-producing sources in the human body, this chapter will explore the different variables to understand the complex odorant picture comprising human odor. Over the past few years, there has been increased interest in the definition and evaluation of human odor due to its expanded applications within the cosmetic/hygiene industry, textile manufacturing, the medical field, and forensics. Human odor has

DOI: 10.1201/9781003141426-2

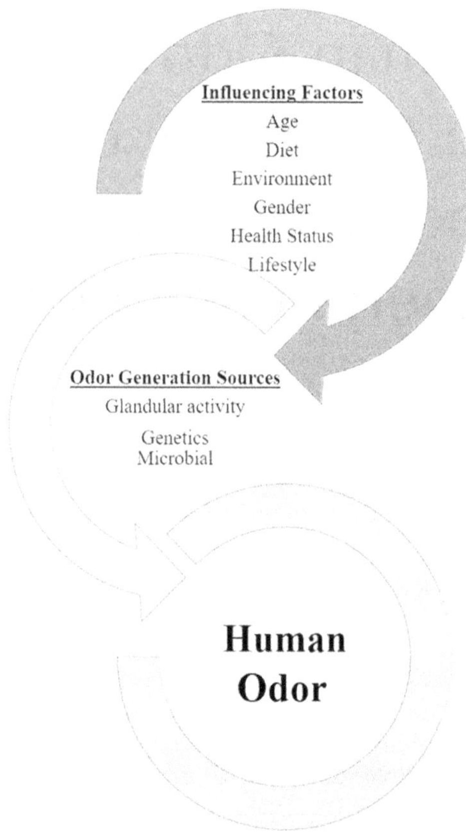

FIGURE 2.1 Schematic overview of human odor production mechanisms and variables.

witnessed novel advances with respect to analytical chemistry perspectives as it relates to different sampling and detection approaches, with the ultimate objective of determining a more exact definition of its chemical composition. This chapter provides recent developments and analytical trend highlights with respect to body odor studies. Some of the topics being addressed have been previously discussed by the author (Prada, Curran, and Furton 2015), and this chapter aims to provide a coherent framework of human odor in today's dynamic research arena.

2.2 PHYSIOLOGICAL ASSESSMENT

2.2.1 Human Skin Structure

Human skin is key in the generation and accumulation of body odor. It serves as a strategic boundary between the internal and external interface of the body with the environment, ultimately serving as a barrier to factors such as light, temperature, and chemical and biological contamination. It is also a region of rich metabolic activity, as seen by glandular secretions of biochemical importance, including lipids, peptides, proteins, nucleic acids, and volatile organic compounds (VOCs). Furthermore, skin serves as an important environment for microbial communities that reflect health conditions and also emit distinctive metabolites from microbial interactions (Duffy and Morrin 2019). Skin varies within a population in terms of thickness and composition due to specific age and body location. Normal skin pH ranges from 4 to 6, whereas internal pH in the internal body cavity ranges from 7 to 9 (Honari, Andersen, and Maibach 2017). As can be observed in Figure 2.2 below, the three layers in human skin are the epidermis, dermis, and hypodermis. The

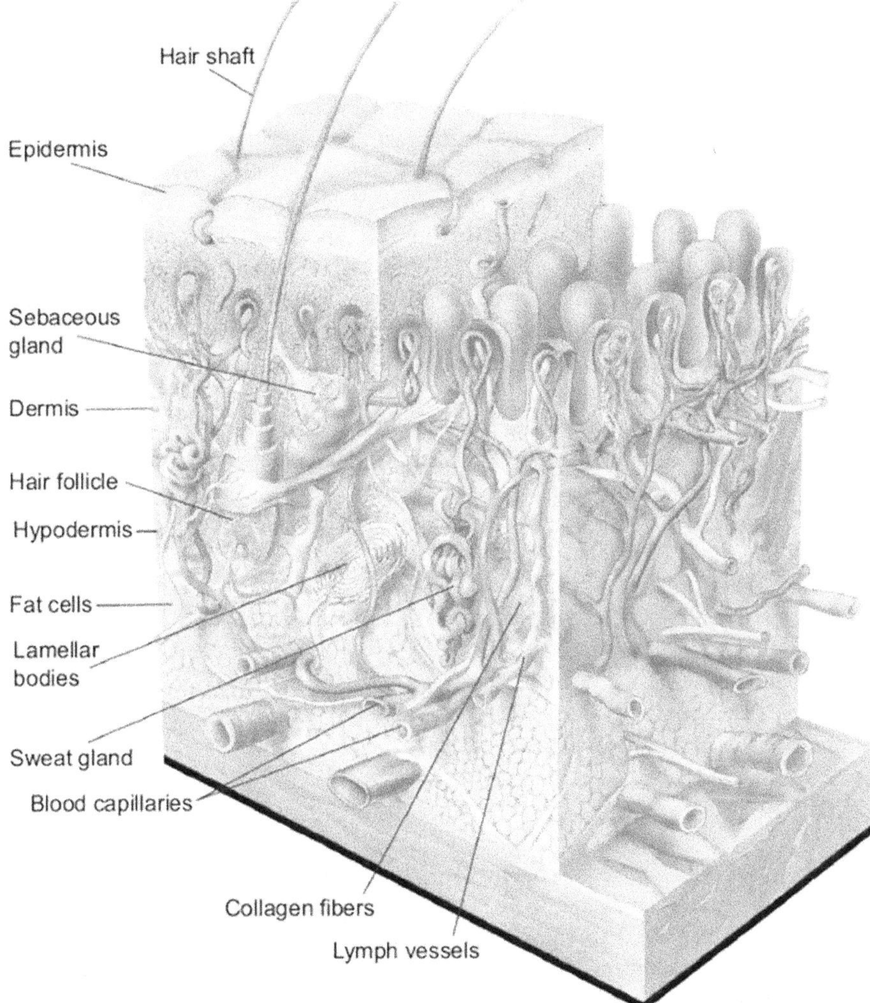

FIGURE 2.2 Schematic of human skin structure. (From Wikipedia: The Free Encyclopedia. Wikipedia Foundation, Inc. Available at https://web.archive.org/web/20090114060549/http://training.seer.cancer.gov/ss_module14_melanoma/images/illu_skin01.jpg, Last edited on November 29, 2021.)

epidermis is the outermost layer of the human skin organ, and it has about 0.5–1 mm of thickness, which varies according to body region. There are three types of cells in this layer – keratinocytes, melanocytes, and Langerhans cells. The keratinocytes are the most abundant type of cells in the epidermal layer. These cells give the skin surface its strength and flexibility (Schaefer and Redelmeier 1996). Melanocytes are found in the bottom layer of the epidermis and are responsible in the generation of the skin's main pigment, melanin. Langerhans cells are yet another type of specialized cells of the body's immune response that act in response to foreign substances that encounter the skin (Jablonski 2006). The stratum corneum is the outer layer of the epidermis, serving as the major barrier with respect to water loss and entry of exogenous matter. Commonly, this barrier is depicted as a brick-and-mortar-like framework where the bricks are represented by nonviable keratinocytes (also known as corneocytes), which are embedded in lipid membranes. As these bridges degrade, there is space created within the stratum corneum that allows for penetration processes (Menon and Elisa 1997; Michaels, Chandrasekaran, and Shaw 1975). Skin absorption is dependent on body region and the particular individual; the variation is correlated to lipid composition and stratum corneum thickness. Different routes have been described for this penetration to occur and include intercellular (chemicals passing through the lipid matrix), intracellular (chemicals passing through both the lipid matrix and corneocytes), through skin appendages, and via mechanical methods (i.e., ablation, needles) (Elias 2004; Prausnitz, Mitragotri, and Langer 2004).

The dermis is the second layer of the skin. Important cells in this layer of skin are the fibroblasts, which are rich in collagen and produce the tensile strength of skin. This layer contains a complex blood vessel network, sweat glands, tissue fibers, and hair follicles. The dermal layer is responsible for thermal regulation, repair, and nutrition processes (Schaefer and Redelmeier 1996). The deepest layer is the hypodermis. Main cells found in this section are the adipocytes, or fat cells. The key physiological function is to connect the skin with bone and muscle. Hence, it is an important layer for energy storage and metabolism. This layer contains intracellular fat droplets that act as a chemical repository that can permeate through the stratum corneum. Thus, compounds that can penetrate this layer can therefore distribute throughout the body (Noble 1992).

2.2.2 Glandular Activity

Human skin plays a key role in the regulation of body temperature and, as such, has dedicated secretion glands that include the eccrine, apocrine, and sebaceous glands to aid in this function. The combined secretions of these glands contribute to the generation and constitution of a subject's characteristic body odor signature. The distribution and function of each glandular group can be summarized in Table 2.1. Eccrine glandular secretions produce a watery perspiration to assist in core body temperature cooling, primarily consisting of water, with traces of salt, proteins, amino acids, lactate, bicarbonate, antimicrobial peptides, urea, and ammonia, to name a few. This is not intended to be an exhaustive list but includes some of the more commonly researched substances reported in the sweat diagnostic literature

TABLE 2.1
Distribution and Function of Human Secretion Glands (Data from: Prada et al. 2015)

	Apocrine Glands	Eccrine Glands	Sebaceous Glands
Body Distribution	Situated at the base of hair follicles, particularly in axilla, areolae of nipples, periumbilical, and genitals	Cover entire body, rich in forehead, axilla, hand palms, and soles of feet	Abundant in face and scalp, upper torso, and pubic area
Main Function	Response to emotional stimuli (e.g., pain, fear, anxiety)	Thermoregulation	Secretions from gland (sebum) spread over skin surface and mix
Secretion Characteristics	Lipids, proteins, steroids; no apocrine function in prepubertal children or the elderly	99% water content, traces of salts, urea, ammonia, amino acids, and proteins	Sebum formation varies with age and subject

(Baker 2019; Baker and Wolfe 2020; Groscurth 2002; Sato et al. 1989). The eccrine gland activity is overseen by neural stimulation guided by nerve fibers located around the gland. Thus, sweating rates are dependent on skin temperature, although it is sometimes related to a nervous response. Sweat composition varies according to specific body location and across subjects; however, it is generally characterized as a colorless and odorless acidic fluid with up to 99% water content, with additional electrolytes such as NaCl, K^+, and $HCO3^-$ (Tobin 2006).

Although apocrine glands exist from birth, their activity is marked by puberty onset. They are restricted to hairy body areas as their ducts exit toward skin layers via hair follicles. The secretion in apocrine glands is characterized by a higher viscosity than eccrine secretions, due to increased concentration of fatty acids and other compounds. Apocrine glands do not respond to thermoregulation functions but are associated with emotional stimuli, such as sexual arousal, pain, and anxiety. Bacterial interactions at the skin surface ultimately convert apocrine secretions into odorous molecules (Jablonski 2006).

The sebaceous glands are associated with hair follicles, with the excretion of sebum being a primary function for this glandular group. Normal sebum production is at a rate of 0.3 mg per 10 cm^3 per hour. It has been hypothesized that sebum is a barrier compound that slows the production of odorous molecules. Sebum secretion is highly dependent on an individual's age, diet, and gender. Males typically produce more sebum per cm^2 than women, presumably due to distinctive hormonal levels between the sexes (Giacomoni, Mammone, and Teri 2009; Jablonski 2006). Thus, it can be said that the functions of all three glandular groups within the human body play an essential role in the regulation of skin emission activity that ultimately yields to odorant compounds via bacterial interactions in the environment and the subsequent development of distinctive body odor within an individual.

2.3 A MICROBIOLOGICAL CONTEXT

Human skin is equipped with glands that not only help in thermoregulation but also in the production of secretions of vital importance for skin health and biochemistry. This diverse chemical landscape on the skin surface is critical for microbial activity. The composition of the skin microbiota is dependent on factors such as body location, skin pH, moisture level, and nutrient availability. Thus, occluded regions with limited surrounding air, such as the axillae and groin, tend to favor microbial activity, whereas regions of low water content (forearms) have reduced microbial density. Furthermore, external factors, such as beauty and hygiene products, can further affect microbial composition (Costello et al. 2009; Grice and Segre 2011; Holland and Bojar 2002; Turnbaugh et al. 2007). There are four main types of bacterial flora that have been characterized from human skin: *Corynebacterium, Micrococcus, Propionibacterium* (recently renamed *Cutibacterium*), and *Staphylococcus* (Smallegange, Verhulst, and Takken 2011). Recent studies have focused efforts on understanding the topographical distribution and composition of these microbial communities on the skin by 3D mapping of mass spectral data along with other techniques, such as 16S rDNA profiling. Seminal results have provided foundational concepts exhibiting the intertwined relationship between skin hygiene, microbiota, and the environment. From the topographical study, it was noted that the family *Staphylococcaceae* was found in typically moist areas (feet, under breast, neck). *Propionibacterium* was mainly linked to sebaceous regions, such as the scalp, face, and torso. The genus *Corynebacterium* was commonly identified in the head, groin, and toes of the subjects (Bouslimani et al. 2015). These studies highlighted the crucial role of the skin in the diversity of cutaneous microbiota. Skin microbiota composition and density can be said to be unique to the individual, being determined by the subject and its corresponding genotype (Dethlefsen et al. 2007). Temporal fluctuations can occur with respect to an individual's microbial community structure, and these can be attributed to intrinsic changes, such as nutrition, or external variations, such as climate, clothing, and beauty products (Gao et al. 2010).

The apocrine gland (located mainly in groin and axilla regions) forms odorless precursor molecules that yield to characteristic malodor formation when there is an interaction with cutaneous microbial flora (Kippenberger et al. 2012; Rennie et al. 2007). Thus, much of the research in this area has stemmed from the industry for deodorant and hygiene product development. Deodorant products employ antimicrobial agents that are also enzyme inhibitors. The notion is that by lowering bacterial numbers in the axilla, the synthesis of odorous compounds can be decreased (Dayan, Wertz, and Drake 2007). In terms of specific bacterial interactions, it has been shown that aerobic Corynebacteria metabolize odorless steroids, producing 16-androstenes (5a-androstenol,5a-androstenone) with a pungent musk- and urine-like odor (Austin and Ellis 2003). Axillary androstenone levels are much higher in men than in women, partly due to the larger size of apocrine glands in males compared to females, resulting in higher male odor levels (Labows, McGinley, and Kligman 1982). Other studies have shown that skin microbiota, such as *Staphylococcus,* can convert branched amino acids, such as leucine, into odorous molecules, such as short-chained (C_4-C_5) methylated volatile fatty acids such as isovaleric acid. Other cutaneous bacterial flora, such as *Corynebacterium,* have also been reported to transform fatty acids into axillary malodor constituents. Thus, the products of the biotransformation process done by *Corynebacterium* yield common malodor products (James, Hyliands, and Johnston 2004). Studies from odor emanations from the human axilla have also shed light on the role of volatile C6-C11 straight-chain, branched, and unsaturated acids as major contributors to axilla malodor, with (E)-3-methyl-2-hexenoic acid (3M2H) being a major contributor to this odor formation (Natsch et al. 2003; Zeng et al. 1991). Studies have depicted that 3M2H is secreted into the apocrine gland, bound to two carrier proteins known as the apocrine secretion odor binding proteins (ASOB1 and ASOB2). The hypothesis for this mechanism is that due to the capability of these proteins to carry apocrine secretion, individuals secreting a high quantity of ASOB2 have a higher capability of stronger underarm odor production (Jacoby et al. 2004; Spielman et al. 1998). Thus, differences in terms of gender with respect to odor intensity have no relationship with odorant amounts or protein structure but rather with the availability of non-odorous precursor materials in apocrine secretions, which male individuals secrete more easily (Spielman et al. 1998). Another mechanism for odor generation with respect to microbial contexts lies in the presence of N-acyl-glutamine-aminoacylasein bacterial flora, which release odorous volatile carboxylic acids, such as 3-hydroxy-3-methylhexanoic acid (Natsch et al. 2006; Natsch, Schmid, and Flachsmann 2004). Thus, there is a relationship between physiological glandular odorless secretions and the interaction of these with cutaneous microbial flora that ultimately yield proposed mechanisms for odor generation.

2.4 GENETIC SOURCES

The major histocompatibility complex (MHC) is the most diverse part of the genome; in humans this genetic coding region is referred to as the human leukocyte antigen (HLA). MHC molecules have two categories, Class I and Class II. Class I molecules are present in all nucleated cells within the body, and their purpose is the processing and recognition by T-cells of any foreign antigen. Class II MHC molecules are present on certain lymphocytes, and their purpose is to present antigens for recognition with other cells to increase immune response (Eggert et al. 1999). Thus, the HLA locus effectively represents an immunological "identity" of an individual. Seminal work in this area has been conducted with mice animal models exhibiting odor preference of conspecifics, which differed in MHC genes (Yamazaki et al. 1976). This foundational work caused an eruption of research in this area, investigating the MHC complex as a source of an individual "odortype" (Havlicek and Roberts 2009; Kwak et al. 2010). Of special interest in the research community is the evaluation of MHC function with respect to mate selection or kin recognition (Yamazaki and Beauchamp 2007). The hypothetical mechanism proposed is that due to immunological individuality defined by the MHC, the allele expression creates a specific individual odor signature, in both biological body fluids and general body odor (Yamazaki et al. 1978). A recent study performed a comprehensive meta-analysis to investigate available evidence and study design approaches with respect to mate selection and MHC dissimilarity. Recommendations were the need of larger sampling sizes, geographical sites, and even cultural diversity in order to fully understand this complex interaction between a "genetic" odortype and mating preferences (Havlíček, Winternitz, and Roberts 2020). All these studies are based on the premise that there is a specific odor compound along with a corresponding combination of MHC genes. The question yet to be answered, however, is the exact mechanism that leads the MHC complex to express a particular odor "fingerprint".

As depicted in Figure 2.3, several hypotheses have been suggested as to how the MHC genes influence body odor generation and potentially yield a distinctive "odorprint". Given that MHC molecules are present in specimens such as sweat and urine, one hypothesis suggests the odor generation is created via the specimen-MHC interaction. Another hypothesis states that MHC molecules bind to allele-specific peptides and that resulting metabolites (i.e., carboxylic acids) yield a particular odor signature (Brennan and Kendrick 2006). A third hypothesis suggests that the MHC genes affect odor by influencing microbial flora populations. A study reared congenic rodents in germ-free conditions, and results highlighted that the MHC interacted with bacteria to give specific urine odors (Brown 1995). Another proposed mechanism suggests that MHC molecules bind to volatiles and thus carry these volatiles to scent glands. This approach has been investigated by correlating human skin volatiles and mosquito attractiveness to MHC genes (Verhulst et al. 2013). A final hypothesis has been that the MHC region influences odor generation at a cellular level. Thus, it is proposed that MHC allele combinations impact metabolic pathways at the intracellular level, which influences the generated volatile

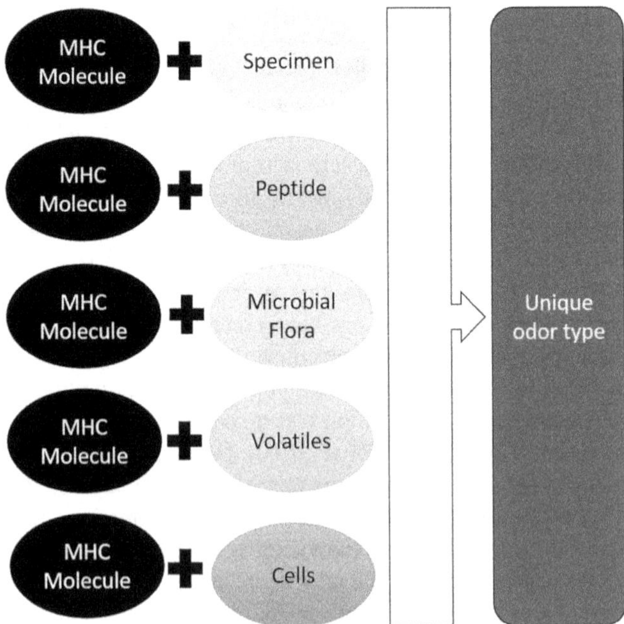

FIGURE 2.3 Proposed mechanisms for MHC gene odor generation.

odor chemicals (Aksenov et al. 2012). It can be said that all these mechanisms propose an odor signature that is a degradation product from MHC gene molecules that interact with peptides, specimens, volatiles, cells, or even microbes to generate a particular odortype.

Recently, research has also focused on the gene ABCC11 (MRP8), which is crucial for the formation of the characteristic axillary odor, otherwise known as a medical condition axillary osmidrosis (strong odor and profuse axilla sweating) (Toyoda et al. 2016). Studies show that ABCC11 is expressed and localized in apocrine sweat glands. Furthermore, results indicated a key function of ABCC11 in the secretion of odorants and their precursors from apocrine sweat glands (Martin et al. 2010).

Thus, from a genetic body odor perspective, there are a number of attractive areas for future study to understand the dynamic relationship between targeted genes, their expressions in different populations, and their subsequent regulation of odorant production.

2.5 HUMAN ODOR – A VOLATOLOMIC PERSPECTIVE

Body odors can be described as a compilation of hundreds of odorous VOCs emitted from various parts of the body via complex metabolic mechanisms. Figure 2.4 shows some of the major sources of body VOCs, including (but not limited to): scalp, breath, saliva, sweat, blood, skin, feet, back, urine, feces, and vaginal secretions. The usefulness of understanding the human volatilome has been observed in disease diagnostics (Broza et al. 2015; Buljubasic and Buchbauer 2014;

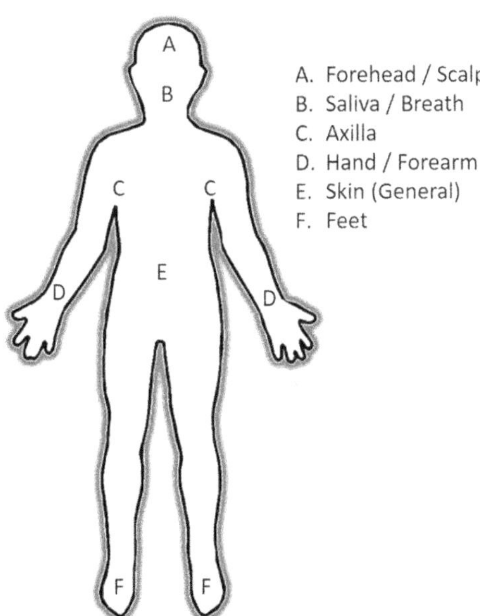

FIGURE 2.4 Body VOC Sources.

Prugnolle et al. 2009; Olsson et al. 2014; Shirasu and Touhara 2011) and forensic applications (Prada and Furton 2008, 2012). In terms of textile-odor evaluations, it is also crucial to understand body odor as human wearing is the pathway for odor transfer in real-life applications (McQueen and Vaezafshar 2020). The generation of body odor within an individual, along with all its related production variables, come into play with odor formation upon textile use and subsequent laundering processes. It is known that human odor, even though originating from the same person, has an instrumentally distinctive odor profile depending on textile material utilized (Prada, Curran, and Furton 2011). Although the scope of this chapter is not to investigate the textile-odor formation mechanism, it is important to understand that the development of such a process is largely dependent on the VOC landscape of the human subject prior to any textile contact.

Recently, a compendium of VOCs from the human body of healthy individuals was published, reporting a total of 1840 compounds from breath, saliva, blood, milk, skin secretions, urine, and feces (de Lacy Costello et al. 2014). Figure 2.5 provides the percentage breakdown of VOCs in the different body areas, with breath and skin emanations being the highest contributors with respect to reported VOCs. The renaissance of research in human odor VOCs has depicted a highly diverse composition to include carboxylic acids, short chain alcohols, ketones, aldehydes, and short and long chain hydrocarbons (Pandey and Kim 2011). The complexity of the human body odor profile can be spoken about in three terms coined in previous work by the author: the "primary odor" of an individual is those constituents that

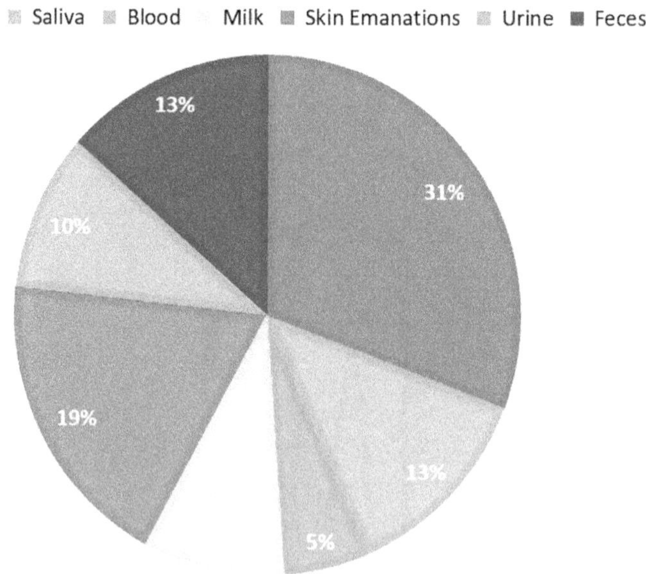

FIGURE 2.5 Body Part VOC Distribution. (Data from: de Lacy Costello et al. 2014.)

come from within and are stable over time regardless of diet or environmental factors; the "secondary odor" is constituents that also come from within and are present due to diet and environmental factors; and the "tertiary odor" contains constituents that are present because they were applied from the outside (e.g., lotions, soaps, perfumes) (Curran et al. 2005). Thus, it is important to highlight that while analytical methods have advanced our understanding of the human volatilome, comparison of all this foundational work needs to be taken with caution as differences in sampling approaches and body regions produce distinctive odor signatures that can yield high variation and divergences (Brown et al. 2013).

2.5.1 Sampling Preconcentration and Extraction Techniques

Due to the close contact of textiles with the human skin, a focused evaluation will be performed on analytical samplings and techniques that have geared efforts toward skin emanations. The study of skin VOCs requires suitable sampling approaches and preconcentration techniques to obtain target odors from the matrix (skin region) that can be adequately detected with analytical systems (Kataoka et al. 2013). The challenge also lies in the low concentrations of VOCs in skin emanations, and thus the need for a robust and high degree of analytical sensitivity (Duffy and Morrin 2019). A thorough review of sampling methods has been provided by Dormont et al. (2013). These typically include solvent extraction procedures (Gallagher et al. 2008) and dynamic headspace absorption onto a variety of sorbents via contact or headspace modes (use of glass beads, gauze pads, worn clothing, polydimethylsiloxane [PDMS] membranes) (Bernier et al. 2000; Jiang et al. 2013; Riazanskaia et al. 2008). Once collected, trapped volatiles can be recovered from these phases for analysis via thermal desorption, solvent extraction, or collection onto adsorbent traps like Tenax by employing an airflow process. One sampling challenge with these methods, however, is that there might be contamination and low volatile yields even before analytical detection. For example, it has been noted that while some of these materials are biologically sterile, this sterilization does not equate to an analytically clean background, thus introducing the potential of exogenous compounds into the collected sample (Prada, Curran, and Furton 2010). Another issue with skin sampling procedures is not only the approach for collection but also the skin treatment process for the individuals being tested. There are studies that limit or restrict food intake and use of cosmetic/hygiene products, and even require participants to engage in physical activity prior to collection procedures (see reviews, Duffy and Morrin 2019; Dormont, Bessiere, and Cohuet 2013). Recent literature has showcased the utilization of a more contemporary methodology – solid phase microextraction (SPME). SPME is a very simple, solventless, and sensitive extraction technique that allows the extraction of volatiles onto coated fibers of variable chemical composition, followed by direct desorption into a gas chromatography (GC) injector (Prada and Furton 2012). SPME can be used directly on the sample or can be collected in the headspace above the sample. For example, recent literature has used SPME fibers directly over the skin surface to collect feet odor volatiles, yielding comparative results with headspace sampling (Dormont et al. 2013). Some inherent drawbacks of the SPME technique are matrix effects, low storage stability, dependence of extraction efficiency on environmental sampling conditions, and competitive sorption within fibers (Kataoka et al. 2013). An example of various SPME skin sampling designs is summarized in Figure 2.6 below.

2.5.2 Analytical Detection

Analytical techniques for the characterization of human body VOCs have been reviewed in the literature (Duffy and Morrin 2019; Kataoka et al. 2013). Instrumentation for their analysis is commonly conducted with gas chromatography coupled with mass spectrometry (GC-MS), the gold standard detection system for VOC odor profile analysis due to its capacity for simultaneous analysis of a large number of compounds, coupled with built-in identification mechanisms, as seen with mass spectral libraries. GC is also advantageous as the odor components released from the human body are mostly VOCs, which have a high vapor pressure and thus are ideal for GC separation. A new analytical trend has been observed with the introduction of multidimensional chromatographic systems. Studies have begun implementing comprehensive two-dimensional gas chromatographic time-of-flight mass spectrometry (GCxGC-TOFMS) for the analysis of emissions from ankles, wrists, and hands (Cuzuel et al. 2018; Dolezal et al. 2017; Roodt et al. 2018). Direct MS methods, such as selected ion flow tube mass spectrometry (SIFT-MS), proton transfer reaction mass spectrometry (PTR-MS), membrane inlet mass spectrometry (MIMS), and secondary electrospray ionization mass spectrometry (SESI-MS),

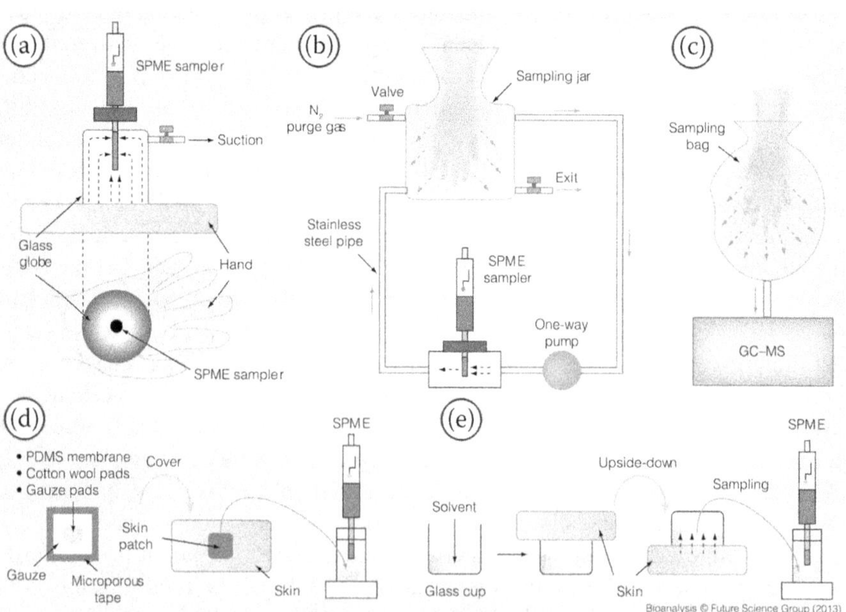

FIGURE 2.6 Examples of SPME Skin Emission Sampling Designs. (a) Direct SPME sampling in sealed glass globes, (b) Direct SPME with airflow sampling chambers, (c) Direct sampling bag, (d) skin patch sampling, (e) Liquid sampling in glass cups; SPME: Solid-phase microextraction. (Source: Reprinted from Kataoka et al. 2013, with permission.)

have also been used. GC has also been coupled with ion mobility spectrometry (IMS), as well as the use of electronic noses (e-nose), for skin VOC detection (see reviews, Pandey and Kim 2011; Duffy and Morrin 2019). In terms of biological detection, canines have been a standard for detection of humans in a range of law enforcement and forensic applications (Agapiou et al. 2015; Prada, Curran, and Furton 2015). Due to the volatility of these personal odor signatures, canines represent a dynamic detection system, with low threshold characteristics, amenity to work in high contamination, and ease of being deployable to operational settings. Although it is still unknown which volatile odor markers are specific for canine use in human odor detection, it has been established that there is enough variation within an individual's odor profile for canines to not only detect human odor but also to perform discrimination on odor samples (Prada, Curran, and Furton 2015).

2.6 CONCLUSIONS

In this chapter, there is a combined perspective on different aspects and impacting factors on the definition and generation of human odor. There is a clear lack of understanding of how these variables interconnect to yield individual odor signatures across a population. Not only are there several physiological, microbial, and genetic factors guiding odor production, but also there are external factors at the interface of these physiological traits. In addition, it is important to learn more about the complex odor picture of the human body, as it is dynamic with constant changes due to human growth, environment, and lifestyle habits. Studies have identified key odor volatiles using a number of sampling approaches, along with analytical detection, GC-MS being the gold standard. It would be interesting to evaluate potential pitfalls in sampling methods and detection systems by performing different techniques on the same samples. Although these results offer useful and important findings, the conclusions must be taken with caution before generalizations can be made. As new technological advances reach the laboratory, higher resolution and information-rich data can be produced as that observed with multidimensional chromatographic separation developments. There is a need for larger scale studies focusing both on the composition and on the quantity of the volatilome across the entire human body. Research must push forward and consider the inter-individual differences as well as different analytical techniques, sampling approaches, environmental conditions, and specific population traits that can offer more insights into the different aspects interacting and impacting the human odor formation process. These insights are much needed to open opportunities for the understanding of odor formation and how it merges to outside interactions, as seen in the textile-user interface.

REFERENCES

Agapiou, A., Amman, A., Mochalski, P., and Statheropoulos, M. 2015. Trace detection of endogenous human volatile organic compounds for search, rescue and emergency applications. *Trends in Analytical Chemistry*, 66: 158–175.

Aksenov, A.A., Gojova, A., Zhao, W., Morgan, J.T., Sankaran, S., Sandrock, C.E., and Davis, C.E. 2012. Characterization of volatile organic compounds in human leukocyte antigen heterologous expression systems: A cell's "Chemical Odor Fingerprint" *ChemBioChem*, 13: 1053–1059.

Austin, C., and Ellis, J. 2003. Microbial pathways leading to steroidal malodour in the axilla *Journal of Steroid Biochemistry and Molecular Biology*, 87: 105–110.

Baker, L.B. 2019. Physiology of sweat gland function: The roles of sweating and sweat composition in human health. *Temperature*. doi: 10.1080/23328940.2019.1632145

Baker, L.B., and Wolfe, A.S. 2020. Physiological mechanisms determining eccrine sweat composition. *European Journal of Applied Physiology*, 120: 719–752.

Bernier, U.R., Kline, D.L., Barnard, D.R., Schreck, C.E., and Yost, R.A. 2000. Analysis of human skin emanations by gas chromatography/mass spectrometry. 2. Identification of volatile compounds that are candidate attractants for the yellow fever mosquito (Aedes aegypti). *Analytical Chemistry*, 72: 747e756.

Bouslimani, A., Porto, C., Rath, C.M., Wang, M., Guo, Y., Gonzalez, A., Berg-Lyon, D., Ackermann, G., Moeller Christensen, G.J., Nakatsuji, T., Zhang, L., Borkowski, A.W., Meehan, M.J., Dorrestein, K., Gallo, R.L., Bandeira, N., Knight, R., Alexandrov, T., and Dorrestein, P.C. 2015. Molecular cartography of the human skin surface in 3D. *PNAS*, 112: E2120–E2129. doi: 10.1073/pnas.1424409112

Brennan, P.A., and Kendrick, K.M. 2006. Mammalian social odours: Attraction and individual recognition. *Philosophical Transactions of the Royal Society B: Biological Sciences 36l*: 2061–2078.

Brown, B.E. 1995. What is the role of the immune system in determining individually distinct body odours? *International Journal of Immunopharmacology*, *17*: 655–661.

Brown, J.S., Prada, P.A., Curran, A.M., and Furton, K.G. 2013. Applicability of emanating volatile organic compounds from various forensic specimens for individual differentiation. *Forensic Science International*, 226: 173–182.

Broza, Y.Y., Mochalski, P., Ruzsanyi, V., Amann, A., and Haick, H. 2015. Hybrid volatolomics and disease detection. *Angewandte Chemie*, 54: 2–15.

Buljubasic, F., and Buchbauer, G. 2014. The scent of human diseases: A review on specific volatile organic compounds as diagnostic biomarkers. *Flavour and Fragance Journal*, 30: 5–25.

Costello, E.K., Lauber, C.L., Hamady, M., Fierer, N., Gordon, J.I., and Knight, R. 2009. Bacterial community variation in human body habitats across space and time. *Science*, 326(5960): 1694–1697.

Curran, A.M., Rabin, S.I., Prada, P.A., and Furton, K.G. 2005. Comparison of the volatile organic compounds present in human odor using SPME-GC/MS. *Journal of Chemical Ecology*, 31(7): 1613–1625.

Cuzuel, V., Leconte, R., Cognon, G., Thiebaut, D., and Vial, J. et al. 2018. Human odor and forensics: Towards Bayesian suspect identification using GC × GC–MS characterization of hand odor *Journal of Chromatography B* 1092: 379–385.

Dayan, N., Wertz, P., and Drake, D. 2007. Antimicrobial compounds (AMC's) as inhibitors of bacterial growth causing sweat malodor. *Journal of Cosmetic Science*, 58: 184–185.

de Lacy Costello, B., Amann, A., Al-Kateb, H., Flynn, C., Filipiak, W., Khalid, T., Osborne D., and Ratcliffe, N.M. 2014. A review of the volatiles from the healthy human body. *Journal of Breath Research*, 8: 014001. doi:10.1088/1752-7155/8/1/ 014001.

Dethlefsen, L., McFall-Ngai, M., and Relman, D.A. 2007. An ecological and evolutionary perspective on human–microbe mutualism and disease. *Nature*, 449: 811–817.

Dolezal, P., Kyjakova, P., Valterova, I., and Urban, S. 2017. Qualitative analyses of less-volatile organic molecules from female skin scents by comprehensive two dimensional gas chromatography–time of flight mass spectrometry. *Journal of Chromatography B* 1505: 77–86.

Dormont, L., Bessiere, J.M., and Cohuet, A. 2013. Human skin volatiles: A review. *Journal of Chemical Ecology* 39: 569–578.

Dormont, L., Bessière, J.M., Mc Key, D., and Cohuet, A. 2013. New methods for field collection of human skin volatiles and perspectives for their application in the chemical ecology of human/pathogen/vector interactions. *Journal of Experimental Biology* doi: 1 0.1242/jeb.085936.

Duffy, E., and Morrin, A. 2019. Endogenous and microbial volatile organic compounds in cutaneous health and disease. *Trends in Analytical Chemistry*, 111: 163–172.

Eggert, F., Luszyk, D., Haberkorn, K., Wobst, B., Vostrowsky, O., Westphal, E., Bestmann, H.J., Muller-Ruchholtz, W., and Ferstl, R. 1999. The major histocompatibility complex and the chemosensory signaling of individuality in humans. *Genetica*, 104: 265–273.

Elias, P.M. 2004. The epidermal permeability barrier: From the early days at Harvard to emerging concepts. *Journal of Investigative Dermatology* 122(2): xxxvi–xxxix.

Gallagher, M., Wysocki, C.J., Leyden, J.J., Spielman, A.I., Sun, X., and Preti, G. 2008. Analyses of volatile organic compounds from human skin. *British Journal of Dermatology*, 159: 780e791. doi: 10.1111/j.1365-2133.2008.08748.x

Gao, Z., Perez-Perez, G.I., Chen, Y., and Blaser, M.J. 2010. Quantitation of major human cutaneous bacterial and fungal populations. *Journal of Clinical Microbiology*, 48: 3575–3581.

Giacomoni, P.U., Mammone, T., and Teri, M. 2009. Gender-linked differences in human skin. *Journal of Dermatological Science*, 55: 144–149.

Grice, E.A., and Segre, J.A. 2011. The skin microbiome. *Nature Reviews Microbiology*, 9(4): 244–253.

Groscurth, P. 2002. Anatomy of sweat glands. *Current Problems in Dermatology*. doi: 10.1159/000060678

Havlicek, J., and Roberts, S.C. 2009. MHC-correlated mate choice in humans: A review. *Psychoneuroendocrinology*, 34: 497–512.

Havlíček, J., Winternitz, J., and Roberts, S.C. 2020. Major histocompatibility complex-associated odour preferences and human mate choice: Near and far horizons. *Philosophical Transactions of the Royal Society B*, 375: 20190260. doi: 10.1098/rstb.2019.0260

Holland, K.T., and Bojar, R.A. 2002. Cosmetics: What is their influence on the skin microflora? *American Journal of Clinical Dermatology*, 3(7): 445–449.

Honari, G., Andersen, R., and Maibach, H.L. 2017. *Sensitive Skin Syndrome*. 2nd ed. CRC Press.

Jablonski, N.G. 2006. *Skin: A Natural History*. University of California Press.

Jacoby, R.B., Brahms, J.C., Ansari, S.A., and Mattai, J. 2004. Detection and quantification of apocrine secreted odor-binding protein on intact human axillary skin. *International Journal of Cosmetic Science*, 26: 37–46.

James A.G., Hyliands D., and Johnston H. 2004. Generation of volatile fatty acids by axillary bacteria. *International Journal of Cosmetic Science*, 26: 149–156.

Jiang R., Cudjoe, E., Bojko B., Abaffy T., and Pawliszyn J. 2013. A non-invasive method for in vivo skin volatile compounds sampling. *Analytica Chimica Acta*, 804: 111e119. doi: 10.1016/j.aca.2013.09.056

Kataoka, H., Saito K., Kato H., and Masuda K. 2013. Noninvasive analysis of volatile biomarkers in human emanations for health and early disease diagnosis. *Bioanalysis*, *5*(11): 1443–1459.

Kippenberger S., Havlicek J., Bernd A., Thaci D., Kaufmann R., and Meissner M. 2012. 'Nosing Around' the human skin: What information is concealed in skin odour? *Experimental Dermatology*, 21: 655–659.

Kwak, J., Willse, A., Preti, G., Yamazaki, K., and Beauchamp, G.K. 2010. In search of the chemical basis for MHC odourtypes. *Proceedings of the Royal Society B*, 277: 2417–2425.

Labows, J.N., McGinley, K.J., and Kligman, A.M. 1982. Perspectives on axillary odor. *Journal of the Society of Cosmetic Chemists*, 1982(34): 193–202.

Martin, A., Saathoff, M., Kuhn, F., Max, H., Terstegen, L., and Natsch, A. 2010. A functional ABCC11 allele is essential in the biochemical formation of human axillary odor. *Journal of Investigative Dermatology*, 130: 529–540.

McQueen, R.H., and Vaezafshar, S. 2020. Odor in textiles: A review of evaluation methods, fabric characteristics, and odor control technologies. *Textile Research Journal*, 90(9–10): 1157–1173.

Menon, G.K., and Elisa, P.M. 1997. Morphologic basis for a pore-pathway in mammalian stratum corneum. *Skin Pharmacology*, 10(5–6): 235–246.

Michaels, A.S., Chandrasekaran, S.K., and Shaw J.E. 1975. Drug permeation through human skin: Theory and in vitro experimental measurements. *AIChE Journal*, 21(5): 985–996.

Natsch, A., Derrer, S., Flachsmann, F., and Schmid, J. 2006. A broad diversity of volatile carboxylic acids, as candidate molecules for the determination of human-body odor-type. *Chemistry & Biodiversity*, 3: 1–19.

Natsch, A., Gfeller, H., Gygax, P., Schmid, J., and Acuna, G. 2003. A specific bacterial aminoacylase cleaves odorant precursors secreted in the human axilla. *The Journal of Biological Chemistry*, 278(8): 5718–5727.

Natsch, A., Schmid, J., and Flachsmann, F. 2004. Identification of odoriferous sulfanylalkanols in human axilla secretions and their formation through cleavage of cysteine precursors by a C-S Lyase isolated from axilla bacteria. *Chemistry & Biodiversity*, 1: 1058–1072.

Noble, W.C. 1992. *The Skin Microflora and Microbial Skin Disease*. Cambridge University Press.

Olsson, M.J., Lundström, J.N., Kimball, B.A., Gordon, A.R., Karshikoff, B., Hosseini, N., Sorjonen, K., Höglund, C.O., Solares, C., Soop, A., Axelsson, J., and Lekander, M. 2014. The scent of disease: Human body odor contains an early chemosensory cue of sickness. *Psychological Science*, 25(3): 817–823.

Pandey, S.K., and Kim, K.H. 2011. Human body-odor components and their determination. *Trends in Analytical Chemistry*, 30(5): 784–796.

Prada, P.A., Curran, A.C., and Furton, K.G. 2015. *Human Scent Evidence*. CRC Press.

Prada, P.A., Curran, A.M., and Furton, K.G. 2010. Comparison of extraction methods for the removal of volatile organic compounds (VOCs) present in sorbents used for human scent evidence collection. *Analytical Methods*, 2: 470–478.

Prada, P.A., Curran, A.M., and Furton, K.G. 2011. The evaluation of human hand odor volatiles on various textiles: A comparison between contact and noncontact sampling methods. *Journal of Forensic Sciences*, 56(4): 866–881.

Prada, P.A., and Furton, K.G. 2008. Human scent detection: A review of its developments and forensic applications. *Rev Cienc Foren*, 1: 81–87.

Prada, P.A., and Furton, K.G. 2012. Recent advances in solid phase microextraction for forensic applications. In J. Pawliszyn (ed.), *Comprehensive Sampling and Sample Preparation. Analytical Techniques for Scientists* (pp. 877–891). Elsevier.

Prausnitz, M.R., Mitragotri, S., and Langer, R. 2004. Current status and future potential of transdermal drug delivery. *Nature Reviews Drug Discovery*, 3(2): 115–124.

Prugnolle, F., Lefevre, T., Renaud, F., Moller, A.P., Misse, D., and Thomas, F. 2009. Infection and body odours: Evolutionary and medical perspectives. *Infection, Genetics and Evolution*, 9: 1006–1009.

Rennie, P.J., Gower, D.B., Holland, K.T., Mallet, A.I., and Watkins, W.J. 2007. The skin microflora and the formation of human axillary odour. *International Journal of Cosmetic Science*, 12: 197–207.

Riazanskaia, S., Blackburn, G., Harker, M., Taylor, D., and Thomas, C.L.P. 2008. The analytical utility of thermally desorbed polydimethylsilicone membranes for in-vivo sampling of volatile organic compounds in and on human skin. *Analyst*, 133: 1020e1027. doi: 10.1039/b802515k.

Roodt, A.P., Naude, Y., Stoltz, A., and Rohwer, E. 2018. Human skin volatiles: Passive sampling and GC × GC-ToFMS analysis as a tool to investigate the skin microbiome and interactions with anthropophilic mosquito disease vectors. *Journal of Chromatography B*, 1097–1098: 83–93.

Sato, K., Kang, W.H., Saga, K., and Sato, K.T. 1989. Biology of sweat glands and their disorders. I. Normal sweat gland function. *Journal of the American Academy of Dermatology*, 20: 713–726.

Schaefer, H., Redelmeier, T.E. 1996. *Skin Barrier Principles of Percutaneous Absorption*. Karger.

Shirasu, M., and Touhara, K. 2011. The scent of disease: Volatile organic compounds of the human body related to disease and disorder. *Journal of Biochemistry*, 150(3): 257–266.

Smallegange, R.C., Verhulst, N.O., and Takken, W. 2011. Sweaty skin: An invitation to bite? *Trends in Parasitology*, 27(4): 143–148.

Spielman, A.I., Sunavala, G., Harmony, J.A.K., Stuart, W.D., Leyden, J.J., Turner, G., Vowels, B.R., Lam, W.C., Yang, S., and Preti, G. 1998. Identification and immunohistochemical localization of protein Precursors to human axillary odors in apocrine glands and secretions. *Archives of Dermatology*, 134: 813–818.

Tobin, D.J. 2006. Biochemistry of human skin – Our brain on the outside. *Chemical Society Reviews*, 35: 52–67.

Toyoda, Y., Gomi, T., Nakagawa, H., Nagakura, M., and Ishikawa, T. 2016. Diagnosis of human axillary osmidrosis by genotyping of the human ABCC11 gene: Clinical practice and basic scientific evidence. *Biomed Research International*. doi: 10.1155/2016/7670483

Turnbaugh, P.J., et al. 2007. The human microbiome project. *Nature*, 449(7164): 804–810.

Verhulst, N.O., Beijleveld, H., Qiu, Y.T., Maliepaard, C., Verduyn, W., Hassnoot, G.W., Claas, F.H.J., Mumm, R., Bouwmeester, H.J., Takken, W., VanLoon, J.J.A., and Smallegange, R.C. 2013 Relation between HLA genes, human skin volatiles and attractiveness of humans to malaria mosquitoes. *Infection, Genetics and Evolution*, 18: 87–93.

Yamazaki, K., and Beauchamp, G.K. 2007. *Genetics of Sexual Differentiation and Sexually Dimorphic Behaviors, Vol. 59*. K. Yamazaki (ed.), Elsevier. pp. 129–145.

Yamazaki, K., Boyse, E.A., Mike, V., Thaler, H.T., Mathieson, B.J., Boyse, E.A., Zoyas, Z.A., and Thomas, L. 1976. Control of mating preferences in mice by genes in the major histocompatibility complex. *Journal of Experimental Medicine*, 144: 1324–1335.

Yamazaki, K., Yamaguchi, M., Andrews, P.W., Peake, B., and Boyse, E.A. 1978. Mating preferences of F2 segregants of crosses between MHC-congenic mouse strains *Immunogenetics*, 6: 253–259.

Zeng, X.N., Leyden, J.J., Lawley, H.J., Sawano, K., Nohara, I., and Preti, G. 1991. Analysis of characteristic odors from human male axillae. *Journal of Chemical Ecology*, 17(7): 1469–1492.

3 Sources and Characterization Approaches of Odour and Odour-Causing Bodily Compounds in Worn Clothing

Mourad Krifa
School of Fashion & Advanced Materials and Liquid Crystal Institute, Kent State University, Kent, Ohio, United Sates

Mathilda Savocchia
School of Fashion, Kent State University, Kent, Ohio, United Sates

CONTENTS

3.1 INTRODUCTION

Cleanliness of clothing and home textiles has throughout history been an essential component of human hygiene, and thus pertains to the "establishment and maintenance of human health" according to Terpstra (1998, 2001, 2003). Textiles constitute a prominent interface between the human body and its environment.

DOI: 10.1201/9781003141426-3

Consequently, maintaining the cleanliness of clothing and other textiles that come in intimate contact with the human body has major human health implications. Historically, clothing items have been laundered for aesthetic purposes to remove visible stains, and to restore the feel and fitness for use through the elimination of odours (Shove 2003; Terpstra 2001). Although laundering practices of people around the world vary significantly due to the influence of social, cultural, and moral norms (Shove 2003; Laitala et al. 2011), the cleanliness of clothing is invariably assessed subjectively by consumers using sensory evaluation of odour (McQueen et al. 2007).

When in contact with the human body, clothing is exposed to both visible and invisible contaminants consisting of stains and solid particles from the surrounding environment, and of organic substances from perspiration, skin shedding, and sebum (Bowers and Chantrey 1969; Chi and Obendorf 1998; Chung and Seok 2012; Terpstra 2001). Human sebum, a sebaceous secretion of the skin, is a major constituent of organic soils found in worn clothing (Bowers and Chantrey 1969; Chi and Obendorf 1998; Chung and Seok 2012).

This chapter provides a brief review of the sources of the odour-causing bodily compounds found in clothing. It then focuses on the measurement techniques used to detect and characterize odour with consideration of standardization efforts. We review both direct (olfactometry) and indirect (analytical and instrumental) methods to detect and characterize odour. We examine issues relating to odour sampling in the various measurement methods and focus on research conducted in the textile field.

3.2 ODOUR-CAUSING BODILY COMPOUNDS

The human skin is one of the major sources of laundry soils (Bowers and Chantrey 1969; Chi and Obendorf 1998; Chung and Seok 2012; Laughlin and Gold 1990). Microscopic analyses of unwashed clothing have shown large amounts of skin lipids found in the inter-fibre capillaries of yarns (Obendorf and Klemash 1982; Obendorf and Webb, 1987). Human sebum is a sebaceous secretion of the skin and is a mixture of unique lipids, notably including squalene, wax esters, and fatty acids (Picardo et al. 2009; Stefaniak et al. 2010). Those substances, particularly squalene, oleic acid, linoleic acid, and triolein, have been shown to be abundant contributors to worn garment soil retention (Bowers and Chantrey 1969; Chi and Obendorf 1998; Chung and Seok 2012). They also serve as a nourishing medium on which microorganisms thrive, leading to the development of malodour (Chung and Seok 2012; McQueen et al. 2007, 2013). For instance, it was shown that *Staphylococcus epidermidis* had a higher growth rate on cotton fabric soiled with triolein (Chung and Seok 2012), a major component of human sebum (Wertz 2009).

In addition to sebum, clothing and other textiles that come in contact with the skin absorb substances caused by perspiration. Perspiration is a key biological function performed by the skin to regulate the body core temperature (Thieme et al. 2003; Wilke et al. 2007). In addition to the thermoregulatory response initiated by the body, stress and anxiety can also cause perspiration or emotional sweating. Dietary factors, such as consumption of spicy food, can also induce sweating, known as gustatory sweating (Wilke et al. 2007).

Human perspiration is a complex mixture of secretions by the eccrine and apocrine sweat glands (Chen et al. 2020; Thieme et al. 2003; Wilke et al. 2007, 2009). A third category of sweat glands, the apoeccrine glands, has been identified in the literature (Sato et al. 1989; Wilke et al. 2007, 2009), and are presumed to develop during puberty from eccrine glands as their number decreases with age. However, it has been difficult to differentiate between apocrine and apoeccrine sweat (Wilke et al. 2007), and several histological studies have failed to show evidence of apoeccrine glands (Bovell et al. 2007, 2011; Wilke et al. 2008).

The eccrine glands are distributed throughout the body and produce perspiration consisting of an aqueous electrolyte solution with small amounts of metabolites, including ammonia, lactic acid, and urea (Sato et al. 1989; Thieme et al. 2003). Eccrine sweat is shown to result in a body odour that is impacted by diet, hydration, and metabolic rate (Chen et al. 2020; Munk et al. 2000). In addition, eccrine sweat provides moisture for bacterial activity (Munk et al. 2000).

The apocrine glands are present in the hairy regions of the body (axillae and groin), are dormant at birth, and become active at puberty (Chen and Jain 2010). Similar to sebaceous glands, apocrine glands secrete into the hair follicle; thus, apocrine sweat may be mixed with sebum (Wilke et al. 2009). Apocrine secretions consist of an oily and odourless (in a pure state) substance that contains proteins, lipids, steroids, and some amount of sebum (Chen et al. 2020; Munk et al. 2000). However, they become odorous due to bacterial activity (Chen et al. 2010; Munk et al. 2000).

Overall, secretions from all three skin glands (apocrine, eccrine, and sebaceous) contribute to odour-causing compounds by promoting the action of cutaneous microorganisms (Chen and Jain 2010; Munk et al. 2000). However, the action of sebum and apocrine sweat is predominant in the axillae and other hairy regions of the body where the three glands are present (Chung and Seok 2012; Munk et al. 2000). Mechanical friction and pressure between the skin and clothing result in the transfer of those secretions onto the fabric, thus supporting the growth of microorganisms that cause odour on the clothing itself (Sanders et al. 2021).

3.3 ODOUR DETECTION AND CHARACTERIZATION

Odour measurement has long been a common practice in many industries. The food and beverage industry, cosmetic and hygiene industry, and the air and water purification industries, have long faced the need for odour detection, measurement, and remediation technologies (Bockreis and Jager 1999; Hobbs et al. 1995; Mackay et al. 1961; Schilling et al. 2009). In the textile industry, characterizing odour constitutes a major component of assessing cleanliness and laundering performance, alongside colour (or whiteness) and visible stains (ASTM 2005).

The primary mechanism to measure odour is through olfactometry using sensory analysis, where the human nose constitutes the "measurement device" (Domingues et al. 2016; Gouronnec and Tomasso 2000). Odour is the result of complex interactions between the olfactory system and diverse chemical substances. The complexity and uniqueness of the olfactory system compared to the other human senses makes the task of directly measuring odours using sensors or analytical techniques difficult (Gouronnec and Tomasso 2000; McGinley et al. 2000). Nevertheless, chemical and physicochemical analytical methods have been developed to indirectly assess odour by detecting the substances that cause it, e.g., sebum (Krifa 2018; Krifa et al. 2019; Obendorf and Klemash 1982; Obendorf et al. 1983) and bacterial activity (McQueen et al. 2013; Obendorf et al. 2007), or the odorous volatile compounds (McQueen et al. 2007, 2008, 2007; Yuwono and Lammers 2004).

Although olfactometry provides an understanding of the human response to odour, physicochemical methods allow instrumental identification and quantification of the composition of the odorous compounds. When the odour emanates from complex mixtures, relating the chemical composition to the resulting human sensation is a challenge (Capelli et al. 2013; Kim and Park 2008; Littarru 2007; Stuetz et al. 1999). However, knowledge of the chemical composition of odorous compounds and of the biological and physicochemical mechanisms resulting in odours is critical for control and remediation.

3.3.1 Olfactometry – Sensory Methods

Sensory methods rely on the sense of smell of human assessors and are used to rate odour according to multiple parameters relevant to the perception of smell. Depending on the area of application, different odour parameters can be measured using sensory approaches. For instance, the fragrance and flavour industry developed parameters such as "long-lastingness", i.e., durability of fragrance perception after its application; "substantivity", i.e., persistence on laundry after washing and drying; and "diffusivity", or the potential for perception at a distance from the source (Gygax and Koch 2001).

More broadly adopted sensory measures include odour concentration, intensity, and character or descriptive profile (Gouronnec and Tomasso 2000; McGinley et al. 2000; McGinley and McGinley 2017). Two of the primary ratings, odour concentration and odour intensity, are quantitative in nature because they correspond to scales that measure the relative strength of the odour attributes (Gouronnec and Tomasso 2000; McGinley et al. 2000; McGinley and McGinley 2017). Quantitative measurements are performed by odour assessors trained and tested for olfactory perception according to strict guidelines formalized by standardization organizations such as the American Society for Testing and Materials, or ASTM (ASTM Committee E-18 on Sensory Evaluation of Materials and Products 1981; ASTM 2014; ISO 2012), and ISO (ISO 2012).

According to McGinley and McGinley (2017) and to Gouronnec and Tomasso (2000), odour concentration corresponds to the number of dilutions required to reduce an odour to its detection threshold. The threshold corresponds to the lowest concentration that still allows the detection of the odour. Higher concentrations (stronger odours) require more dilution to reach the perception or detection threshold. Gouronnec and Tomasso (2000) describe the perception threshold as "the concentration of odorous molecules in air so that the probability of perception is 50%".

Standardized procedures based on varying the dilution of the odorous sample with odourless neutral air using a device called an olfactometer are referred to as "dynamic olfactometry" in the European standard EN 13725 (CEN 2007) and the Australian Standard AS/NZS 4323:3:2001 (Standards Australia 2001), or as "forced-choice ascending concentration" methods in the American standard ASTM E679-19 (ASTM 2019) and in the international standard ISO 13301 (ISO 2018). The "forced choice" method refers to the fact that odour thresholds are established by having the assessors sniff the odour sample at a given dilution with air, along with two blank samples of air with no odorous compounds. The odour perception threshold is thus determined at the point where the assessor is able to discriminate between the odorous sample and the blanks (ASTM 2019; ISO 2018).

Odour intensity, also referred to as suprathreshold intensity (ASTM 2018), is defined as the "magnitude of perceived sensation" beyond the detection threshold (Gostelow et al. 2001). Intensity can be rated based on a subjective ordinal scale, such as faint, moderate, and strong (Gostelow et al. 2001). However, standardized methods in the United States (ASTM 2018) and in Europe (AFNOR 1996) preconize the "olfactory matching" of the sample odour intensity to a calibrated Standard Odour Intensity Referencing Scale (OIRS), using a known gas at varied concentrations (Gouronnec and Tomasso 2000; McGinley et al. 2000; McGinley and McGinley 2017). The American standard (ASTM E544) stipulates the use of n-butanol with a range of 8 known concentrations from 12 ppm to 1550 ppm (McGinley and McGinley 2017). The French standard NF X43–103 (AFNOR 1996) uses known concentrations of either butanol or pyridine (Gouronnec and Tomasso 2000).

One prominent challenge in olfactometry is that the quantification of the parameters is impacted by the variability of human olfaction (Capelli et al. 2013; van Harreveld et al. 1999). This difficulty is managed through standardization and strict criteria for selecting and training assessors. For instance, both standards cited above require the panel of assessors to consist of individuals who are preselected based on sensitivity to some reference compounds (e.g., isovaleric acid), then formally trained and "calibrated" according to strict guidelines (ASTM Committee E-18 on Sensory Evaluation of Materials and Products 1981; ASTM 2014; AFNOR 1996; Bratcher 2013; ISO 2012).

Reinbach et al. (2011) used a "Sniffin' Stick" test to select normosmic panellists, i.e., with normal smell perception. The researchers then performed an odour intensity ranking test using a range of dilutions of 1-butanol, the same odorous reference used in the standards ASTM E544 or NF X43–103. Prescreened and trained panellists are required in test methods, such as the standards cited above, where the objective is to discriminate between odours or describe odours based on sensory properties and their intensities (Bratcher 2013).

In addition to the quantifiable olfactometry parameters (concentration and intensity), there is a class of qualitative parameters consisting of a descriptive profiling of the odour. Depending on the sources and standard practices, odour qualitative testing can be purely hedonic or affective (Bratcher 2013; Gouronnec and Tomasso 2000). For instance, Gouronnec and Tomasso (2000) define one such measure as odour annoyance, i.e., the level of discomfort caused by the odour. The measure is performed by asking a panel to rate the odour on a 5-level ordinal scale ranging from "not annoyed" to "extremely annoyed". Affective measures are typically conducted to gage the environmental disturbances caused by odour within a particular population, or the acceptance and preference of a product by a target group of consumers (Bratcher 2013). Therefore, unlike the quantitative measures discussed above, affective measures are assessed using a panel of volunteers from the local population or target consumer segment without the need for specific training or selection criteria (Gouronnec and Tomasso 2000).

More wholistic descriptive profiling is practiced using standard descriptors that classify the odour both based on its character and its hedonic connotation (ASTM 1992; Fisher et al. 2018; McGinley and McGinley 2017). ASTM has published an "Atlas of Odor Character Profiles" for use in a broad range of fields, including food and beverage science, indoor air quality, cosmetics and hygiene, and medicine (ASTM 1992). A commonly used tool in odour character profiling is the 'odour descriptor wheel' (Dalton et al. 2011; Fisher et al. 2018; McGinley et al. 2000; McGinley and McGinley 2017; Suffet et al. 2004). An odour descriptor wheel depends on the specific field of application and can be customized with a set of odour descriptors that match its needs. There are, however, major categories of descriptors that are common to most applications. Figure 3.1 depicts an illustration

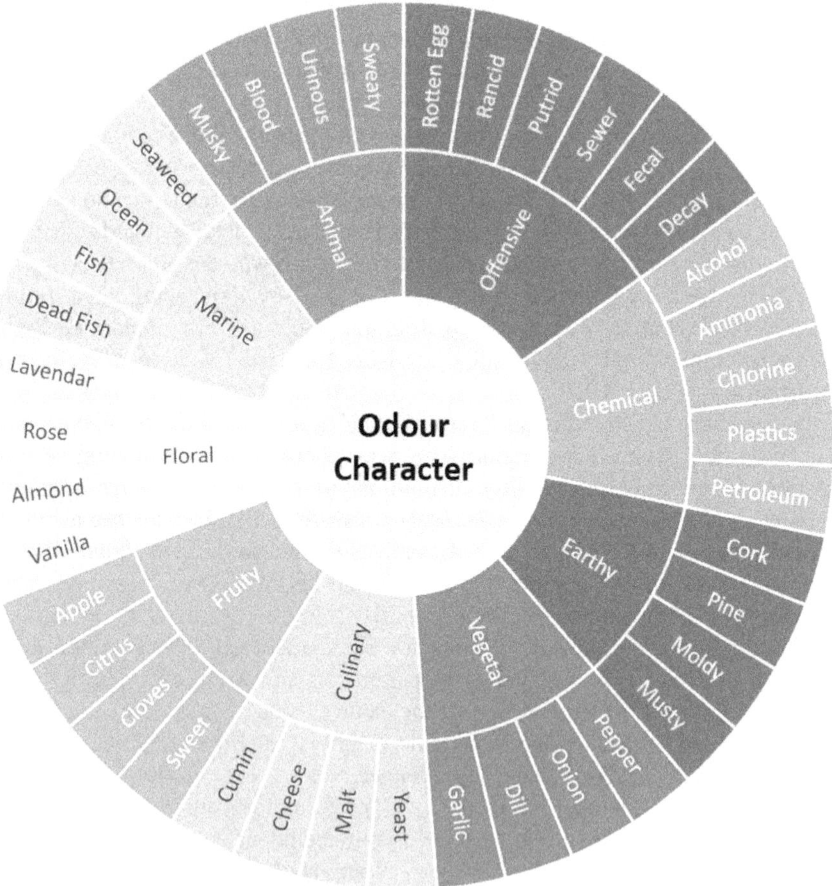

FIGURE 3.1 Example of an Odour Wheel. (Based on Dalton et al. 2011; Fisher et al. 2018; McGinley et al. 2000; McGinley and McGinley 2017; Suffet et al. 2004).

of an odour descriptor wheel with some of those major descriptor categories compiled based on multiple sources (Dalton et al. 2011; Fisher et al. 2018; McGinley et al. 2000; McGinley and McGinley 2017; Suffet et al. 2004). Some of the major categories of descriptors shown in Figure 3.1 include ‘offensive odours’, ‘animal odours’, and ‘earthy odours’. Within each category, more detailed descriptors are used to provide an accurate depiction of the odour character (ASTM 1992). Unlike the purely hedonic assessments done by consumer panels, standardized wholistic odour descriptor wheel characterizations rely on preselected and trained assessors (ASTM 2014; ISO 2012).

3.3.1.1 Sensory Odour Measurement in Textile Research

Because of the complexity and subtility of the primary odorants in human body odours, the reliance on sensory analysis by pre-elected human assessors remains predominant when characterizing odour on textiles (Chung and Seok 2012; Hasegawa et al. 2004; McQueen et al. 2007; Natsch et al. 2006; Stapleton et al. 2013). Odour descriptors classified under the “Animal” category in Figure 3.1 are highly relevant to the area of textiles exposed to odorous bodily compounds. Indeed, some of the substances caused by apocrine and sebaceous gland secretions and their interaction with microorganisms present on the skin correspond to odour characters referenced in ASTM E1207-14 “Standard Guide for Sensory Evaluation of Axillary Deodorancy” (ASTM 2014). Among those, *5-alpha-androst-16-en-one,* which is a result of bacterial action on apocrine secretion, confers a “urinous” character to human axillary odour (ASTM 2014; Labows and Kligman 1982). Another result of skin bacterial action on apocrine gland secretions, *5-alpha-androst-16-en-3-alpha-ol*, confers a “musky” character to the axillary odour (ASTM 2014; Labows and Kligman 1982). Finally, the ASTM standard attributes the “sweaty, acid” character of axillary odour to the *isovaleric acid (3-methylbutnoic acid)* component (ASTM 2014).

Clothing creates a unique ecosystem for the bodily odour components to transfer and evolve. Indeed, odour is not only transferred to clothing through contact with the body, but there is also evidence that it is intensified to different extents by clothing of different fibre contents (McQueen et al. 2007). It has been shown that fabric properties, such as moisture management, thermal conductivity and air permeability, can affect the clothing-skin microclimate (Guo et al. 2008; Morrissey and Rossi 2013; Zimniewska and Krucinska 2010), thus favouring the growth of different microbial populations (Fowler et al. 2019; Sanders et al. 2021).

Odour transferred to and developing in clothing is referred to as secondary, by opposition to the primary odour on the human body prior to any interaction with the textile structure (Dravnieks et al. 1968). In addition to odour intensity and character, the concept of odour retention in the garments is of critical importance in the textile field (Huisman and Morris 1972; McQueen et al. 2013, 2008, 2007; Mukhtar Abdul-Bari et al. 2020; Wang et al. 2019). Similarly, the inverse phenomenon of odour release from the fabric is of utmost importance because of its dependence on the adherence to the fabric structure of the high-molecular weight, low volatility, and low-water solubility odorous compounds present in sebum and apocrine secretions (Munk et al. 2000; Obendorf et al. 2007). The two phenomena of odour retention and release can present significant challenges for some textile products and applications, such as athletic wear, with the accumulation and persistence of odours (Gocek and Duru 2019).

Numerous studies have used sensory analysis to investigate the retention (and release) of odours in textiles worn in close contact with the human skin, with particular attention to comparing different types of fibre compositions and finishes. For instance, it has been shown that synthetic materials have a higher propensity for odour retention and lead to more pronounced odours than textiles made of natural fibres (Abdul-Bari et al. 2018; Klepp et al. 2016; McQueen et al. 2007; Munk et al. 2000). McQueen et al. (2007) found in a study comparing axillary odour retention of a range of fibres in different knit structures, that polyester resulted in a higher odour intensity than cotton and wool. Abdul-Bari et al. (2018) compared nylon and polyester and found no differences in odour intensity measured by sensory analysis according to ASTM 1207 (ASTM 2014).

Given that synthetic fibres such as polyester and nylon have been dominant in the activewear market (Guruprasad et al. 2015; Krifa and Stewart Stevens 2016), there have been major efforts to impart odour-control technology through antibacterial finishes such as silver nanoparticles (Gao and Cranston 2008; Klepp et al. 2016; Krifa and Prichard 2020). However, sensory analysis of odours observed in athletic wear of various fibre compositions, including polyester with and without odour-control technology, revealed that natural fibres had a lower odour intensity than both treated and untreated polyester samples (Klepp et al. 2016). In this research, Klepp et al. (2016) measured the odour intensity of 13 different types of fabrics commonly used in sportswear, including wool, cotton, polyester, and polyester treated with odour controlling properties. The researchers used a sensory analysis with a consumer panel of untrained individuals to determine odour intensity in these fabrics during different stages of usage (sweaty, aired, and washed). Participants were asked to smell different fabric samples and rank them on a 5-point ordinal scale from "no odour" to "very strong odour". The assessors were also asked whether they would wear the sample or choose to wash it considering the odour intensity they perceived (Klepp et al. 2016), which represents an affective measure similar to the level of annoyance discussed in Gouronnec and Tomasso (2000). A recent study by Wang et al. (2019) explored odour retention in polyester/wool blends. This study used the dynamic olfactometry method for sensory analysis according to the Australia Standard AS/NZS 4323:3:2001 (Standards Australia 2001).

3.3.2 Physicochemical and Instrumental Methods

Many of the studies discussed above sought to complement the direct sensory analysis of odour with indirect objective measures using analytical methods targeting the physical, chemical, or microbiological interactions of the odour-causing compounds (Brattoli et al. 2013; Gostelow et al. 2001; Kim and Park 2008; McQueen et al. 2007; 2008; Munk et al. 2000; Wang et al. 2019). Some of those indirect methods attempted to analytically determine the chemical composition of the odorous volatile compounds (Gostelow et al. 2001; Kim and Park 2008; McQueen et al. 2008, 2007). Other studies attempted to detect and quantify the chemical compounds at the origin of the odour, including apocrine secretions and sebum (Krifa 2018; Krifa et al. 2019; Obendorf and Klemash 1982; Obendorf et al. 1983; Wang et al. 2019). Finally, another research stream focused on microbiological analysis, such as bacterial counts and identification of microorganisms associated with odour formation (Callewaert et al. 2014; McQueen et al. 2007; Stapleton et al. 2013).

3.3.2.1 Gas Chromatography and Mass Spectrometry

In the first category of methods, gas chromatography coupled with mass spectrometry (GC-MS) of the volatile odour effluents represent a common approach (Brattoli et al. 2013; Kim and Park 2008; McQueen et al. 2008; Munk et al. 2000; Prada et al. 2011, 2014; Wang et al. 2019). GC-MS allows determining the chemical composition of the volatile compounds in the odorous mixture and is therefore useful in remediation efforts. The individual compounds in the mixture are separated through elution by GC then conveyed to the MS detector to be chemically identified based on mass spectra (Li 2014; Stapleton et al. 2013). Other types of sensors, such as flame ionization detectors, FID (Takeuchi et al. 2012), or ion mobility spectrometry, IMS (Rudnicka et al. 2010; Ruzsanyi et al. 2012), have also been coupled with GC to detect volatile compounds emanating from the human body.

To enable GC-MS or other sensor analysis of textile odours, it is necessary to extract the odorous VOCs and direct them to the GC column. This can be done by directly drawing the VOC mixture from the sample headspace and injecting it into the GC injector port (McQueen et al. 2008; Miracle et al. 2020; Abdul-Bari et al. 2020). An example of headspace collection referred to as headspace solid phase microextraction (HS-SPME) is depicted in Figure 3.2 (Schmidt and Podmore 2015).

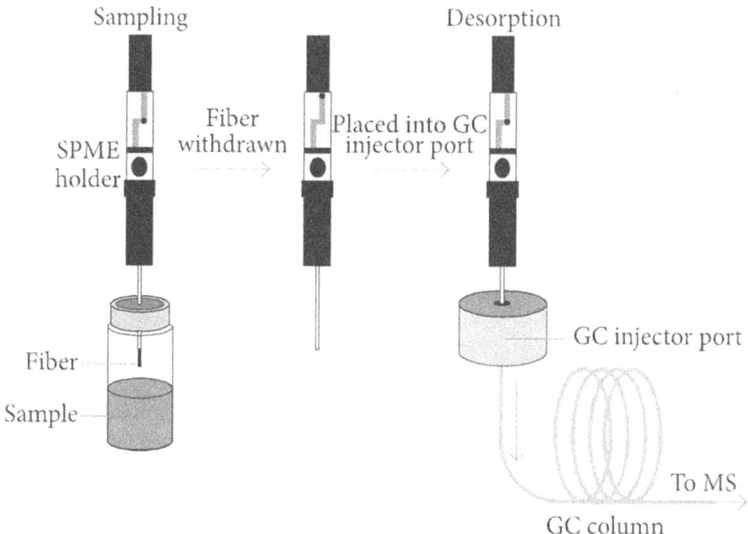

FIGURE 3.2 GC-MS analysis with solid phase headspace microextraction. (Reprinted with permission from Schmidt and Podmore (2015) under the Creative Commons Attribution, License.)

The HS-SPME technique uses a fibre assembly suspended in the headspace above the sample to collect the volatile molecules through adsorption (Schmidt and Podmore 2015). The analytes are then desorbed into the GC injector for separation and subsequent MS analysis (Figure 3.2).

McQueen et al. (2008) used headspace analysis of axillary volatile compounds released from cotton, wool, and polyester fabrics. The analytical method used in this research was proton transfer reaction-mass spectrometry (PTR-MS). The VOCs were conveyed from the sample headspace to the PTR-MS inlet through direct drawing using a heated capillary. The authors compared polyester, cotton, and wool and found results that are concordant with sensory analysis showing a higher odour intensity in polyester (McQueen et al. 2008).

In addition to headspace analysis, other direct extraction methods have been reported as a preparation for GC-MS evaluation. For example, Munk et al. used a multi-step extraction procedure to isolate the volatiles through overnight soaking of the textile swatches in double-distilled diethyl ether or in dichloromethane followed by microdistillation and high-vacuum sublimation (Munk et al. 2000, 2001).

GC-MS offers a powerful tool to instrumentally separate and identify a broad range of odorous compounds. However, the performance of the human nose remains unrivalled from the perspective of the actual smell perception by humans. Therefore, significant research was conducted to couple GC separation with olfactometric detection and characterization of the compounds (GC-O) (Brattoli et al. 2011, 2013; Debonneville et al. 2002; Peres et al. 2013). Debonneville et al. (2002) describe a "multisniffing system" (Figure 3.3) in which a GC is used to separate the analytes within an odorous mixture. The eluate is conveyed through splitters and a connexion box toward multiple sniff ports corresponding to the number of panellists in the experiment (Debonneville et al. 2002).

3.3.2.2 Chemical Sensors – Electronic Noses

Another category of instrumental analysis of odour was made possible by specialized sensors or arrays of sensors referred to as 'electronic noses' (Bockreis and Jager 1999; Cipriano and Capelli 2019; Loutfi et al. 2015; Nicolas et al. 2000). Electronic noses (EN) were the result of research conducted more than four decades ago and aimed at developing machine olfaction to monitor odour in applications such as food processing, water treatment, and environmental pollution (Bartlett and Gardner 2000; Hudon et al. 2000; Macleod et al. 1976; Persaud and Dodd 1982). The early definition of an electronic nose (EN) describes an instrument equipped with an "array of heterogenous electrochemical sensors with partial specificity and a pattern recognition system" (Loutfi et al. 2015). A broad range of gas sensors with different operating mechanisms have been used to enable machine olfaction using EN (Cipriano and Capelli 2019; Hudon et al. 2000; Loutfi et al. 2015; Nicolas et al. 2000). The most common operate based on metal oxide semiconductors, or on conductive polymers and polymer composites (Cipriano and Capelli 2019; Hudon et al. 2000; Loutfi et al. 2015; Nicolas et al. 2000). However, other types of detection mechanisms have been reported, including optical sensors, gas sensitive field effect transistors, and quartz microbalance sensors (Di Natale et al. 1997; Loutfi et al. 2015).

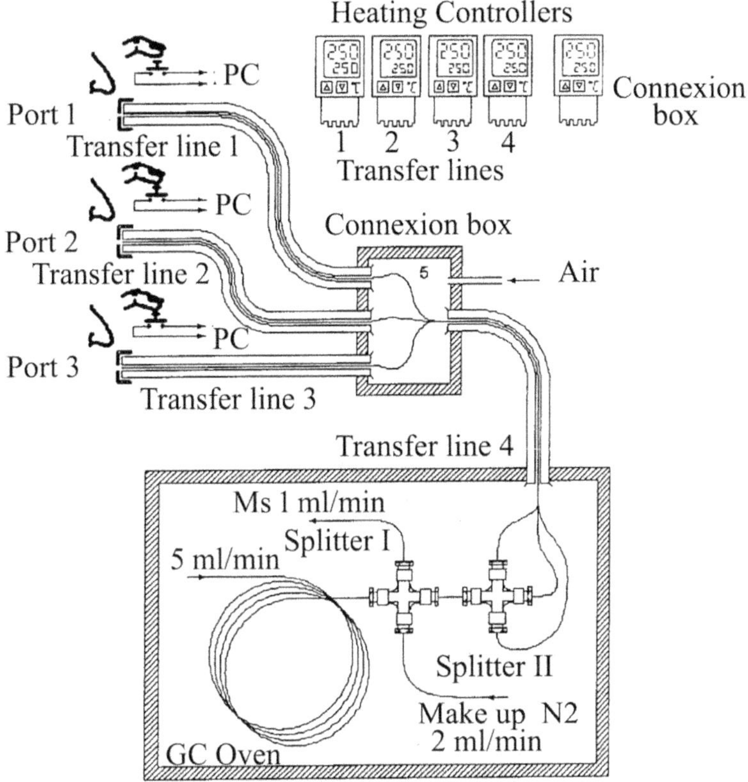

FIGURE 3.3 Schematic diagram of the odour detection system using GC coupled with Olfactometry – multisniffing system GC-O. (Reprinted with permission from Debonneville et al. (2002). Copyright (2002) American Chemical Society.)

EN instruments typically combine multiple sensors calibrated with specificity to capture different odour compounds. Loutfi et al. (2015) reviewed a range of 12 commercial EN models with arrays comprising 2 to 38 metal oxide or conducting polymer sensors. Research continues to explore new types of sensors with operating mechanisms based on nano-enhanced micro-electro-mechanical systems (MEMS), or ultra-fast GC-MS (Loutfi et al. 2015; Nie et al. 2020). The latter principle capitalizes on the capability of GC-MS analysis to separate and identify volatile odorous compounds and on advances made in drastically reducing the duration required to separate the analytes (Mondello et al. 2004). Depending on the specificity of the sensors used, electronic noses can be used to identify not only odours, but also other chemical substances in the mixture (Bockreis and Jager 1999).

Regardless of the sensor array operating mechanism, odorous compound identification and monitoring require processing the signals for pattern recognition and classification using approaches such as principal component analysis and cluster analysis (Loutfi et al. 2015), or artificial intelligence tools such as fuzzy logic and neural networks (Bockreis and Jager 1999; Scott et al. 2006).

The applications most sought through EN include food processing (Baldwin et al. 2011; Di Natale et al. 1997; Deisingh et al. 2004; Loutfi et al. 2015; Mielle 1996), pharmaceutical and medical diagnosis (Baldwin et al. 2011; Chang et al. 2009; Farraia et al. 2019; Gardner et al. 2000; Thaler and Hanson 2005), and real-time environmental odour nuisance monitoring (Cipriano and Capelli 2019; Littarru 2007; Nicolas et al. 2000; Sironi et al. 2007; Stuetz et al. 1999). For example, Nicolas et al. (2000) experimented with EN using tin oxide sensors to monitor malodour in the vicinity of compost facilities, paint shops, and wastewater treatment plants. Littarru (2007) applied both EN and dynamic olfactometry to evaluate the odour nuisance in waste treatment plants and chemical plants. The author found that both methods offer complementary measures, and that the combination of both approaches provides optimal capabilities for the monitoring of odour nuisance (Littarru 2007).

Santos et al. (2010) compared the performance of EN for the detection of aromatic compounds in wine with sensory analysis by a preselected and trained panel of assessors. The electronic nose comprised an array of 16 tin oxide sensors followed by pattern recognition using principal component analysis and neural networks (Santos et al. 2010). The researchers concluded that the sensitivity of the electronic nose instrument compared favourably to the sensory panel and had a detection threshold up to 10 times lower than the human nose. In addition, the electronic nose offered the capability to quantitatively determine the concentrations of the compounds (Santos et al. 2010). In addition to achieving high sensitivity with sensor specificity, another advantage of electronic noses is that once they are calibrated, they can be used to perform odour assessment on a continuous basis at a minimal cost (Hudon et al. 2000).

Attempts have been made to use electronic noses in characterizing human body odour secretions. For example, Gardner et al. (2000) reported on attempts to use electronic noses to diagnose illness from breath samples. Di Natale et al. (2000) developed an electronic nose procedure to evaluate human skin odour. The researchers evaluated EN sensitivity to an important odorous compound of human apocrine secretions, namely, pheromone 5a-androst-16-en-3-one, and reported encouraging results for the assessment of smell and odour in humans (Di Natale et al. 2000).

Haeringer and Goschnick (2008) focused on odour-causing contaminants found in textiles with the goal of reducing the environmental impact of laundry. They used a commercially available EN system based on 38 metal-oxide sensor elements to test the odour caused by cigarette smoke, rotten milk, and human sweat applied to a range of textiles, including cotton, wool, and polyester (Haeringer and Goschnick 2008). The fabric samples were exposed to the odour sources, then placed in a glass cylinder under a constant flow of a carrier gas (Figure 3.4). The gas flow carrying the odour sample was then conveyed to the electronic nose (KAMINA in Figure 3.4), and the resulting signal was processed for pattern recognition and identification of the compounds using the dedicated operating computer (Haeringer and Goschnick 2008). The system successfully distinguished between contaminated and noncontaminated fabric samples. In addition, it was able to distinguish and quantify the levels of the various odours (Haeringer and Goschnick 2008).

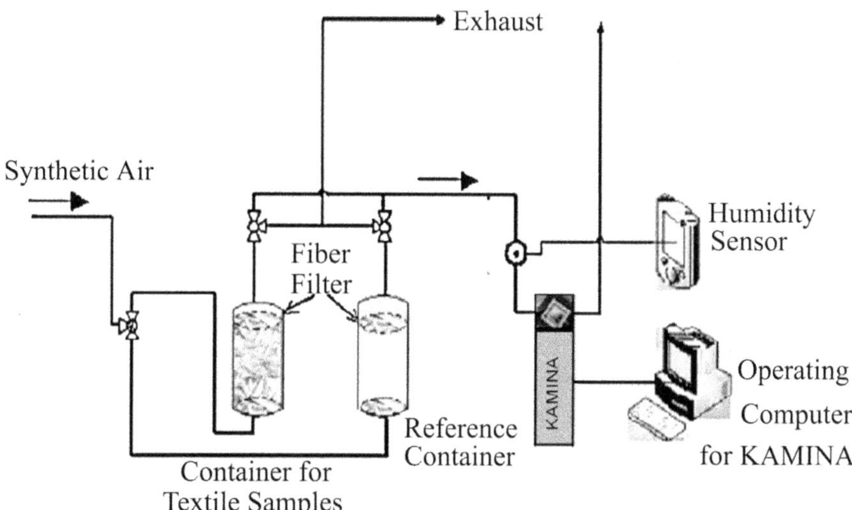

FIGURE 3.4 Schematic diagram of the odor detection system using an electronic nose (KAMINA). (Reprinted with permission from Haeringer and Goschnick (2008). Copyright (2008) Elsevier.)

Eza et al. (2012) used both olfactometry according to ISO 8589 (ISO 2007) and an electronic nose based on an array of metal oxide sensors to test the effectiveness of different odour-control fabric treatments with activated carbon against onion odour used as a proxy for body odour (Amato 2009; Eza et al. 2012).

More recently, Shakoorjavan et al. (2016) used a single metal oxide sensor but with temperature modulation of the semiconductor surface to generate a virtual sensor array for the measurement of fragrance on cotton and polyester fabric. The authors examined the correlation between the response of the virtual sensor array and the results of sensory analysis by human assessors. They found a highly significant correlation between the two methods and concluded that the single-sensor device modulated as a virtual sensor array provides an effective alternative to olfactometry (Shakoorjavan et al. 2016). Another recent study by Asadi Farad et al. (2018) used a similar electronic nose to examine the relationship between odour intensity and antibacterial properties of thyme essential oil applied to cotton fabric. The authors found a good correlation between the electronic nose measurement of odour intensity and the antibacterial effectiveness of the essential oil (AsadiFard et al. 2018).

Although most applications of electronic nose instruments were relevant to the food industry and to environmental odour nuisance monitoring, there appears to be emerging research interest in textile applications, as seen in the studies reviewed above. Recently, electronic nose odour evaluation was included in the ISO standard 17299-5 for the determination of the deodorant properties of textiles (ISO 2014). The standard specifies the use of arrays of metal-oxide semiconductor sensors as the EN operating mechanism for the detection of complex odours in textiles (ISO 2014).

ISO 17299-5 is not the only standard focusing on electronic noses. Notable standardization efforts have been undertaken to ensure the quality of electronic noses in odour monitoring. In 2012, the national standardization body for the Netherlands (NEN) published NTA 9055 (Netherlands Standardization Institute 2012) to provide specific guidelines for using electronic noses to monitor the composition of the ambient air. More recently, the Association of German Engineers (VDI) published VDI/VDE 3518-3 (VDI Standards 2018) with the goal of creating a set of guidelines for measurements made by electronic noses. Similarly, the Italian standardization body (UNI) created UNI1605848 (UNI, 2019) in response to the growing use of electronic noses in Italy. The Italian standard provides guidelines for calibrating and verifying electronic noses results.

Aside from initiatives by individual countries, the European Committee on Standardization established a working group (CEN TC/264 WG41) with a panel of experts from multiple countries within Europe, tasked with developing metrics and guidelines for use of electronic noses and other sensors in characterizing odour as perceived by humans (Cipriano and Capelli 2019).

3.3.2.3 Indirect Methods Targeting Odour-Causing Compounds

A third category of indirect instrumental methods to measure odour in textiles is based on detecting and characterizing some of the primary chemical compounds at the origin of the odour. Some of those compounds include sebaceous gland secretions, i.e., sebum, and apocrine gland secretions. As mentioned, squalene, oleic acid and triolein are abundant contaminants in worn clothing soils (Bowers and Chantrey 1969; Chi and Obendorf 1998; Chung and Seok 2012; Picardo et al. 2009; Stefaniak et al. 2010) and serve as a nourishing medium for microorganisms causing malodour in clothing (Chung and Seok 2012; McQueen et al. 2007, 2007, 2013). Consequently, some of those compounds have been targeted for specific reactivity and used as potential indirect detection proxies for odours in textiles.

Studies aiming at quantifying sebum are relatively abundant in medical research focusing on dermatology and cosmetics. For instance, Robosky et al. (2008) reported a quantitative method for sebum analysis using nuclear magnetic resonance (NMR). More recently, Ashraf et al. (2011) used infrared and visible absorbance spectroscopy to quantify sebum collected from the skin using Sebutape® patches (Ashraf et al. 2011).

In the textile field, Obendorf et al. used electron microscopy (Obendorf and Klemash 1982) as well as ^{14}C radiotracer analysis (Obendorf et al. 1983) to detect the presence of residual triolein in various fabrics.

One particular constituent of human sebum, squalene, has recently been the focus of significant research because of its reactivity with ozone (Morrison et al. 2021; Wisthaler and Weschler 2010). Ozone is an effective oxidant with unsaturated organic compounds and has been successfully used to control malodour in different industries (Khadre et al. 2001; Omer and Walker 2011; Pandrangi and Morrison 2008; Rice et al. 1981; Wells et al. 2008). Therefore, the kinetics of ozone reaction with odorous compounds, including human skin secretions, was used to detect body odour and personal care products (Corsi et al. 2007; Krifa et al. 2019), and to examine the variability of reaction products from the human skin (Morrison

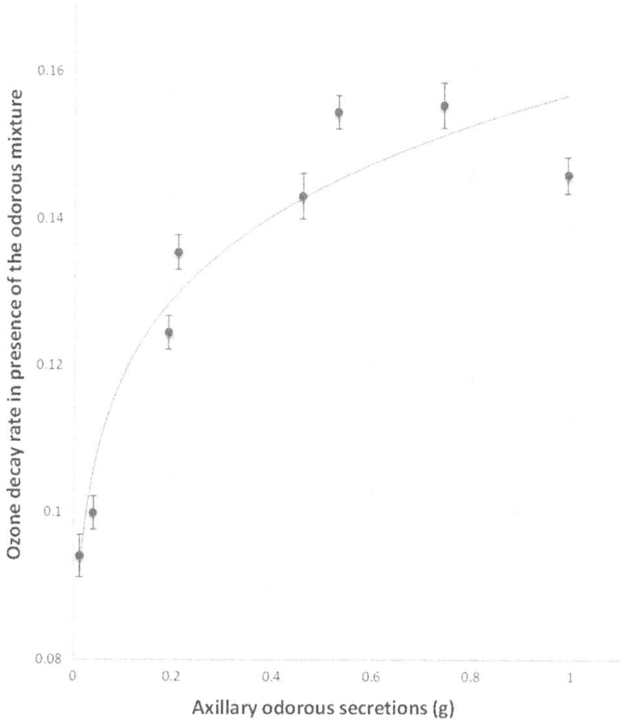

FIGURE 3.5 Relationship between ozone decay rate and the axillary odorous secretion amount on the fabric samples. (From Krifa et al. 2019.)

et al. 2021). Corsi et al. (2007) investigated ozone reactions with terpenes and terpenoids emitted from personal care products, such as cosmetics and fragrances. More recently, Krifa et al. (2018, 2019) conducted ozone decay experiments in the presence of fabric samples soiled with axillary odour-causing secretions collected through wear trials. The results showed a significant nonlinear relationship between axillary perspiration amounts and ozone decay rates in the presence of the contaminated fabric (Krifa et al. 2019) (Figure 3.5). Using this relationship, it appeared feasible to detect and quantify the presence of odorous soils on garments using the variation of ozone decay rates (Krifa 2018; Krifa et al. 2019).

One important consideration in research aimed at characterizing odour in clothing is the sampling approach, i.e., the method adopted to transfer the odour from its human body source to the fabric medium of interest. Most research studies reviewed above rely on human wear trials to transfer the odorous mixtures from the body to the fabric sample. For instance, in some studies, sample fabric swatches were sewn to the axillary underarms of T-shirts worn by a human subject during physical exercise sessions of varied intensity (Abdul-Bari et al. 2018; Krifa et al. 2019; McQueen et al. 2013).

Odour sampling done through contact and noncontact sampling methods was evaluated by Prada et al. (2011). For forensic evidence purposes, human body odours from six volunteers were sampled by contact sampling methods with the help of absorbent textiles and noncontact sampling methods through air flow systems while the odour components were transferred to a Scent Transfer Unit-100 for odour collection (Prada et al. 2011).

Just as the selection of sensory panel assessors is the subject of several standards and guidelines (ASTM Committee E-18 on Sensory Evaluation of Materials and Products 1981; ASTM 2014; AFNOR 1996; Bratcher 2013), so is the selection of odour-producing individuals for bodily odours such as axillary emanations (ASTM 2014). Wear trials aimed at collecting odorous samples are typically preceded by a screening phase to ensure that the participants generate sufficient odour intensity (Abdul-Bari et al. 2018).

3.4 CONCLUSION

Smell is one of the most complex human senses as it depends on unique interactions between the olfactory system and diverse chemical substances. The olfactory system works as a defence mechanism for the human body against malodours and irritants. Substances transferred to clothing through direct contact with the body, including sebum, eccrine, and apocrine gland secretions, constitute a prominent source of malodour. The primary mechanism of odour formation due to those substances is related to the bacterial activity they enable. Odour detection and characterization is achieved through sensory analysis using human assessors, or through instrumental and physicochemical methods based on a variety of principles. Sensory analysis or olfactometry measures both quantitative and qualitative parameters, allowing the characterization of odour concentration, intensity, and descriptive character. There have been many standardization efforts relevant to odour measurement by major organizations. Due to the complex and subjective nature of the sense of smell, sensory methods to measure odour have been extensively specified through a multitude of standards by major organizations in many countries. Part of the standard procedures specified in many of those guidelines pertain to the screening and training of panelists who serve as assessors in the sensory analysis, in addition to general considerations related to the test methods and reference materials. One consideration that is specific to odour measurement in textiles is related to the selection of odour-producing individuals who volunteer in wear trials (ASTM 2014).

In addition to sensory analysis, a range of instrumental methods have been documented in research about odour detection. GC-MS is the most common among those methods. GC-MS allows identifying individual compounds in odorous mixtures. Some of the shortcomings of GC-MS include the lack of information on the odour's impact or olfactory perception. A development that proved effective in addressing this shortcoming is GC-O.

Electronic noses constitute another class of instruments developed to measure odour. They consist of arrays of sensors with specificity to certain odorous compounds and have been tested for several decades in many industries, including environmental pollution monitoring, pharmaceutical and medical diagnostics, and the food industry. Recent attempts made to use electronic noses to measure odour in textiles have been reported with success. The approach has been included in ISO

17299-5 for the determination of the deodorant properties of textiles (ISO 2014). Electronic noses have also been the subject of extensive standardization efforts in ambient odour and air quality monitoring.

Recent research has attempted to indirectly assess odour based on the chemical reactivity of the bodily compounds that cause it. For instance, sebum and its constituents (e.g., squalene) have been detected and quantified based on ozone decay rates in the presence of fabric samples containing varied amounts of axillary secretions.

REFERENCES

Abdul-Bari, M.M., McQueen, R.H., Nguyen, H., Wismer, W.V., Paulina de la Mata, A., and Harynuk, J.J. 2018. Synthetic clothing and the problem with odor: Comparison of nylon and polyester fabrics. *Clothing and Textiles Research Journal*, 36(4): 251–266. 10.1177/0887302x18772099.

Abdul Bari, M. M., McQueen, R.H., Paulina de la Mata, A., Batcheller, J.C., and Harynuk, J.J. 2020. Retention and release of odorants in cotton and polyester fabrics following multiple soil/wash procedures. *Textile Research Journal*, 0(0): 0040517520914411. 10.1177/0040517520914411.

AFNOR. 1996. *Qualité de l'air, Mesurage olfactométriques, Mesurage de l'odeur d'un effluent gazeux, Méthodes supraliminaires*. NF X 43-103. French Standard AFNOR.

Amato, I. 2009. You smell. *Chemical & Engineering News*, 87(41): 50–54.

AsadiFard, P., Shakoorjavan, S., and Akbari, S. 2018. The relationship between odour intensity and antibacterial durability of encapsulated thyme essential oil by PPI dendrimer on cotton fabrics. *The Journal of The Textile Institute*, 109(6): 832–841. 10.1080/00405000.2017.1376820.

Ashraf, Z., Pasha, U., Greenstone, V., Akbar, J., Apenbrinck, E., Foulks, G.N., and Borchman, D. 2011. Quantification of human sebum on skin and human meibum on the eye lid margin using Sebutape®, spectroscopy and chemical analysis. *Current Eye Research*, 36(6): 553–562. 10.3109/02713683.2011.574331.

ASTM. 1992. *Atlas of Odor Character Profiles*. DS61-EB. American Society for Testing and Materials.

ASTM. 2005. *D2960 - 05 Standard Test Method of Controlled Laundering Test Using Naturally Soiled Fabrics and Household Appliances*. ASTM International.

ASTM. 2014. *E1207 – 14 – Standard Guide for Sensory Evaluation of Axillary Deodorancy*. American Society for Testing and Materials.

ASTM. 2018. *Standard Practice for Referencing Suprathreshold Odor Intensity*. E544-18. American Society for Testing and Materials.

ASTM. 2019. *Standard Practice for Determination of Odor and Taste Thresholds By a Forced-Choice Ascending Concentration Series Method of Limits*. E679-19. ASTM International.

ASTM Committee E-18 on Sensory Evaluation of Materials and Products. 1981. *Guidelines for the Selection and Training of Sensory Panel Members*, 1–1. ASTM International.

Baldwin, E.A., Bai, J., Plotto, A., and Dea S. 2011. Electronic noses and tongues: Applications for the food and pharmaceutical industries. *Sensors*, 11(5): 4744–4766.

Bartlett, J.W., and Gardner, N.P. 2000. Electronic noses. Principles and applications. *Measurement Science and Technology*, 11(7): 1087–1087. 10.1088/0957-0233/11/7/702.

Bockreis, A., and Jager, J. 1999. Odour monitoring by the combination of sensors and neural networks. *Environmental Modelling & Software*, 14(5): 421–426. 10.1016/s1364-8152(98)00105-4.

Bowers, C.A., and Chantrey, G. 1969. Factors controlling the soiling of white polyester cotton fabrics part I: Laboratory studies. *Textile Research Journal,* 39(1): 1–11. 10.1177/004051756903900101.

Bratcher, C.L. 2013. Trained sensory panels. In C.R. Kerth (ed.). *The Science of Meat Quality* (pp. 207–213), https://doi.org/10.1002/9781118530726.ch11

Brattoli, M., Cisternino, E., Dambruoso, P.R., De Gennaro, G., Giungato, P., Mazzone, A., Palmisani, J., and Tutino, M. 2013. Gas chromatography analysis with olfactometric detection (GC-O) as a useful methodology for chemical characterization of odorous compounds. *Sensors*, 13(12): 16759–16800.

Brattoli, M., De Gennaro, G., De Pinto, V., Loiotile, A.D., Lovascio, S., and Penza, M. 2011. Odour detection methods: Olfactometry and chemical sensors. *Sensors,* 11(5): 5290–5322.

Bovell, D.L., Corbett, A.D., Holmes, S., MacDonald, A., and Harker, M. 2007. The absence of apoeccrine glands in the human axilla has disease pathogenetic implications, including axillary hyperhidrosis. *British Journal of Dermatology*, 156(6): 1278–1286. https://doi.org/10.1111/j.1365-2133.2007.07917.x.

Bovell, D.L., MacDonald, A., Meyer, B.A., Corbett, A.D., MacLaren, W.M., Holmes, S.L., and Harker, M. 2011. The secretory clear cell of the eccrine sweat gland as the probable source of excess sweat production in hyperhidrosis. *Experimental Dermatology*, 20(12): 1017–1020. https://doi.org/10.1111/j.1600-0625.2011.01361.x.

Callewaert, C., De Maeseneire, E., Kerckhof, F.-M., Verliefde, A., Van de Wiele, T., and Boon, N. 2014. Microbial odor profile of polyester and cotton clothes after a fitness session. *Applied and Environmental Microbiology*, 80(21): 6611. 10.1128/AEM.01422-14.

Capelli, L., Sironi, S., and Del Rosso, R. 2013. Odor sampling: Techniques and strategies for the estimation of odor emission rates from different source types. *Sensors,* 13(1): 938–955.

CEN. 2007. *Air Quality—Determination of Odour Concentration by Dynamic Olfactometry.* EN 13725:2003. European Committee for Standardization.

Chang, H.C., Kish, L.B., King, M.D., and Kwan, C. 2009. Fluctuation-enhanced sensing of bacterium odors. *Sensors and Actuators B: Chemical,* 142(2): 429–434. 10.1016/j.snb.2009.04.005.

Chen, J.C., and Jain, S.A. 2010. Chapter 104 - principles of skin grafts. In J. Weinzweig (ed.), *Plastic Surgery Secrets Plus, 2nd ed.* (pp. 677–683). Mosby.

Chen, Y.-L., Kuan, W.-H., and Liu, C.-L. 2020. Comparative study of the composition of sweat from eccrine and apocrine sweat glands during exercise and in heat. *International Journal of Environmental Research and Public Health,* 17(10): 3377.

Chi, Y.-S., and Obendorf, S.K. 1998. Aging of oily soils on textile materials: A literature review. *Journal of Surfactants and Detergents,* 1(3): 407–418. 10.1007/s11743-998-0044-0.

Chung, H., and Seok, H.J. 2012. Populations of malodor-forming bacteria and identification of volatile components in triolein-soiled cotton fabric. *Fibers and Polymers,* 13(6): 740–747. 10.1007/s12221-012-0740-5.

Cipriano, D., and Capelli, L. 2019. Evolution of electronic noses from research objects to engineered environmental odour monitoring systems: A review of standardization approaches. *Biosensors* 9(2): 75.

Corsi, R.L., Siegel, J., Karamalegos, A., Simon, H., and Morrison, G.C. 2007. Personal reactive clouds: Introducing the concept of near-head chemistry. *Atmospheric Environment,* 41(15): 3161–3165. 10.1016/j.atmosenv.2006.07.054.

Dalton, P., Caraway, E.A., Gibb, H., and Fulcher, K. 2011. A multi-year field olfactometry study near a concentrated animal feeding operation. *Journal of the Air & Waste Management Association,* 61(12): 1398–1408. 10.1080/10473289.2011.624256.

Debonneville, C., Orsier, B., Flament, I., and Chaintreau, A. 2002. Improved hardware and software for quick gas chromatography–olfactometry using CHARM and GC-"SNIF" analysis. *Analytical Chemistry,* 74 (10): 2345–2351. 10.1021/ac011064a.

Deisingh, A.K., Stone, D.C., and Thompson, M. 2004. Applications of electronic noses and tongues in food analysis. *International Journal of Food Science & Technology,* 39(6): 587–604. 10.1111/j.1365-2621.2004.00821.x.

Di Natale, C., Macagnano, A., Davide, F., D'Amico, A., Paolesse, R., Boschi, T., Faccio, M., and Ferri, G. 1997. An electronic nose for food analysis. *Sensors and Actuators B: Chemical,* 44(1): 521–526. 10.1016/S0925-4005(97)00175-5.

Di Natale, C., Macagnano, A., Paolesse, R., Tarizzo, E., Mantini, A., and D'Amico, A. 2000. Human skin odor analysis by means of an electronic nose. *Sensors and Actuators B: Chemical,* 65 (1–3): 216–219. 10.1016/s0925-4005(99)00313-5.

Domingues, R.R., Arias, R., Vera, L., Villatoro, C., and Ceballos, V. 2016. Sensory analysis of the foot deodorisation efficiency of a commercial product. *Chemical Engineering Transactions*, 54: 325–330. 10.3303/CET1654055.

Dravnieks, A., Krotoszynski, B.K., Lieb, W.E., and Jungermann, E. 1968. Influence of an antibacterial soap on various effluents from Axillae. *Journal of the Society of Cosmetic Chemists,* 19: 611–626.

Eza, T.S.M., Ahmad, W.Y.W., Omar, K., and Ahmad, M.N. 2012. The activated carbon anti-odour treated fabrics by coating and pigment printing. 2012 IEEE Symposium on Business, Engineering and Industrial Applications, 23–26 Sept. 2012.

Farraia, M.V., Rufo, J.C., Paciência, I., Mendes, F., Delgado, L., and Moreira, A. 2019. The electronic nose technology in clinical diagnosis: A systematic review. *Porto Biomedical Journal,* 4(4): e42. 10.1097/j.pbj.0000000000000042.

Fisher, R.M., Barczak, R.J., "Mel" Suffet, I.H., Hayes, J.E., and Stuetz, R.M. 2018. Framework for the use of odour wheels to manage odours throughout wastewater biosolids processing. *Science of The Total Environment,* 634: 214–223. 10.1016/j.scitotenv.2018.03.352.

Fowler, J.F.J., Fowler, L.M., and Lorenz, D. 2019. Effects of merino wool on atopic dermatitis using clinical, quality of life, and physiological outcome measures. *Dermatitis*, 30(3): 198–206. https://doi.org/10.1097/der.0000000000000449.

Gao, Y., and Cranston, R. 2008. Recent advances in antimicrobial treatments of textiles. *Textile Research Journal,* 78(1): 60–72. 10.1177/0040517507082332.

Gardner, J.W., Shin, H.W., and Hines, E.L. 2000. An electronic nose system to diagnose illness. *Sensors and Actuators B: Chemical,* 70(1): 19–24. 10.1016/S0925-4005(00)00548-7.

Gocek, İ., and Duru, S.C. 2019. Investigating the effects of wicking and antibacterial finishing treatments on some comfort characteristics of Meryl skinlife for seamless activewear/sportswear. *Journal of Engineered Fibers and Fabrics* 14: 1558925019852790. 10.1177/1558925019852790.

Gostelow, P., Parsons, S.A., and Stuetz, R.M. 2001. Odour measurements for sewage treatment works. *Water Research*, 35(3): 579–597. 10.1016/S0043-1354(00)00313-4.

Gouronnec, A.M., and Tomasso, V. 2000. Measurement of odours by sensory analysis or "olfactometry". *Analusis*, 28(3): 188–199. 10.1051/analusis:2000280188.

Guo, Y., Li, Y., Tokura, H., Wong, T., Chung, J., Wong, A.S.W., Gohel, M.D., and Leung, P.H.M. 2008. Impact of fabric moisture transport properties on physiological responses when wearing protective clothing. *Textile Research Journal*, 78(12): 1057–1069. 10.1177/0040517508090496.

Guruprasad, R., Vivekanandan, V., and Chattopadhyay, S. 2015. The use of cotton as a sportswear material: A critical analysis. *Cotton Research Journal*, 7(1): 61–64.

Gygax, H., and Koch, H. 2001. The measurement of odours. *CHIMIA International Journal for Chemistry*, 55(5): 401–405.

Haeringer, D., and Goschnick, J. 2008. Characterization of smelling contaminations on textiles using a gradient microarray as an electronic nose. *Sensors and Actuators B: Chemical*, 132(2): 644–649. 10.1016/j.snb.2008.01.068.

Hasegawa, Y., Yabuki, M., and Matsukane, M. 2004. Identification of new odoriferous compounds in human axillary sweat. *Chemistry & Biodiversity*, 1(12): 2042–2050. 10.1002/cbdv.200490157.

Hobbs, P.J., Misselbrook, T.H., and Pain, B.F. 1995. Assessment of odours from livestock wastes by a photoionization detector, an electronic nose, olfactometry and gas chromatography-mass spectrometry. *Journal of Agricultural Engineering Research*, 60(2): 137–144. 10.1006/jaer.1995.1007.

Hudon, G., Guy, C., and Hermia, J. 2000. Measurement of odor intensity by an electronic nose. *Journal of the Air & Waste Management Association (1995)*, 50(10): 1750–1758.

Huisman, M.A., and Morris, M.A. 1972. Oily soil retention by abraded durable press fabrics. *Textile Research Journal*, 42(1): 30–32. 10.1177/004051757204200106.

ISO. 2007. *Sensory analysis — General guidelines for the design of test rooms*. 8589. International Organization for Standardization.

ISO. 2012. *Sensory analysis — General guidelines for the selection, training and monitoring of selected assessors and expert sensory assessors*. 8586. International Organization for Standardization.

ISO. 2014. *Textiles — Determination of deodorant property — Part 5: Metal-oxide semi-conductor sensor method*. 17299-5. International Organization for Standardization.

ISO. 2018. *Sensory Analysis - Methodology - General Guidance for Measuring Odour, Flavour and Taste Detection Thresholds by a Three Alternative Forced Choice (3AFC) Procedure*. 13301. ASTM International.

Khadre, M.A., Yousef, A.E., and Kim, J.G.. 2001. Microbiological aspects of ozone applications in food: A review. *Journal of Food Science,* 66 (9): 1242–1252. 10.1111/j.1365-2621.2001.tb15196.x.

Kim, K.-H., and S.-Y. Park. 2008. A comparative analysis of malodor samples between direct (olfactometry) and indirect (instrumental) methods. *Atmospheric Environment,* 42(20): 5061–5070. 10.1016/j.atmosenv.2008.02.017.

Klepp, I.G., M. Buck, K. Laitala, and M. Kjeldsberg. 2016. What's the problem? Odor-control and the smell of sweat in sportswear. *Fashion Practice,* 8(2): 296–317. 10.1080/17569370.2016.1215117.

Krifa, M. 2018. Objective quantitative assessment of worn clothing soils. International Textile and Apparel Association (ITAA) Annual Conference, Cleveland, OH, November, 6–9

Krifa, M., and Stewart Stevens, S. 2016. Cotton utilization in conventional and non-conventional textiles—a statistical review. *Agricultural Sciences,* 7: 747–758. 10.4236/as.2016.710069.

Krifa, M., and Prichard, C. 2020. Nanotechnology in textile and apparel research – an overview of technologies and processes. *The Journal of The Textile Institute,* 111(12): 1778–1793. 10.1080/00405000.2020.1721696.

Krifa, M., Rajaganesh, S., and Fahy, W. 2019. Perspectives on textile cleanliness – detecting human sebum residues on worn clothing. *Textile Research Journal.* 10.1177/0040517519855323.

Labows, J.N., and Kligman, A.M. 1982. Perspectives on axillary odor. *Journal of the Society of Cosmetic Chemists,* 34: 193–202.

Laitala, K., Boks, C., and Klepp, I.G. 2011. Potential for environmental improvements in laundering. *International Journal of Consumer Studies,* 35(2): 254–264. 10.1111/j.1470-6431.2010.00968.x.

Laughlin, J., and Gold, R.E. 1990. Methyl parathion residue removal from protective apparel fabric soiled with synthetic sebum or vegetable oil. *Archives of Environmental Contamination and Toxicology,* 19(2): 205–213. 10.1007/bf01056088.

Li, S. 2014. Recent developments in human odor detection technologies. *J Forensic Sci Criminol,* 1(1): 1–12. 10.15744/2348-9804.1.S104.

Littarru, P. 2007. Environmental odours assessment from waste treatment plants: Dynamic olfactometry in combination with sensorial analysers "electronic noses". *Waste Management,* 27(2): 302–309. 10.1016/j.wasman.2006.03.011.

Loutfi, A., Coradeschi, S., Mani, G.K., Shankar, P., and Rayappan, J.B.B. 2015. Electronic noses for food quality: A review. *Journal of Food Engineering,* 144: 103–111. 10.1016/j.jfoodeng.2014.07.019.

Mackay, D.A.M., Lang, D.A., and Berdick, M. 1961. Objective measurement of odor. ionization detection of food volatiles. *Analytical Chemistry,* 33(10): 1369–1374. 10.1021/ac60178a027.

Macleod, W.D., Green, D.E., and Seet, E. 1976. Automated analysis of phencyclidine in urine by probability based matching GC/MS. *Clinical Toxicology,* 9(4): 561–572. 10.3109/15563657608988158.

McGinley, C.M., McGinley, M.A., and McGinley, D.L. 2000. "Odor Basics", understanding and using odor testing. The 22nd Annual Hawaii Water Environment Association Conference, Honolulu, Hawaii, 6–7 June.

McGinley, M.A., and McGinley, C.M. 2017. Methods for odor evaluation of textiles and other materials. American Association of Textile Chemists and Colorists International Conference Wilmington, North Carolina, USA 28–30 March 2017.

McQueen, R.H., Keelan, M., Xu, Y., and Mah, T. 2013. In vivo assessment of odour retention in an antimicrobial silver chloride-treated polyester textile. *Journal of the Textile Institute,* 104(1): 108–117. 10.1080/00405000.2012.697623.

McQueen, R.H., Laing R.M., Brooks H.J.L., and Niven B.E.. 2007. Odor intensity in apparel fabrics and the link with bacterial populations. *Textile Research Journal,* 77(7): 449–456. 10.1177/0040517507074816.

McQueen, R.H., Laing, R.M., Delahunty, C.M., Brooks, H.J.L., and Niven, B.E. 2008. Retention of axillary odour on apparel fabrics. *Journal of the Textile Institute,* 99(6): 515–523. 10.1080/00405000701659774.

McQueen, R.H., Laing, R.M., Wilson, C.A., Niven, B.E., and Delahunty, C.M. 2007. Odor retention on apparel fabrics: Development of test methods for sensory detection. *Textile Research Journal,* 77(9): 645–652. 10.1177/0040517507078792.

Mielle, P. 1996. 'Electronic noses': Towards the objective instrumental characterization of food aroma. *Trends in Food Science & Technology,* 7(12): 432–438. 10.1016/S0924-2244(96)10045-5.

Miracle, G.S., Randall, S.L., Liu, Z., Brogden, D.W., Ketcha, M.M., Good, D.A., Johnson, M.B., Stenger, P.C., Hertz, P.R., and Meli, F. 2020. Copper chelants and antioxidants in laundry detergent formulations reduce formation of malodor molecules on fabrics. *Journal of Surfactants and Detergents,* 23(6): 1125–1134. 10.1002/jsde.12467.

Mondello, L., Shellie, R. , Casilli, A., Tranchida, P.Q., Marriott, P., and Dugo, G. 2004. Ultra-fast essential oil characterization by capillary GC on a 50 μm ID column. *Journal of Separation Science,* 27 (9): 699–702. 10.1002/jssc.200301602.

Morrison, G.C., Eftekhari, A., Majluf, F., and Krechmer, J.E. 2021. Yields and variability of ozone reaction products from human skin. *Environmental Science & Technology,* 55(1): 179–187. 10.1021/acs.est.0c05262.

Morrissey, M.P., and Rossi, R.M. 2013. Clothing systems for outdoor activities. *Textile Progress,* 45(2-3): 145–181. 10.1080/00405167.2013.845540.

Munk, S., Johansen, C., Stahnke, L.H., and Adler-Nissen, J. 2001. Microbial survival and odor in laundry. *Journal of Surfactants and Detergents,* 4(4): 385–394. 10.1007/s11743-001-0192-2.

Munk, S., Münch, P., Stahnke, L., Adler-Nissen, J., and Schieberle, P. 2000. Primary odorants of laundry soiled with sweat/sebum: Influence of lipase on the odor profile. *Journal of Surfactants and Detergents,* 3 (4): 505–515. 10.1007/s11743-000-0150-z.

Natsch, A., Derrer, S., Flachsmann, F., and Schmid, J. 2006. A broad diversity of volatile carboxylic acids, released by a bacterial aminoacylase from axilla secretions, as candidate molecules for the determination of human-body odor type. *Chemistry & Biodiversity,* 3 (1): 1–20. 10.1002/cbdv.200690015.

Netherlands Standardization Institute. 2012. *Air quality - Electronic air monitoring - Odour (nuisance) and safety*. NEN NTA 9055. Netherlands Standardization Institute (NEN).

Nicolas, J., Romain, A.C., Wiertz, V., Maternova, J., and André, P.h. 2000. Using the classification model of an electronic nose to assign unknown malodours to environmental sources and to monitor them continuously. *Sensors and Actuators B: Chemical,* 69(3): 366–371. 10.1016/S0925-4005(00)00487-1.

Nie, J.-Y., Li, R., Jiang, Z.-T., Wang, Y., Tan, J., Tang, S.-H., and Zhang, Y. 2020. Antioxidant activity screening and chemical constituents of the essential oil from rosemary by ultra-fast GC electronic nose coupled with chemical methodology. *Journal of the Science of Food and Agriculture,* 100(8): 3481–3487. 10.1002/jsfa.10388.

Obendorf, S.K., Kim, J., and Koniz, R.F. 2007. Measurement of odor development due to bacterial action on antimicrobial polyester fabrics. *AATCC Review,* 7(7): 35–40.

Obendorf, S.K., and Klemash, N.A. 1982. Electron microscopical analysis of oily soil penetration into cotton and polyester/cotton fabrics. *Textile Research Journal*, 52(7): 434–442. 10.1177/004051758205200702.

Obendorf, S.K., and Webb, J.J. 1987. Detergency study: Distribution of natural soils on shirt collars. *Textile Research Journal*, 57(10): 557–563. https://doi.org/10.1177/004051758705701001.

Obendorf, S.K., Namasté, Y.M.N., and Durnam, D.J. 1983. A microscopical study of residual oily soil distribution on fabrics of varying fiber content. *Textile Research Journal,* 53(6): 375–383. 10.1177/004051758305300609.

Omer, A.R., and Walker, P.M. 2011. Treatment of swine slurry by an ozone treatment system to reduce odor. *Journal of Environmental Protection*, 2(7): 867–872. doi: https://doi.org/10.4236/jep.2011.27098

Pandrangi, L.S., and Morrison, G.C. 2008. Ozone interactions with human hair: Ozone uptake rates and product formation. *Atmospheric Environment,* 42(20): 5079–5089. 10.1016/j.atmosenv.2008.02.009.

Peres, F., Jeleń, H.H., Majcher, M.M., Arraias, M., Martins, L.L., and Ferreira-Dias, S. 2013. Characterization of aroma compounds in Portuguese extra virgin olive oils from Galega Vulgar and Cobrançosa cultivars using GC–O and GC×GC–ToFMS. *Food Research International,* 54(2): 1979–1986. 10.1016/j.foodres.2013.06.015.

Persaud, K., and Dodd, G. 1982. Analysis of discrimination mechanisms in the mammalian olfactory system using a model nose. *Nature,* 299(5881): 352–355. 10.1038/299352a0.

Picardo, M., Ottaviani, M., Camera, E., and Mastrofrancesco, A. 2009. Sebaceous gland lipids. *Dermato-endocrinology,* 1(2): 68–71.

Prada, P.A., Curran, A.M., and Furton, K.G.. 2011. The evaluation of human hand odor volatiles on various textiles: A comparison between contact and noncontact sampling methods*,†. *Journal of Forensic Sciences,* 56(4): 866–881. 10.1111/j.1556-4029.2011.01762.x.

Prada, P.A., Curran, A.M., and Furton, K.G. 2014. Characteristic human scent compounds trapped on natural and synthetic fabrics as analyzed by SPME-GC/MS. *Journal of Forensic Science & Criminology,* 1(1): 1–10. 10.15744/2348-9804.1.S101.

Reinbach, H.C., Allesen-Holm, B., Kristoffersson, L., and Bredie, W.L.P. 2011. Development of a sensory test method for odor measurement in a package headspace. *Journal of Sensory Studies,* 26(2): 118–127. 10.1111/j.1745-459X.2011.00328.x.

Rice, R.G., Michael Robson, C., Wade Miller, G., and Hill, A.G. 1981. Uses of ozone in drinking water treatment. *Journal (American Water Works Association),* 73(1): 44–57. 10.2307/41270552.

Robosky, L.C., Wade, K., Woolson, D., Baker, J.D., Manning, M.L., Gage, D.A., and Reily, M.D. 2008. Quantitative evaluation of sebum lipid components with nuclear magnetic resonance. *Journal of Lipid Research,* 49(3): 686–692. 10.1194/jlr.D700035-JLR200.

Rudnicka, J., Mochalski, P., Agapiou, A., Statheropoulos, M., Amann, A., and Buszewski, B. 2010. Application of ion mobility spectrometry for the detection of human urine. *Analytical and Bioanalytical Chemistry*, 398(5): 2031–2038. 10.1007/s00216-010-4147-0.

Ruzsanyi, V., Mochalski, P., Schmid, A., Wiesenhofer, H., Klieber, M., Hinterhuber, H., and Amann, A. 2012. Ion mobility spectrometry for detection of skin volatiles. *Journal of Chromatography B*, 911: 84–92. 10.1016/j.jchromb.2012.10.028.

Sanders, D., Grunden, A., and Dunn, R.R. 2021. A review of clothing microbiology: the history of clothing and the role of microbes in textiles. *Biology Letters*, 17(1): 20200700. 10.1098/rsbl.2020.0700.

Santos, J.P., Lozano, J., Aleixandre, M., Arroyo, T., Cabellos, J.M., Gil, M., and del Carmen Horrillo, M. 2010. Threshold detection of aromatic compounds in wine with an electronic nose and a human sensory panel. *Talanta* 80(5): 1899–1906. 10.1016/j.talanta.2009.10.041.

Sato, K., Kang, W.H., Saga, K., and Sato, K.T. 1989. Biology of sweat glands and their disorders. I. Normal sweat gland function. *Journal of the American Academy of Dermatology,* 20(4): 537–563. 10.1016/s0190-9622(89)70063-3.

Schilling, B., Kaiser, R., Natsch, A., and Gautschi, M. 2009. Investigation of odors in the fragrance industry. *Chemoecology,* 20(2): 135–147. 10.1007/s00049-009-0035-5.

Schmidt, K., and Podmore, I. 2015. Current challenges in volatile organic compounds analysis as potential biomarkers of cancer. *Journal of Biomarkers*, 2015: 981458. 10.1155/2015/981458.

Scott, S.M., James, D., and Ali, Z. 2006. Data analysis for electronic nose systems. *Microchimica Acta*, 156(3): 183–207. 10.1007/s00604-006-0623-9.

Shakoorjavan, S., Akbari, S., Haghighat Kish, M., and Akbari, M. 2016. Correlation of sensory analysis with a virtual sensor array data for odour diagnosis of fragrant fabrics. *Measurement,* 90: 396–403. 10.1016/j.measurement.2016.04.074.

Shove, E. 2003. Converging conventions of comfort, cleanliness and convenience. *Journal of Consumer Policy*, 26(4): 395–418. 10.1023/a:1026362829781.

Sironi, S., Capelli, L., Céntola, P., Del Rosso, R., and Il Grande, M. 2007. Continuous monitoring of odours from a composting plant using electronic noses. *Waste Management,* 27(3): 389–397. 10.1016/j.wasman.2006.01.029.

Standards Australia. 2001. *Stationary source emissions - Part 3: Determination of odour concentration by dynamic olfactometry*. AS/NZS 4323.3-2001. Standards Australia.

Stapleton, K., Hill, K., Day, K., Perry, J.D., and Dean, J.R. 2013. The potential impact of washing machines on laundry malodour generation. *Letters in Applied Microbiology,* 56(4): 299–306. 10.1111/lam.12050.

Stefaniak, A.B., Harvey, C.J., and Wertz, P.W. 2010. Formulation and stability of a novel artificial sebum under conditions of storage and use. *International Journal of Cosmetic Science,* 32(5): 347–355. 10.1111/j.1468-2494.2010.00561.x.

Stuetz, R.M., Fenner, R.A., and Engin, G. 1999. Assessment of odours from sewage treatment works by an electronic nose, H2S analysis and olfactometry. *Water Research,* 33(2): 453–461. 10.1016/S0043-1354(98)00246-2.

Suffet, I.H., Burlingame, G.A., Rosenfeld, P.E., and Bruchet, A. 2004. The value of an odor-quality-wheel classification scheme for wastewater treatment plants. *Water Science and Technology,* 50(4): 25–32. 10.2166/wst.2004.0211.

Takeuchi, K., Hasegawa, Y., Ishida, H., and Kashiwagi, M. 2012. Identification of novel malodour compounds in laundry. *Flavour and Fragrance Journal,* 27(1): 89–94. 10.1002/ffj.2088.

Terpstra, M.J. 1998. Domestic and institutional hygiene in relation to sustainability. Historical, social and environmental implications. *International Biodeterioration & Biodegradation,* 41(3–4): 169–175. 10.1016/s0964-8305(98)00017-1.

Terpstra, M.J. 2001. The correlation between sustainable development and home hygiene. *American Journal of Infection Control,* 29(4): 211–217. 10.1067/mic.2001.115684.

Terpstra, M.J. 2003. The validity of domestic hygiene indicators in the scope of social and technological changes. *International Biodeterioration & Biodegradation,* 51(4): 233–238. 10.1016/s0964-8305(03)00037-4.

Thaler, E.R., and Hanson, C.W. 2005. Medical applications of electronic nose technology. *Expert Review of Medical Devices,* 2(5): 559–566. 10.1586/17434440.2.5.559.

Thieme, D., Anielski, P., Grosse, J., Sachs, H., and Mueller, R.K. 2003. Identification of anabolic steroids in serum, urine, sweat and hair: Comparison of metabolic patterns. *AnalyticaChimica Acta,* 483(1): 299–306. 10.1016/S0003-2670(02)01604-5.

UNI. 2019. UNI1605848: Emissioni e qualità dell'aria - Determinazione degli odori tramite IOMS (Instrumental Odour Monitoring Systems). In *UNI Standards* (Vol. 1605848).

van Harreveld, A.P., Heeres, P., and Harssema, H. 1999. A review of 20 years of standardization of odor concentration measurement by dynamic olfactometry in Europe. *Journal of the Air & Waste Management Association,* 49(6): 705–715. 10.1080/10473289.1999.11499900.

VDI Standards. 2018. *Multigas Sensors - Odour-related Measurements With Electronic Noses and their Testing*. VDI/VDE 3518. VDI (Association of German Engineers).

Wang, J., Lu, X., Wang, J., and Wang, X. 2019. Quantitative and sensory evaluation of odor retention on polyester/wool blends. *Textile Research Journal,* 89(13): 2729–2738. 10.1177/0040517518801183.

Wells, J.R., Morrison, G.C., Coleman, B.K., Spicer, C., and Dean, S.W. 2008. Kinetics and reaction products of ozone and surface-bound squalene. *Journal of ASTM International,* 5(7). 10.1520/jai101629.

Wertz, P.W. 2009. Human synthetic sebum formulation and stability under conditions of use and storage. *International Journal of Cosmetic Science,* 31(1): 21–25.

Wilke, K., Martin, A., Terstegen, L., and Biel, S.S. 2007. A short history of sweat gland biology. *International Journal of Cosmetic Science,* 29(3): 169–179. 10.1111/j.1467-2494.2007.00387.x.

Wilke, K., Martin, A., Terstegen, L., and Biel, S.S. 2009. Neurobiology of skin appendages: Eccrine, apocrine, and apoeccrine sweat glands. In R.D. Granstein, and T.A. Luger (eds.), *Neuroimmunology of the Skin* (pp. 167–175). Springer Berlin Heidelberg.

Wisthaler, A., and Weschler, C.J. 2010. Reactions of ozone with human skin lipids: Sources of carbonyls, dicarbonyls, and hydroxycarbonyls in indoor air. *Proceedings of the National Academy of Sciences,* 107(15): 6568–6575. 10.1073/pnas.0904498106.

Yuwono, A.S., and Lammers, P.S. 2004. Odor pollution in the environment and the detection instrumentation. *Agricultural Engineering International: CIGR Journal*, VI, 1–33.

Zimniewska, M., and Krucinska, I. 2010. The effect of raw material composition of clothes on selected physiological parameters of human organism. *Journal of the Textile Institute,* 101(2): 154–164.

4 Axillary Odour Formation and Retention: The Role of Textiles Fibers and Fabric Structure

R. Rathinamoorthy
Department of Fashion Technology, PSG College of Technology, Coimbatore, India

G. Thilagavathi
Department of Textile Technology, PSG College of Technology, Coimbatore, India

CONTENTS

4.1 INTRODUCTION TO AXILLA ODOUR

Odour generation on the human body occurs mainly due to microbial biotransformation of sweat into odour-causing molecules. Numerous microbiological studies have been performed in the previous decades relating to the bacterial population on body parts, especially the axilla. The axilla region is one of the dominant odour-creating sites in the human body. The most common reason for axilla-odour generation is the apocrine sweat glands. Though segregated sweat appears to be odourless, skin-born bacteria develops specific odour through the generation of volatile organic compounds (VOCs). Different VOCs are derived from the different parts of the body based on the available sweat glands, namely eccrine, sebaceous, and apocrine. This develops different VOC profiles and different odours at different parts of the body (Labows et al. 1982). In later years, research results showed a significant contribution of textile materials to increasing body odour. Shelley et al. (1953) reported that textiles contribute to body odour by transferring body secretions, skin debris, and bacteria from the body to the fabric

DOI: 10.1201/9781003141426-4

surface. The textile material is often referred to as a secondary odour source by Dravnieks et al. (1968) as it constantly is in contact with the axilla and other parts of the skin. Basically, the textile fibers provide initial adherence to the bacteria based on the type of material and bacteria (Teufel et al. 2010). In addition to that, the textiles also provide a suitable microclimate to the bacteria to grow, with skin temperature and moisture developed due to sweat. The air-permeable characteristics of the textile is an additional feature that creates a microenvironment more suitable for bacterial growth. The body secretions and different finishes used in the textiles serve as a nutrient for bacterial growth (Gao and Cranston 2008). Though the body odour developed on the skin, the textile material worn next to the skin plays an important role in retaining or enhancing the odour. Hence, this chapter details the role of textile material on body odour formation and propagation. The first part of this chapter details the odour formation mechanism and the role of responsible microorganisms in the axilla region of the human body. The second part of this chapter reviews the impact of fiber types on odour formation in textiles, whereas the last part of this chapter details the effect of structure on odour formation.

4.2 BACTERIA AND ODOUR FORMATION

Axillary odour formation is mainly attributed to skin-resident microorganisms. These bacterial strains convert odourless secretion into odourous liquid (Shelley et al. 1953). Generally, gram-positive bacterial strains, such as *Staphylococcus, Micrococcus, Corynebacterium,* and *Propionibacterium,* are reported as skin-resident bacteria (Leyden et al. 1981). Researchers studied the relationship between bacterial populations and axilla body odour formation. The highest bacterial densities are most commonly found in the axilla region of the body. Based on a study conducted among 229 human subjects, the results were quantitatively correlated and reported no difference among the left and right sides of the axilla and gender. They also reported bacterial densities of 500,000/cm^2 and 1,000,000/cm^2, respectively, for the aerobic and anaerobic organisms (Leyden et al. 1981). *Carynebacteria* were found to develop pungent, apocrine odour, and the faint acid non-apocrine odour was mainly developed by *staphylococci* stains. Leyden et al. (1981) confirmed these findings through *In-vivo* experiments, where apocrine sweat samples were used to incubate the above-mentioned bacteria. Taylor et al. (2003) reported that the axilla was mainly dominated by *staphylococcus* species or aerobic *coryneforms* (predominantly *Corynebacterium* species), along with a few *propionibacterium* species. They compared the individual subject odour intensities with the microbial count in their axilla. The results were highly significant with respect to *corynebacterium* species and *micrococcus* species. However, due to the lower density of the *micrococcus* species in axilla, their contribution was not significant. They also reported no association of *staphylococci* and *propionibacteria* with odour formation.

The structure of the odour-causing precursor for body odour was first reported by Natsch et al. (2003). They analysed acid generation from the axilla and the reaction mechanism with a glutamine residue in fresh axilla secretions. They reported that the acids found in the human axilla mainly belonged to the *Corynebacteria* strains. Similar research identified that *Propionibacteria* and *staphylococci* can ferment glycerol and lactic acid to the short-chain (C2–C3) VFAs, acetic, and propionic acid. *Staphylococci* stains were noted as the main microorganism for converting the branched aliphatic amino acids, such as leucine, to highly odourous short-chain (C4–C5) methyl-branched VFAs, such as isovaleric acid. These components were mainly associated with the acidic note of axilla malodour (James et al. 2004). The molecular basis of the human body was noted as the main reason for odour formation. Out of several researched bacterial strains, aerobic *Corynebacteria* was found to be the major source of strong axillar odour generation versus the other skin resident strains. Detailed analysis of the different methods of body odour formation was performed by Fredrich et al. (2012). They reported that steroid derivatives, short volatile branched-chain fatty acids, and sulphonyl alkanols are the major odour contributor in human sweat. Detailed representations of various odour compounds reported to form axillary sweat, along with their odourless precursor, are provided in Table 4.1 (Fredrich et al. 2012).

The development of short-chain fatty acids and medium-chain fatty acids on the skin by the biotransformation of sweat by microorganism is noted as the main reason for odour formation. However, the detailed mechanism of the malodor formation was not known until the last decade. James et al. (2004) reported a new pathway for the generation of malodor from axilla. In their research, they analysed the generation of volatile fatty acids from skin lipids. Their result reported partial catabolism of longer fatty acids by a new *Corynebacterium* genus, which is denoted as *corynebacteria* (A) by the researchers. These subgroups play a vital role in converting the long-chain fatty acids into short- and medium-chain volatile fatty acids (C2–C11) in the axilla. Similarly, the research findings also showed that *staphylococci* and *propionibacteria* strains are capable of converting the glycerol, lactic acid, and andamino acids to short-chain (C2–C5) volatile fatty acids through metabolism. In the meantime, the kinetic data of the research indicated that the contribution of these stains is comparatively less than *corynebacteria* (A). The odour formation mechanism proposed by James et al. (2004) is reported in Figure 4.1.

4.3 ROLE OF TEXTILE FIBERS IN ODOUR FORMATION

The physical structure and chemical nature of textile fibers differs greatly. Changes occur due to their different origins and manufacturing methods. Generally, the textile material acts as a carrier to propagate the odour due to its moisture sorption characteristics. Every individual might have felt this in their daily routine. It is also apparent that we are comfortable with natural fibers due to their lower odour intensity compared to synthetic textiles. Stench formation in clothing is a complex phenomenon. Several parameters are involved to predict the role of textiles in odour generation. The preliminary studies on odour retention characteristics of textile fibers were analysed by McQueen et al. (2007). They took cotton, polyester, and wool yarn for analysis. The selected fiber types were made into fabric and attached to the axilla region of a T-shirt for a wear trial. The subjects were allowed to wear the swatches for two days, and the odour analysis was performed in two different periods, namely for 7 and 28 days. The subjective analysis results revealed that

TABLE 4.1
Structure of odour components identified from sweat, its precursor, and the enzymes and microbes involved in the bio-transformation of the precursor molecules (BCCA, branched-chain amino acid) (Reprinted with permission from Fredrich et al. 2012)

Odour compound	Odour precursor	Enzyme	Organism
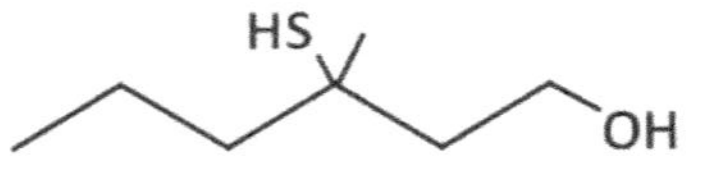 3-Methyl-3-sulfanylhexan-1-ol 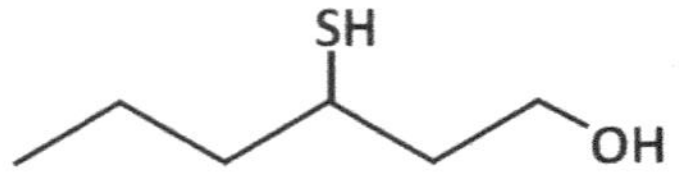 3-Sulfanylhexan-1-ol	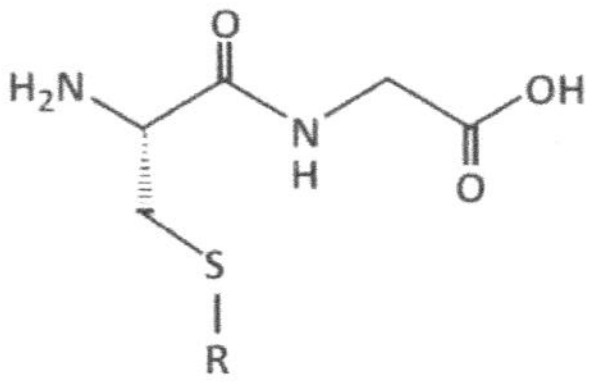 Glycylcysteinyl-*S*-conjugate	TpdA dipeptidase AecDC-S lyase	*Corynebacterium* spp. *Staphylococcus* spp.
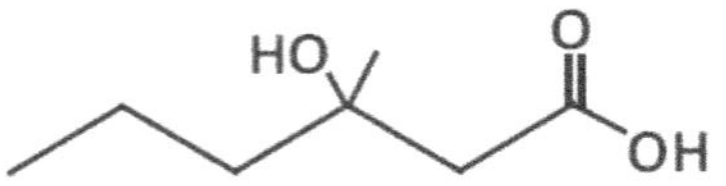 3-Methyl-3-hydroxy hexanoic acid 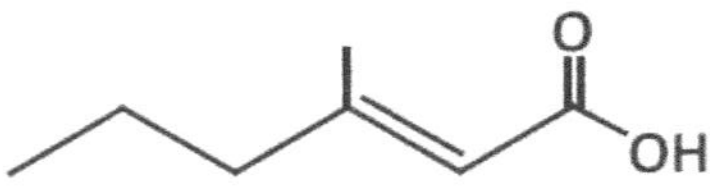 3-Methyl-2-hexenoic acid	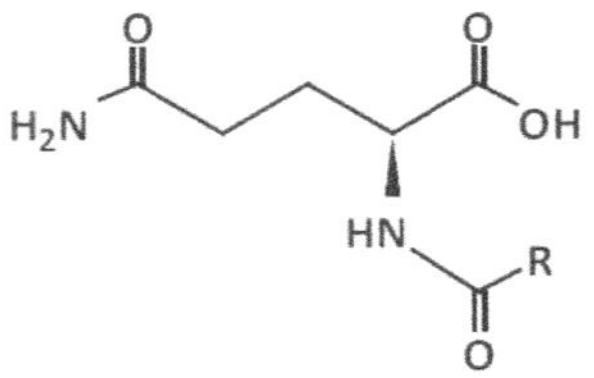 Glutaminyl-conjugate	AgaA*N*α-acylglutamine aminoacylase	*Corynebacterium* spp.

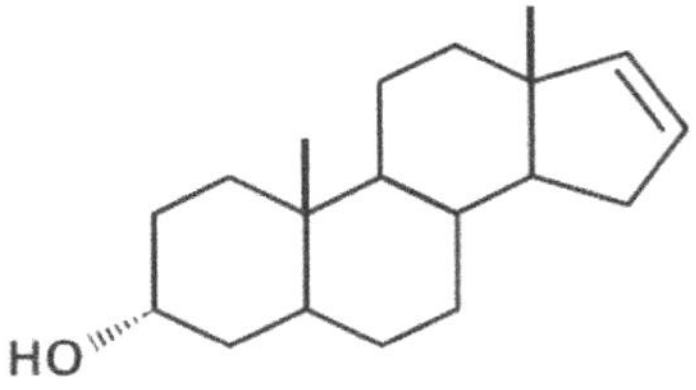 HO; 5α-Androst-16-en-3α-ol	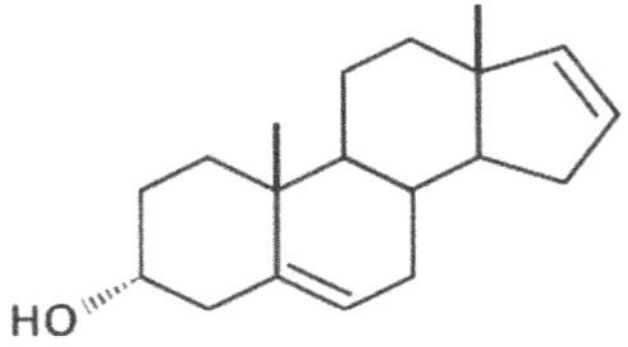 HO; Androsta-5,16-dien-3α-ol	4,5-,or 5α-Reductase5α(β)-Sterol dehydrogenase Steroid 4,5-isomerase	*Corynebacterium* spp. *Micrococcus* spp.
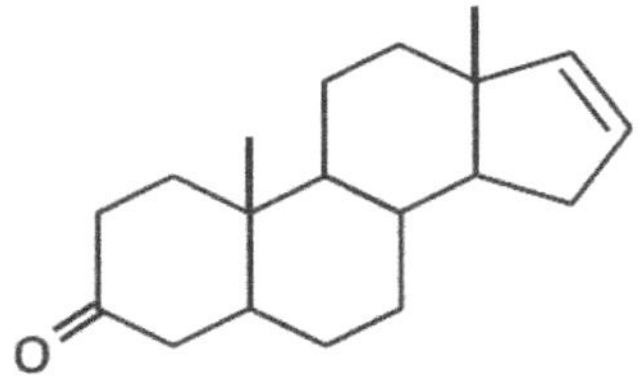 O; 5α-Androst-16-en-3-one	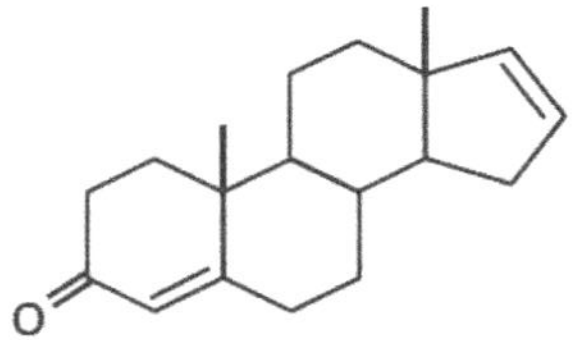 O; Androst-14,6-dien-3-one		
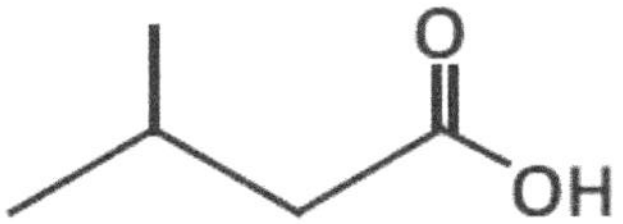 O OH; Isovaleric acid	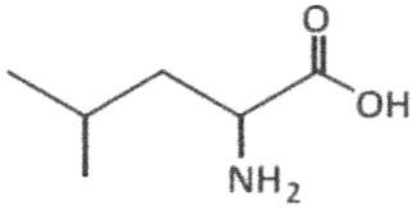 O OH NH_2; Leucine and branched-chain amino acids (BCCA)	BCCA aminotransferase BCCA dehydrogenase	*Staphylococcus* spp.*Corynebacterium* spp.
Volatile fatty acids (VFA)		FadD, FadE, FadB, FadA (β-oxidation enzymes)	Lipid-catabolizing *Corynebacterium* spp.

(Continued)

TABLE 4.1 (Continued)
Structure of odour components identified from sweat, its precursor, and the enzymes and microbes involved in the biotransformation of the precursor molecules (BCCA, branched-chain amino acid) (Reprinted with permission from Fredrich et al. 2012)

Odour compound	Odour precursor	Enzyme	Organism
	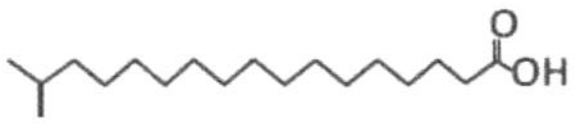 O OH Isostearic acid and methyl-branched fatty acids		
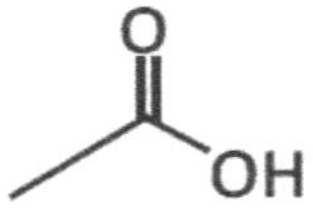 O OH Acetic acid	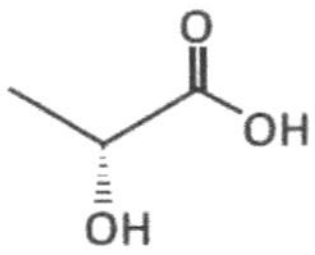 O OH OH Lactic acid	Alternative fermentation pathway enzymes (Ldh, AckA) Wood–Werkman cycle enzymes	Facultative anaerobic *Staphylococcus* spp. Microaerophilic *Propionibacterium* spp.
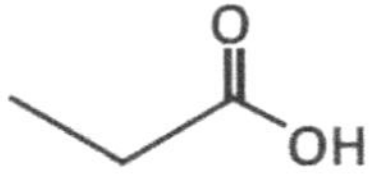 O OH Propionic acid	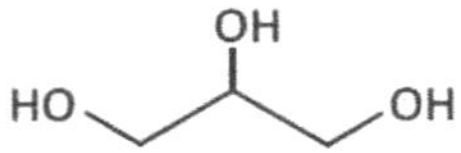 OH HO OH Glycerol		

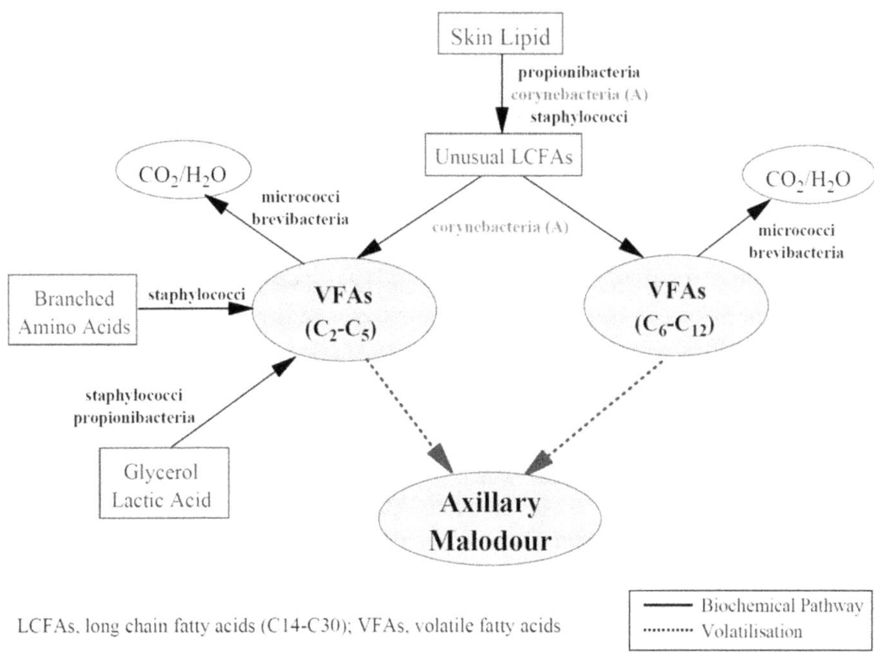

FIGURE 4.1 The formation and utilization of volatile fatty acids by axilla bacterial as reported by James et al. (2004). (Reprinted with permission.)

polyester fabric showed a higher odour intensity than the remaining two fabrics, even after different storage times, namely 7 and 28 days with a rating (mean) 71.8 ± 22.9. Compared to polyester, cotton and wool fabric showed reduced odour intensities. The subjective mean intensity rating of 33.8 ± 11.4 was obtained for cotton, followed by wool fabric, with a rating of 24.4 ± 4.7. To confirm the reason for the higher odour intensity, they analysed the bacterial population in the worn fabric. The bacterial count was measured on days 1, 7, and 28 of removal. The results showed a significant reduction in the bacterial count in subsequent days. With respect to cotton, there is no difference noted between day 1 and 7, but on day 28 significant reduction in bacterial population was noted. The least changes were noted in the wool fabric. These unexpected results of wool fabric contradict the odour intensity of the respective fabrics (McQueen et al. 2007). The researchers concluded that the difference between odour intensity and bacterial population was mainly attributed to the chemical structure and physical morphology of the respective fibers. In the next work, with a similar experimental setup, the research group tried to analyse the influence of fiber type with objective measurement methods over the conventional subjective analysis (McQueen et al. 2008).

Further, McQueen et al. (2008) used polyester, cotton, and wool fabrics for a wear trial and evaluated odour intensity. The worn fabrics were measured for their VOCs using proton-transfer-reaction mass spectrometry (PTR-MS) analysis. Though the technique will not provide any direct measurement of the odour-causing compounds, this method is used to compare increment in volatile groups after a wear trial by comparing the sample with the control sample. The results of the research revealed that the worn fabric showed 18 mass variations (increments due to soiling with odour components) compared to the control sample by representing the changes in the odourous component in the fabric after the wear trial. The *In-vivo* analysis results showed that odour intensity of the apparel differs based on the fiber type of the textile material. It was noted that fiber type has a strong correlation with the odour intensity of the textiles. Out of three fibers tested, polyester fiber was noted to have higher odour intensity, followed by cotton and wool, similar to previous studies (McQueen et al. 2007). The higher odour formation characteristics of the polyester fiber are related to its lower moisture regain properties. Bowers and Chantrey (1969) reported that moisture regains of the textile fiber had an inverse relationship with the oily soil sorption. Above 4% of the moisture regain the fibers did not have much difference in the soiling characteristics. With this prelude, McQueen et al. (2008) plotted the moisture regain of the cotton, wool, and polyester against the odour intensity of the textile. The results were in line with the findings of Bowers and Chantrey (1969), as the odour difference of cotton and wool slightly differed (with moisture regain above 4%), and polyester had significantly higher intensity. They reported a negative correlation between the moisture regain of the fiber and odour intensity (McQueen et al. 2008). Further, the fabric samples were also analysed for their odour retention characteristic. The fabrics were stored for 7 days after the wear trial. In the case of cotton and wool, the odour intensity reduced with an increase in the storage time, but for the polyester, an increase in the odour intensity was noted. Based on these results, the researcher proposed two mechanisms for the higher odour formation in the polyester fabric:

i. First, the strong oleophilic nature of the polyester fabric readily attracts oily substances. This enhances the deposits of long-chain fatty acids from the sweat on the surface of the fiber. As the oily substance is present on the surface of the fiber and is readily available for biotransformation by the microorganism, these substances act as a precursor for odour formation. However, in the case of cotton and wool fiber, the oily substance from the sweat can penetrate inside the structure. This restricts the use of this absorbed oily substance for odour generation as the bacteria cannot access these precursors inside the structure (Obendorf and Namaste 1983; Obendorf and Webb 1987).
ii. Secondly, cotton and wool fabric acts as an absorbing material and keeps the skin dry during sweating. Due to its hydrophobic nature, polyester allows skin to remain moist and hot. This provides a suitable microclimate for bacterial growth; this, in turn, increases the metabolic activity and creates odourous substances as a result (McQueen et al. 2008).

Though these findings showed a reason for higher odour generation during the wear time of polyester over cotton and wool, it does not provide any insight about the odour increment during the seven-day storage after the reduction of the bacterial population. Mcqueen et al. (2008) explained this phenomenon with different parameters. The bacterial strains have the capacity to use one carbon source over another source when multiple resources are accessible. In the case of polyester, this behavior and odour formation mainly depends on the versatility of the bacteria over the quantity or number of bacteria in the body part. Based on this theory, it is reported that during longer storage time, exhaustion of available carbon sources might be the major reason for the bacterial reduction, but at the same time, a particular group of strains may thrive by using available secondary carbon sources and produce the short-chain VOCs. This is why the polyester fabric showed an increased odour intensity after seven days, even though the bacterial population reduced (McQueen et al. 2008).

In similar research, Callewaert et al. (2014) analysed odour formation capabilities of cotton and polyester fabric after a fitness session. They collected worn T-shirt samples from 26 subjects after an hour bicycle spinning session. The subjective analysis results were conducted to rate the different odour parameters, namely hedonic value, intensity, musty, ammonia, strongness, sweatiness, and sourness. All the characteristics were rated on a 0 (*no odour*) to 10 (*very strong/intolerable*) scale, except for hedonic value. The hedonic value was rated from –4 (*very unpleasant*) to 0 (*neutral*), and +4 (*very pleasant*). The results of the sniffing test showed that out of cotton and polyester samples used in this study, polyester clothing had a higher average hedonic value of -2.04 ± 0.90, and the cotton averaged -0.61 ± 1.08. In addition to that, the results also indicated that the polyester fabric was more intense, was mustier, had more ammonia, was stronger, was sweatier, and was sourer than the cotton fabric used. They have reported the poor moisture absorption characteristics of the polyester as the main reason for higher odour intensity and unpleasant odour. The odour propagation in the cotton-based textiles was restricted as it also absorbs and traps the odour molecules inside the structure, as with moisture.

Secondly, there is selective bacterial growth in the polyester textile. Though they obtained 91 isolates, all the textile samples showed a higher population of *Micrococcus* spp. Out of seven tested types of textiles, a maximum growth of (10^7 CFU per cm^2) *Micrococcus* spp. was found in polyester fabric. The selective growth of *Micrococcus* spp. in the polyester fabric may be due to the quorum-sensing nature of the strains and also the favourable aerobic growth conditions on polyester fabric. Though *corynebacteria* is one of the odour-causing bacterial strains, *Micrococcus* spp. is also well known for its odour generation ability. *Micrococcus* spp generally converts the saturated, monounsaturated, and methylbranched fatty acids into malodor compounds (James et al. 2004). In the meantime, a few quantities of *Micrococcus* spp are found in the axillary skin region. This can be interpreted that the aerobic nature of the *Micrococcus* spp. species prefers textile material as a place for its metabolism, whereas *corynebacteria* prefer the anaerobic environment in the skin (Marples and McGinley 1974). This is the main reason for the odour difference between the skin (axilla) and textiles (Callewaert et al. 2014). These findings also provide insight into the findings of McQueen (2007, 2008), who reported higher odour intensity in polyester textiles with a longer storage time of 28 days, even after removal from the wearer. Figure 4.2 depicts the odour formation mechanism of synthetic and natural fibers.

The strong affinity of polyester towards the odour was confirmed by another researcher who analysed the ability of the laundry process to remove malodour from the textile (Munk et al. 2001). They reported that the hydrophobic nature of

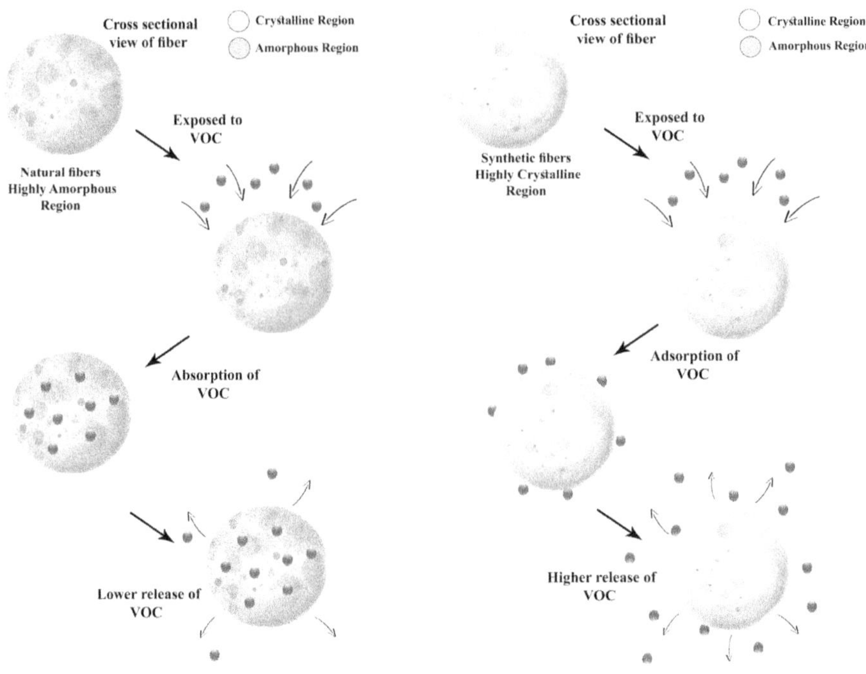

FIGURE 4.2 Odour formation mechanism in synthetic and natural fibers.

polyester is the main reason for its higher aromatic soiling. The washing process did not remove the odourant from polyester textile, and the odour profile of the laundered and unlaundered fabrics was the same, whereas in cotton there was a significant removal of odour from the textile after the first laundry. In odour component analysis, they noted that though both cotton and polyester possessed 30 active odourants, out of this, polyester textile showed a higher number of odour components, with higher flavor dilution (FD) chromatogram numbers. This represents the complex odour profile of the polyester textile. Specifically, octanal, (*E*)-2-octenal,methional, (*Z*)-2-nonenal, (*E*)-2-nonenal, (*E,E*)-2,4-nonadienal,(*E,Z*)-2,4-decadienal, (*E*)-4,5-epoxy-(*E*)-2-decenal, 2-Methoxybenzaldehyde were noted in polyester as the prevalent components. The main reason was noted as the polyester's higher adhesion of lipophilic component. Further, it is also reported that the textiles dried immediately or shorter time showed lower bacterial count and odour (cotton) than the textile dried after an extended period (Munk et al. 2001).

Other research work analysed the odour formation characteristics of axilla-worn 100% cotton, viscose, linen, cotton/polyester blend, 100% polyester, and nylon fabrics. After wear trail analysis, the fabric swatches were subjectively evaluated for their odour intensity ranges. The results reported that out of selected fabric types, a higher odour intensity rating of 3.83 and 4.17 (out of 5) was noted for polyester fabric, respectively, for sedentary and nonsedentary workers. Nylon was noted as the fiber with the second most odour intensity with a rating of 3.5 and 3.47, respectively, for sedentary and nonsedentary workers. The overall odour formation characteristics noted in the sequence of Polyester > Nylon > Cotton/polyester > Linen > Viscose > Cotton. They reported that the least odour intensity with the cellulose-based fibers in the test, least on cotton and viscose followed by linen. The least odour formation in cotton and viscose is attributed to the higher moisture absorption nature of those fibers. These fibers absorb sweat and secretions from the skin effectively; hence, the axilla remains dry. This restricts microbial growth and reduces odour formation in the axilla and textile. They also pointed out the higher water vapour transmission ability of the cotton and viscose fabric structure as potential reasons for higher air transportation. These characters significantly reduced the moist nature of the textile and hence the odour intensity. In the case of linen, a slightly higher intensity was reported than the cotton and viscose. This is again due to its poorer hygroscopic nature than cotton and viscose. Further, the thickness difference among the fabrics is also reported as one of the reasons for the higher odour intensity. This was also reflected in the water vapour transmission character of the linen fabric. In the case of synthetic, it was reported that with the higher oleophilic nature of the polyester and nylon, along with poor moisture management ability, the axilla remains wet longer than natural fibers. This increases the metabolism of skin bacteria and therefore the odour intensity. The bacterial isolation results showed that fabrics mostly consist of *corynebacterium* Sp., *Bacillus* Sp., *Pseudomonas* Sp., and *Staphylococcus* Sp., for all fabrics invariantly (Rathinamoorthy et al. 2014a). While comparing the odour formation characteristics of nylon and polyester, Abdul-Bari et al. (2018) reported that there was no significant difference between nylon and polyester. They have analysed the odour intensity after storing the sample swatches at room temperature and also at freezing point. Similarly, they also analysed odour intensity rates of these two fibers *in vitro* against isovaleric acid and 2-nonenal. The result showed a higher odour absorption rate of nylon over polyester. They reported that a higher absorption of odourous molecules in nylon is associated with the presence of polar groups and higher moisture regain of the nylon. The results of odour intensity percentage for nylon against reported odourant showed significantly lower than polyester. At the same time, the wear trial analysis did not show any difference in the odour intensity among the two fabrics. However, upon storage, odour intensity in nylon increases rapidly at room temperature; this was not the case with polyester (Abdul-Bari et al. 2018).

In order to analyse the effect of the laundry process, McQueen et al. (2014) used worn cotton and polyester fabric for sensory analysis. The results reported higher odour intensity of polyester fabric than cotton. Further, they also evaluated the influence of laundry on odour removal. After the laundry process, the odour-causing volatile organic acids and carboxylic acids were removed more effectively in the case of cotton fabric than the polyester fabric. Thus, the polyester remained with higher odour intensity even after the laundry. It was reported that the moisture regain and oleophilic nature of the polyester were the main reasons. Wang et al. (2019) analysed the effect of blending different percentages of wool with polyester on odour sorption and emission characteristics of the textiles. In initial studies, Wang et al. (2019) evaluated the sorption and emission characteristics of ammonia, acetic acid, and butyric acid with 100% polyester and 10%, 20%, 40%, and 50%

wool blends. The results showed that in all the cases 100% wool fabric showed a higher sorption percentage, whereas, 100% polyester showed the least absorption. They also reported that successive addition of wool % in the fabric significantly increased the sorption ability of the fabric. By comparing the results of blend proportion and their impact on sorption the study concluded that an addition of 20% wool in the polyester fabric significantly reduced the odour emission intensity of the polyester fabric. Further, to evaluate the real-time effect they conducted a wear trial analysis for 20% wool and 80% polyester fabric. The results revealed that 100% polyester fabric showed an odour intensity ranging from 49 to 97 odour units (ou). However, in the case of 20% wool added fabric, the odour value noted less than 16 ou, which represents no odour. The presence of acidic and ammonia groups in the wool provides more sorption sites to absorb volatile components. The acidic and peptide sites created bonds with odourous molecules and reduced the emission (Wang et al. 2019). Additionally, increments in the sorption of acetic acid were noted, with the increment in the relative humidity due to its lower glass transition temperature, whereas the higher Tg of polyester is not influenced by the moisture (Pucher 1971). Hence, the higher hygroscopic nature of wool increased the sorption of oily substance and showed a lower odour emission than polyester (Wang et al. 2019).

PTR-MS analysis performed to evaluate the adsorption and emission capabilities of cotton, polyester, and wool fiber against nonanal, 2-propanethiol, dimethyl disulfide (DMDS), and benzaldehyde. Their results confirmed that the sorption characteristics of fiber strongly depend on its chemical nature. Three different mechanisms were proposed for the selected three fibers.

1. Wool showed a higher absorption of odourous components but a lower release rate, so it is noted as a low odour fabric. This is mainly attributed to the presence of the larger amount of polar and nonpolar amino acid in its structure. These arrangements strongly interact with various oily substances through hydrogen bonds, sulfur–sulphur forces, COOH-NH_2 links, and p–hydrogen interactions. They reported that wool absorbed almost 40–95% of all volatile components.
2. Polyester has a high adsorption characteristic of sulfur compounds and a constant release over time. The compounds interacted with medium-strength noncovalent forces (dipole-dipole force, hydrogen–sulfur bonding), which causes a sustained release of components over time during storage.
3. Cotton fiber has a lower initial absorption and lower release of odour substance on longer storage. However, an initial burst of odour from the cotton was reported. Though cotton is made of cellulose and readily creates hydrogen bonds, the presence of noncellulosic substances reduces the absorption compared to wool and polyester. Other weaker bonds' (e.g., sulfur–hydrogen) interaction with VOCs was noted as the reason for the initial (fast) release of absorbed components (Richter et al. 2018; Yao et al. 2015).

Similar research by Hammer et al. (2013) evaluated odour intensity through olfactometry and reported a less sweat odour intensity rating for wool, a medium rating for cotton, and a higher level of emission from polyester fabric. They mentioned that the higher odour release from polyester was attributed to the lower sorption capabilities of the polyester fabric compared to the other two fabrics. Higher crystallinity and lower moisture absorption capacities of the polyester were noted as the main reason for the poor sorption characteristics. On further research, the results showed a negative correlation between the isovaleric acid retention and fabric odour intensity in all the fabrics. Though polyester retained less isovaleric acid than cotton, it showed a higher odour emission 3hours and 20hours after the first hour (Hammer et al. 2013). These findings were in line with the previous research reports by McQueen et al. (2007, 2008) and Munk et al. (2001). From the analysis, it can be noted that cotton, wool, and other natural fibers possessed lower odour release characteristics due to their chemical structure, which shows higher sorption of odourous compounds. Out of researched fibers, lower odour release and higher sorption were noted with wool followed by cotton. The higher release was noted with nylon and polyester fiber types. A negative correlation between moisture regain of fibers and odour intensity was noted. Synthetic textiles are more odourous due to their hydrophobic nature compared to natural fibers. There is no significant difference noted between nylon and polyester with respect to odour intensity. Higher oleophilic nature of polyester aids in more sorption of volatile organic compounds and oily substances. There is a selective growth of microorganisms (quorum sensing) and selective sorption of sulfur compounds (with poor bond strength) noted on the surface and identified as a source of continuous odour release from synthetic textiles. Table 4.2 summarises the complete findings concerning textile fibers.

4.4 ROLE OF TEXTILE STRUCTURES IN ODOUR FORMATION

Next to the textile fibers, the structure of the textile material was also found to have a significant impact on the odour formation characteristics of the fabrics. As the physical properties of the textile strongly influence the microclimate inside the apparel, they also have a strong influence on odour formation. Though the odour sorption characteristics of textiles depend on fiber type, the structural parameter decides the moisture on the skin. In general, a loose and thinner structure will provide enough ventilation for the skin so that it can remain dry as much as possible. Similarly, loose structures and textures of knitted fabric facilitate the odour substances to reach active adsorption sites and increase odour sorption. A few researchers evaluated the impact of various structural parameters with odour intensity. McQueen et al. (2007) compared the single jersey, 1X1 rib, and interlock fabrics of cotton, wool, and polyester fabric to analyse the effect of structural parameters on odour formation. After the wear trial, the odour intensity of the fabric was analysed subjectively, and ratings were provided. The results showed that the odour formation in the same fiber type differed significantly among the different structures. In polyester fabric, out of all the structures tested, single jersey fabric was noted to have lower odour intensity than the heavy structures. They reported that due to higher density, the fabric with a thick structure resulted in more bacteria generation, which subsequently created a higher odour. In contrast, the lighter fabric allowed

TABLE 4.2

The role of textile fiber type on odour formation – summary of research findings

S.No	Type of Fibers	Odour intensity ranking	Findings	Reference
1.	Cotton, Polyester	Polyester > Cotton	• The higher adhesion ability of lipophilic component on the surface created more odour in polyester. • Due to its hydrophobic nature, polyester keeps axilla wet for a longer time and hence encourages bacterial metabolism more than cotton.	Munk et al. (2001)
2.	Cotton, Polyester, Wool	Polyester > Cotton > Wool	A negative correlation exists between moisture regain of the fibers and odour intensity.	McQueen et al. (2007)
3.	Cotton, Polyester, Wool	Polyester > Cotton > Wool	• The oleophilic nature of the polyester increases the adsorption of VOCs. • Due to the hydrophobic nature, the changes in microclimate between textile and skin increases the bacterial growth.	McQueen et al. (2008)
4.	Cotton, Polyester, Wool	Polyester releases more odour.	• A negative correlation exists between isoveletic acid sorption and odour release. • Higher odour emission was noted, even after 20 hours (after wear trial).	Hammer et al. (2013)
5.	Cotton, Polyester	Polyester > Cotton	• Poor moisture absorption increases odour formation. • Selective growth of *Micrococcus* spp., in polyester fabric – due to the quorum sensing nature of the strains.	Callewaert et al. (2014)
6.	Cotton, Polyester	Polyester > Cotton	• Higher odour intensity exists in polyester before and after laundry. • The laundry process did not remove a significant amount of carboxylic acid from the polyester surface. This causes higher odour.	McQueen et al. (2014)

7.	Cotton, Viscose, Linen, Polyester, Cotton/ polyester, Nylon	Polyester > Nylon > Cotton/ polyester > Linen > viscose > cotton	• Poor moisture absorbency of the synthetic fibers and oleophilic characteristics causes higher odour. • Poor moisture handling ability causes favourable axilla conditions for bacterial growth and causes more odour in synthetic textiles.	Rathinamoorthy et al. (2014a)
8.	Polyester, Nylon	Polyester = Nylon	• Similar odour intensity was found in the wear trial. • Higher odour reduction rate was noted for nylon over polyester.	Abdul-Bari et al. (2018)
9.	Wool, Polyester	Polyester> Wool	• Higher sorption in wool fiber was due to the presence of acidic and peptide sites. • Higher moisture regains of wool also impacts the higher sorption of odourous component into the fiber. • Polyester was noted as a higher odour-emitting fiber.	Wang et al. (2019)
10.	Cotton, Polyester, Wool	Polyester > Cotton > Wool	• Wool showed higher sorption, and cotton showed medium sorption of VOCs and released a lot fewer VOCs. • Sulfur compounds were highly adsorbed by polyester, and consistent higher release was noted with time – due to weaker bonds like sulfur–hydrogen interaction.	Richter et al. (2018); Yao et al. (2015)

more moisture transmission through thinner structure, so the skin remained dry, which resulted in less odour. A similar effect was also noted in the case of other fiber types, like cotton and wool, used in this study.

A similar study evaluated odour generation in cotton fabric with different knit structures, single jersey, rib, and interlock (Rathinamoorthy and Thilagavathi 2014b). Selected fabrics were used for a wear trial and analysed for odour intensity subjectively. The results showed that increment in fabric thickness and grams/square meter positively increased the odour intensity of the material. Out of the three structures analysed, rib structure was noted to be the thickest one, with higher grams per square meter value. This was followed by interlock and single jersey material. The same trend was noted in the odour intensity results, with higher intensity in rib, an average odour intensity rating of 4.84 out of 5, followed by interlock (4.8), and single jersey (3.8), with statistical significance ($p < 0.05$). They reported that the difference was mainly attributed to the structural differences of the fabric (Rathinamoorthy and Thilagavathi 2014b). The structures of the selected fabrics are provided in Figure 4.3.

In their findings, researchers reported that the higher odour intensity of the rib fabric is associated with poor sweat transmission ability from the skin compared to other structures. They also reported a higher thickness of the rib material compared to other structures selected for the study. The increment in the fabric thickness results in air entrapment, and thus it restricts the heat and water vapour transmission. Pac et al. (2001) also reported that the cool feeling in interlock fabric is due to its higher thermal absorptivity nature. They reported that these differences are mainly attributed to the structural difference among the fabric (Pac et al. 2001). The increment in the thickness also influenced the thermal resistance of the fabric. It is a measure of the ability of the fabric's heat transfer from one side to the other side. There is a strong difference among the structures of the same fiber reported in the previous research works (Oğlakcioğlu and Marmarali 2007). Rathinamoorthy and Thilagavathi (2014b) compared the water vapour permeability values of the three different structures of the same fiber and reported that the thinner fabric can effectively transfer body sweat to the atmosphere, so the skin remains dry compared to thicker fabric. Further, this higher water vapour transmission ability of the single jersey fabric resulted in lower odour generation than other structures. Further, to

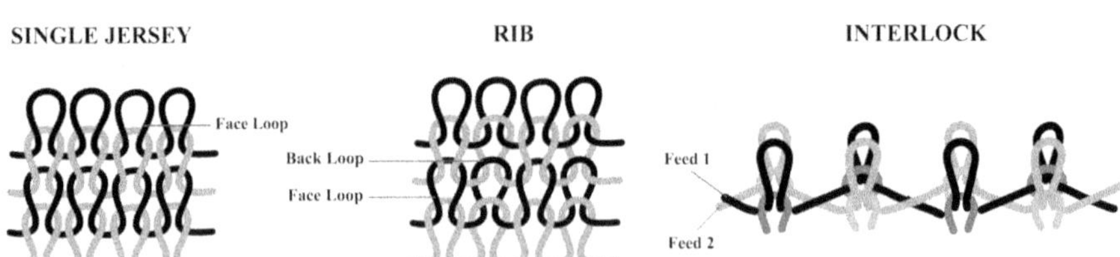

FIGURE 4.3 Common knitted structures used in apparels.

add, Rathinamoorthy and Thilagavathi (2014b) analysed the bacterial population and presence of different bacterial strains in the fabric. In findings, they reported that most of the bacterial popualtions in fabric consisted of *Bacillus* Sp., *Pseudomonas* Sp., and *Staphylococcus* Sp., irrespective of the fabric structure types. These bacterial strains were able to produce different short-chain carboxylic acids like butyric acid, acetic acid, and valeric acid. The development of these precursors were the main source of odour formation in the fabric. This statement was validated by a previous research work, where an analysis was performed on apocrine sweat collected from sterile axilla. They collected apocrine sweat and incubated it for 14 days at room temperature and noted that the sweat did not generate any odour. At the same time, tubes containing apocrine sweat from "unsterile" axilla developed a strong odour within six hours, which became very strong at 24 hours (Shelley et al. 1953). These findings confirm the role of bacteria in odour generation in the axilla and further on clothing.

These findings were also correlated with the PTR-MS results, where there was an increment in the mass components with respect to the produced carboxylic acids. The study also analysed the interrelationship between the selected fabric parameters and odour formation. For bacterial isolation results, a further confirmation of more bacterial colonies was found in a unit area of the sample, as shown in Figure 4.4. It was noted that the thicker rib fabric had more bacterial strains, followed by interlock and single jersey fabric. (Rathinamoorthy and Thilagavathi 2014c). The results reported that this higher population of bacteria in the worn textiles was mainly due to the axilla environment. The thicker rib fabric with higher grams per square meter values keeps the axilla moist and damp due to its poor moisture vapour and air transmission rate. Hence, the microclimate generated provides a suitable environment for bacterial growth compared to the other two structures discussed. The results depicted the correlation between the thickness, mass per square meter with the fabric's moisture management capabilities and therefore the odour formation. In their research, a higher correlation of 73% ($r^2 = 0.73$) was noted between the odour and thickness, whereas in between the mass per square meter and odour it is 57%. Concerning the fabric parameters and odour formation, they also analysed the interrelationship between the grams per square meter (GSM) of the knitted fabric,

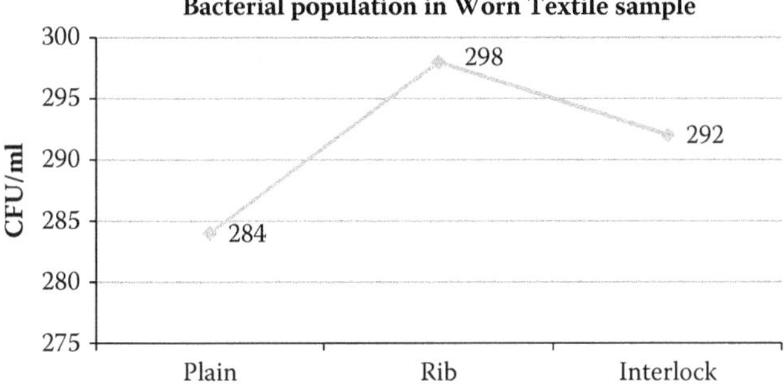

FIGURE 4.4 Bacterial population analysis in the axilla-worn fabric sample with different structure.

loop density, and fabric thickness with the odour intensity rating. The statistical analysis results showed that the increment in fabric GSM directly increased the odour intensity due to its poor moisture handling ability, as discussed earlier. Similarly, the thickness of the fabric also positively influenced the odour intensity. This was reported due to the increase in microclimate temperature between the skin and cloth due to the higher thermal resistance of the fabric. However, in the case of loop density, there was very little effect noted on the odour intensity (Rathinamoorthy and Thilagavathi 2014b).

Similar research performed on cotton knitted socks made of different structures and odour generation capacity evaluated against the foot odour (Rathinamoorthy and Thilagavathi 2013). In this study, the most common structures used in socks like single jersey, rib, and terry were selected. After a wear trial, the collected fabric samples were subjectively analysed for their odour generation capacity. The results showed a higher odour formation with respect to the terry structure compared to the other fabric used. The results of the study are provided in Figure 4.5. From the results, it is noted that terry socks had higher odour intensity with an average value of 4.53 on a 5-point scale compared to the rib (4.2) and single jersey (3.5) (Rathinamoorthy and Thilagavathi 2013). These results were in line with the previous findings reported by McQueen et al. (2007) and Rathinamoorthy et al. (2014a, b).

The results of the research also showed that the textile material with higher thickness and grams per square meter had higher odour formation characteristics irrespective of the types of structure. Out of all the fabrics analysed, the single jersey fabric was reported as the lowest odour-generating material by all the researchers. The results were supported with different thermal and moisture characteristics of the fabric. The lower thickness and open structure of the single jersey fabric permitted a higher amount of moisture and heat transfer and kept the skin (axilla) dry. This restricted the bacterial growth compared to the other structure. Hence, the single jersey fabric exhibited a lower odour intensity. In the case of interlock, rib, and terry fabric, the thicker fabric always showed higher odour formation. McQueen et al. (2007) and Rathinamoorthy et al. (2014a, b, c) reported higher odour intensity with the Rib fabric compared to the interlock and single jersey. This is mainly attributed to the compact structure, which results in poor air and moisture vapour transmission from the skin. The restriction creates a hot and

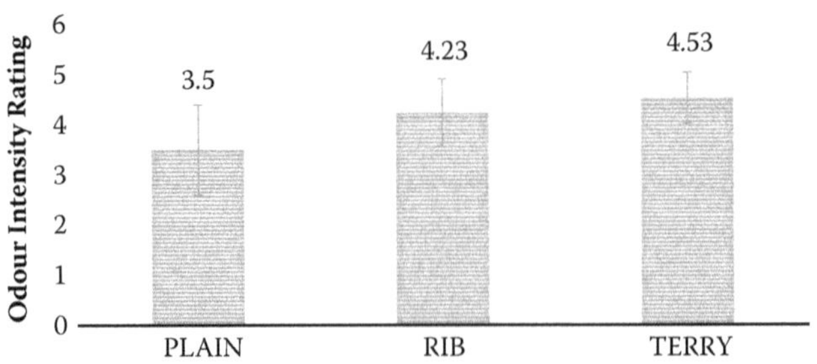

FIGURE 4.5 Average odour intensity value of cotton socks by subjective analysis.

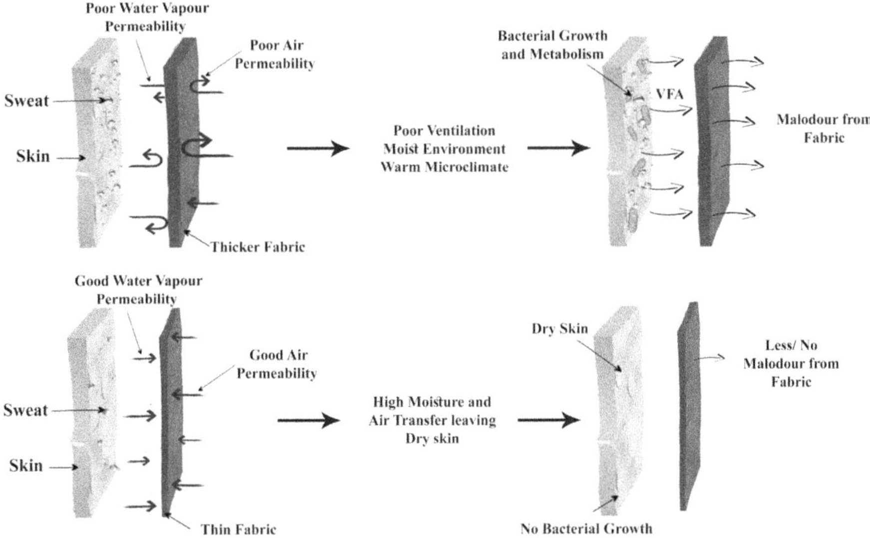

FIGURE 4.6 Role of structural parameters on odour formation in textiles.

wet microclimate suitable for bacterial growth, and this, in turn, creates more odour. Figure 4.6 illustrates the effect of fabric structural parameters on the odour formation characteristics of textiles.

4.5 SUMMARY AND FUTURE SCOPE

This chapter outlines the two different parts of the textile and odour interaction. The first section details the role of textile fibers and odour formation on fabrics. The findings of the analysis revealed that natural fiber (wool, cotton) has lower odour formation than synthetic fiber (nylon, polyester). This is attributed to the hydrophilic nature of the fibers. It is noted that the moisture regain is inversely proportional to odour intensity. In the case of synthetic textiles, due to their hydrophobic nature, sweat is not absorbed and is adsorbed on the surface. This aids the bacterial metabolism and odorous components generated on the surface of the synthetic textile. The oleophilic nature of synthetic textiles like polyester is also noted to be one of the main reasons for higher odour intensity detection. Out of all the synthetic fibers, polyester and nylon were most explored as they were commonly used in apparel. In the case of textile structural parameters, very few research works were performed. The results showed that irrespective of the structures used, the fabric thickness and compactness favoured bacterial growth. The thicker and compact structure restricts airflow and creates a moist surface on the skin. This changes the microclimate to be a suitable place for bacterial metabolic activity. This helps with higher bacterial growth and leads to the development of more odourous substances by biotransformation.

From the results, it is also evident that both *in-vivo* and *in-vitro* analysis produced similar results, either by subjective or objective evaluations. The findings also showed that most of the research works performed on cotton, wool, polyester, and nylon fibers in specifically knit structures. Hence, future works on regenerated synthetic textiles and different synthetic fibers like polyacrylic, polypropylene, etc., must be explored as it is a quite common material to blend with natural fibers. Similarly, the effect of blends has not been explored even though cotton/polyester blends are more common in commercial fabrics. In the case of the fabric structure, though the role of thermal, air, and water vapour resistance has been considered, the direct correlation of these parameters with odour intensity will provide a better understanding to the end-users. The woven structures are quite common in the outerwears; however, no research was found on the woven fabric structural parameters. Exploring these sections will reduce the complexity and enlighten the interaction mechanism of textile and odourours components (volatile fatty acids).

REFERENCES

Abdul-Bari, M.M., McQueen, R.M., Nguyen, R., Wismer, W.V., Paulina de la Mata, A., and James, J.H. 2018. Synthetic clothing and the problem with odor: Comparison of nylon and polyester fabrics. *Clothing and Textiles Research Journal*, 36(4): 251–266.

Bowers C.A., and Chantrey, G. 1969. Factors controlling the soiling of white polyester cotton fabrics: Part I. Laboratory studies. *Textile Research Journal*, 39: 1–11.

Callewaert, C., De Maeseneire, E., Kerckhof, F.M., Verliefde, A., Van de Wiele, T., and Boona, N. 2014. Microbial odor profile of polyester and cotton clothes after a fitness session. *Applied and Environmental Microbiology*, 80(21): 6611–6619.

Dravnieks, A., Krotoszynski, B.K., Lieb W.E., and Jungermann, E. 1968. Influence of an antibacterial soap on various effluents from axillae. *Journal of the Society of Cosmetic Chemists*, 19: 610–626.

Fredrich, E., Barzantny, H., Brune, I., and Tauch, A. 2012. Molecular basis of human body odour formation: Insights deduced from corynebacterial genome sequences. *International Journal of Cosmetic Science*, 34: 2–11.

Gao, Y., and Cranston, R. 2008. Recent advances in antimicrobial treatments of textiles. *Textile Research Journal*, 8(1): 60–72.

Hammer, T.R., Berner-Dannenmann, N., and Hoefer, D. 2013. Quantitative and sensory evaluation of malodour retention of fibre types by use of artificial skin, sweat and radiolabelled isovaleric acid. *Flavour and Fragrance Journal*, 28(4): 238–244. 10.1002/ffj.3134

James, A.G., Casey, J., Hyliands, D., and Mycock, G. 2004. Fatty acid metabolism by cutaneous bacteria and its role in axillary malodour. *World Journal of Microbiology & Biotechnology*, 20: 787–793. 10.1007/s11274-004-5843-8.

Labows, J.B. McGinley, K.J., and Kligman, A.M. 1982. Perspectives on axillary odor. *Journal of the Society of Cosmetic Chemists*, 34: 193–202.

Leyden, J.J., McGinley, K.J., Holzle, E., Labows, J.N., and Kligman, A.M. 1981. The microbiology of the human axilla and its relationship to axillary odour. *The Journal of Investigative Dermatology*, 77: 413–416.

Marples, R.R., and McGinley, K.J. 1974. *Corynebacterium acnes* and other anaerobic diphteroids from human skin. *Journal of Medical Microbiology*, 7: 349–352. 10.1099/00222615-7-3-349.

McQueen, R.H., Laing, R.M., Brooks, H.J.L., and Niven, B.E. 2007. Odor intensity in apparel fabrics and the link with bacterial populations. *Textile Research Journal*, 77(7): 449–456.

McQueen, R.H., Laing, R.M., Delahunty, C.M., Brooks, H.J.L., and Niven, B.E. 2008. Retention of axillary odour on apparel fabrics. *The Journal of The Textile Institute*, 99(6): 515–523, DOI: 10.1080/00405000701659774

McQueen, R.H., Harynuk, J.J., Wendy, V.W., Keelan, M., Xu, Y., and Paulina de la Mata, A. 2014. Axillary odour build-up in knit fabrics following multiple use cycles. *International Journal of Clothing Science and Technology*, 26(4): 274–290. 10.1108/IJCST-05-2013-0064

Munk, S., Johansen, C., Stahnke, L.H., and Adler-Nissen, J. 2001. Microbial survival and odor in laundry. *Journal of Surfactants and Detergents*, 4: 385–394.

Natsch, A., Gfeller, H., Gygax, P., Schmid, J., and Acuna, G. 2003. A specific bacterial aminoacylase cleaves odorant precursors secreted in the human axilla. *Journal of Biological Chemistry*, 278: 5718–5727.

Obendorf, S.K., and Namaste, Y.M.N. 1983. A microscopical study of residual oily soil distribution on fabrics of varying fiber content. *Textile Research Journal*, 53: 375–383.

Obendorf, S.K., and Webb, J.J. 1987. Detergency study: Distribution of natural soils on shirt collars. *Textile Research Journal*, 57: 557–563.

Oğlakcioğlu, N., and Marmarali, A. 2007. Thermal comfort properties of some knitted structures. *FIBRES & TEXTILES in Eastern Europe January/December*, 15(5–6): 64–65.

Pac, M.J., Bueno, M.A., Renner, M., and El Kasmi, S. 2001. Warm-cool feeling relative to tribological properties of fabrics. *Textile Research Journal*, 71(9): 806–812.

Pucher, M. 1971. Sorption of unpleasant odors by wool. *Applied Polymer Symposium*, 18: 347–352.

Rathinamoorthy, R., and Thilagavathi, G. 2013. Studies on human axillary odour retention properties of textile fabrics and development of combined antimicrobial and deodorant finish, University Grants Commission Report. (Unpublished data)

Rathinamoorthy, R., and Thilagavathi, G. 2014b. Knitted fabric specifications and axilla odour retention characteristics. *Indian Journal of Fiber and Textile Research*, 39: 244–253.

Rathinamoorthy, R., and Thilagavathi, G. 2014c. Effect of antimicrobial finish on odour control properties of apparel fabric. *Journal of Textile, Apparel Technology and Management (JTATM)*, 9(1): 1–15.

Rathinamoorthy, R., Thilagavathi, G., Brindha, S., Gayathri, P., Poornakala, N.S., and Pradeep, B. 2014a. Odour control studies on apparel fabric finished with methanol extract of *Terminalia chebula. Fibers and polymers*, 15(8): 1669–1676.

Richter, T.M., Bremer, P. J., Silcock, P., and Laing, R.M. 2018. Textile binding and release of body odor compounds measured by proton transfer reaction – mass spectrometry. *Textile Research Journal*, 88(22): 2559–2567. 10.1177/0040517517725126

Shelley, W.B., Hurley, H.J. Jr., and Nichols, A.C. 1953. Axillary odor: Experimental study of the role of bacteria, apocrine sweat and deodorants. *Archive for Dermatology and Syphilis*, 68: 430–446.

Taylor, D., Daulby, A., Grimshaw, S., James, G., Mercer, J., and Vaziri, S. 2003. Characterization of the microflora of the human axilla. *International Journal of Cosmetic Science*, 25: 137–145.

Teufel, L., Pipal, A., Schuster, K.C., Staudinger, T., and Redl, B. 2010. Material-dependent growth of human skin bacteria on textiles investigated using challenge tests and DNA genotyping. *Journal of Applied Microbiology*, 108(2): 450–461.

Wang, J., Lu, X., Wang, J., and Wang, X. 2019. Quantitative and sensory evaluation of odor retention on polyester/wool blends. *Textile Research Journal*, 89(13): 2729–2738, 10.1177/0040517518801183

Yao, L., Laing, R.M., Bremer, P.J., Silcock, P.J., and Leus, M.J. 2015. Measuring textile adsorption of body odor compounds using proton-transfer- reaction mass spectrometry. *Textile Research Journal*, 85(17): 1817–1826. 10.1177/00405 17515576325.

5 Textile Odour Control by Adsorption

Ayşegül Çetmeli Bakadur
Üniteks Textile R&D Center, İzmir, Turkey

Ahmet Çay
Department of Textile Engineering, Faculty of Engineering, Ege University, Bornova, İzmir, Turkey

CONTENTS

5.1 INTRODUCTION

The formation and release of unpleasant body odour has a negative impact on people's overall comfort and psychology. This effect is much higher in areas where odour formation cannot be prevented, such as gyms or activity-intensive work environments. In addition, excessive odour formation is a factor that increases energy and water consumption due to the increased need for washing and drying of garments. Therefore, the development of functional textile materials that can control the formation of unpleasant odours is an important issue for the textile industry.

Textile and odour interaction is a complex issue due to many parameters, such as fibre type in textile material, finishing processes applied to the material, air/water vapour permeability and wicking property of fabrics, chemical structure and properties of odorous substances, skin structure, and environmental factors. Prevention of sweat evaporation by clothing increases condensation; therefore, this moisture in the microclimate, combined with nutrients, causes the growth of microorganisms and an increase in the potential production of malodours (Le Blan and Vatinel 2018). Therefore, clothes made of breathable fabrics are advantageous for odour control.

Studies have shown that fibre type is one of the most important parameters in terms of odour retention and release. In this context, it has been observed that wool fibres generally have the lowest odour intensity compared to cotton and regenerated

DOI: 10.1201/9781003141426-5

cellulose fibres. Synthetic fibres, such as polyester and polyamide, however, retain the highest odour intensity among all fibres (Abdul-Bari et al. 2018; Hammer, Berner-Dannenmann, and Hoefer 2012; McQueen et al. 2007; Wang et al. 2019). Here, the wicking and drying properties of the fibres and the bacterial/fungal retention of the textile material come into prominence. Laing (2019) stated that the mechanism for odour intensity is not fully understood because there are many factors, such as fabric properties and the interaction between textile material and skin microbiome. On the other hand, in the study by McQueen et al. (2007) with wool, cotton, and polyester, there was not a similar trend for bacterial numbers with respect to fibre type with the resultant odour intensity. It was reported that the number of initial bacteria was similar in all fibres, and the numbers in wool fabrics remained unchanged. However, for polyester fabrics a decrease in bacterial numbers was observed, although they retained the highest odour intensity. In this context, the release properties of the odour compounds retained (adsorbed/absorbed) by the textile material stand out in terms of odour density. The more a fibre retains the odour compound and does not release it into the surrounding environment, the lower the odour density will be. As will be discussed later, this is the principle of odour control by adsorption. The study of Hammer 2012 with isovaleric acid as an odorous substance showed that, three hours after the inoculation of the odour solution, a nearly complete evaporation was observed for polyester fabrics, whereas wool fibres almost completely retained the substance. Fifty percent retention was observed for cotton fibres. Similarly, Richter et al. (2018) reported that wool fibres showed high adsorption capacity and low release for all tested compounds; on the other hand, polyester showed a constant release of the compounds over time. The low odour release in wool fibres might be due to the potential ionic interaction of odorous compounds with functional groups, such as amino and carboxyl (Hammer, Berner-Dannenmann, and Hoefer 2012; Wang et al. 2019). On the other hand, low absorption property and a lack of functional groups that will provide ionic interaction with the odorous compounds might be the reason of the higher odour release for polyester fabrics.

It was observed that the fabric structure has an effect on the odour density to a certain extent. McQueen et al. (2007) investigated that polyester single jersey fabrics had lower odour intensity compared to heavier and thicker interlock and rib fabrics. On the other hand, in wool and cotton fabrics, a significant difference has not been detected since the odour intensity is low.

Although the interaction of textile materials with odour differs, it is not possible to control odour with textile material alone. Especially for polyester fabrics, which are mostly used in sportswear (Shishoo 2015), odour control and prevention are important. In this context, three basic methods stand out in prevention of odour release during use of clothing:

i. Deodorization
ii. Odour masking
iii. Odour trapping

Deodorization is the prevention or control of the growth of malodour-forming microorganisms and can be achieved by biochemical and/or chemical methods. For this purpose, fibres containing antibacterial substances, such as triclosan, silver, copper, chitosan, titanium dioxide, amine-terminated dendritic materials, etc., can be used or antibacterial/antifungal finishing processes can be applied (Klaykruayat, Siralertmukul, and Srikulkit, 2010; Le Blan and Vatinel 2018). Odour masking is the covering and masking of the unpleasant odour created by the body with a pleasant fragrance. For this purpose, there are applications such as aromatic fibres and textile materials containing microcapsules that can release fragrance. The principle of odour trapping is the prevention of the release of unpleasant odours created by the body. In this context, textile materials with odour adsorption can be produced by incorporation of adsorbents such as activated carbons, cyclodextrins (CDs), etc. to textile materials during fiber production or by finishing treatments.

This chapter deals with the textile materials with adsorption properties to control body malodours. In the following sections, some basic information about adsorption is given. Then, the adsorbents with the most potential use in the textile industry for odour control and their textile applications in both commercial and academic terms are mentioned.

5.2 ADSORPTION

Adsorption is defined as "the change in concentration of a given substance at the interface as compared with the neighbouring phases" (Dabrowski 2001). In the adsorption process, atoms, ions, or molecules brought in contact with solid materials are captured on the surface or within the pores (Saha and Grappe 2017). The adsorbed phase, which forms a film on the surface of the materials, is called adsorbate, and the adsorbing phase, on which the adsorbate attaches, is called adsorbent (Ben-Mansour et al. 2016). If the mass transfer occurs from the adsorbent to the adsorbate, this phenomenon is called desorption. It should be noted that the desorption rate is as important as the adsorption capacity in textile odour control.

A heterogeneous porous structure is the common feature of the adsorbents. The adsorption mechanism on solid surfaces includes transport of adsorbate to the external surface of the adsorbent, diffusion of adsorbate through the pores, and finally attachment onto the pore surface (LeMinh et al. 2018), as schematised in Figure 5.1. According to the recommendations of IUPAC, pore sizes of the adsorbents can be classified as macropores (pores with diameters greater than 50 nm), mesopores (pores with diameters between 2 nm and 50 nm), and micropores (pores with diameters less than 2 nm) (Sing et al. 1985). The adsorption capacity of adsorbents is directly related to the distribution of these pores. Macropores are the entrance to adsorption sites, whereas mesopores act as the transportation passage to micropores. The majority of the adsorption takes place in micropores; on the other hand, for larger molecules, mesopores may be the adsorption sites (LeMinh et al. 2018; Ryu et al. 1999).

In general, the adsorption process can be classified into two categories, physical and chemical adsorption. Physical adsorption, or physisorption, is a physical attraction between the adsorbate and adsorbent by van der Waals forces. Due to the interaction through a weak bond, it is a reversible process. On the other hand, chemical adsorption, or chemisorption, is the development of chemical bonds

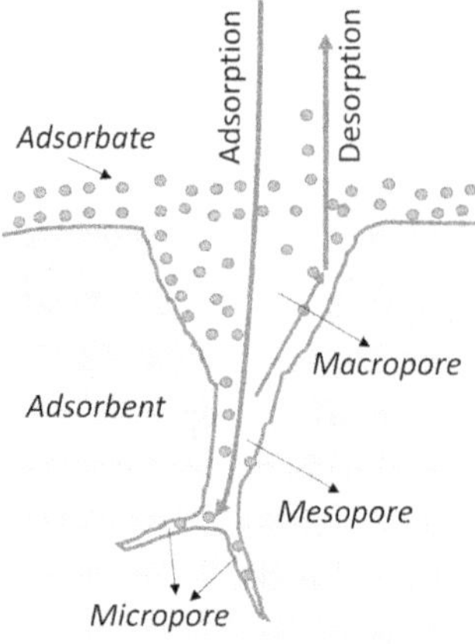

FIGURE 5.1 A schematic of adsorption on solid phase.

between adsorbate and adsorbent. Adsorbed molecules can form multiple layers in physical adsorption, whereas a single layer always occurs in chemical adsorption (Bansal and Goyal 2005; LeMinh et al. 2018; Saha and Grappe 2017; Webb 2003). Under favourable conditions, both physical and chemical adsorption processes can occur simultaneously or alternately (Dabrowski 2001).

The specific surface area, specific pore volume, pore size distribution and surface functional groups are the most important properties that determine the adsorption capacity, as well as pH, temperature, pressure, and contact time. The adsorption behaviours, depending on these factors, are expressed as adsorption isotherms, which is defined as "the relationship between the amount of substance adsorbed by the adsorbent at constant temperature and its equilibrium pressure (for gas adsorption) or concentration (for liquid adsorption)" (Bansal and Goyal 2005; Saha and Grappe 2017). In order to evaluate the adsorption isotherms, different models have been derived. Depending on the properties of the adsorbent and the adsorbed material, adsorption may be suitable for one or more of the isotherm models. Some of the most used adsorption isotherm models are Langmuir, Freundlich, and the Brunauer-Emmett-Teller (BET) isotherms. The Langmuir model, which was first developed to describe gas-solid adsorption onto activated carbons, shows the characteristics of a monolayer and assumes that the adsorption is homogeneous. On the other hand, the Freundlich model is applicable to heterogeneous surfaces and is not restricted to the monolayer formation (Ayawei et al. 2017; Al-Ghouti and Da'ana 2020; Foo and Hameed 2010; Saha and Grappe 2017). The BET equation, which is a special form of the Langmuir model, was developed to derive multilayer adsorption systems and has gained importance, especially in the analysis of the physical adsorption of gases onto porous compounds. It is widely used in the characterization of these adsorbents, such as surface area and pore dimension determination (Al-Ghouti and Da'ana 2020; Bansal and Goyal 2005).

5.3 ADSORBENTS FOR TEXTILE ODOUR CONTROL

In this section, textile applications of adsorbents for malodour adsorption are given, and studies in the literature are also summarised in Table 5.1.

5.3.1 Activated Carbons

Activated carbons are amorphous carbonaceous materials known for their high porosity, large surface area, and high surface reactivity. Because of these features, they are unique and versatile adsorbents used for the removal of malodour, colour, taste, and other inorganic or organic impurities (Bansal and Goyal 2005). Common activated carbons have a specific surface area of 600–1400 m^2/g and pore volume of 0.5–1.4 cm^3/g (Zhu, Shen, and Luo 2020).

Commercial activated carbons are produced from coal, wood, and coconut shells; moreover, they can be produced from a variety of carbonaceous materials, such as industrial and agricultural wastes (Sudaryanto et al. 2006; Timur et al. 2010). The activated carbon production process involves two main steps, (i) carbonisation of the carbonaceous precursor in an inert atmosphere and (i) activation of the carbonised product (Bansal and Goyal 2005), as shown in Figure 5.2. The activation process can be classified into two categories: (i) physical activation and (ii) chemical activation. Activated carbon production by physical activation is a two-step process that involves first carbonization of the raw material, followed by activation at a high temperature in carbon dioxide or a water vapour atmosphere. In chemical activation, carbonisation and activation are applied simultaneously after being impregnated with chemicals, such as zinc chloride, potassium hydroxide, phosphoric acid, etc.

Odorous substances are physically and chemically adsorbed onto activated carbons through Van der Waals, hydrogen, or ionic bonds. The selectivity depends on the pore size and the polarity of the adsorbent (Le Blan and Vatinel 2018). It should be noted that the surface chemistry of activated carbons can significantly affect adsorption performance as well as pore structure.

Commercial activated carbons can be classified as powdered activated carbon, granular activated carbon, and activated carbon fibres. Powdered and granular activated carbons have the highest share in the global activated carbon market (48% powdered, 35% granular, and 17% others) (Chen 2017). The application of powdered activated carbons onto textile materials through conventional finishing processes, such as padding, printing, or coating, might be of interest for the production of functional textile materials with odour adsorption property. For example, Yen et al. (2012) deposited activated carbon produced from bamboo onto a polyester nonwoven fabric using acrylic resin as a binder. Ammonia gas was used for deodorization tests, and a good deodorization effect was indicated, which increased with the increase in activated carbon concentration. In another study, activated carbons produced from coconut and oil palm shells were applied on cotton and polyester fabrics by a printing and coating method created by Eza et al. (2014). They used human olfactometry and an electronic nose to determine odour intensity, and they used onion odour as a substitution for body odour. Both activated carbons were successful in reducing the odour intensity, and the

TABLE 5.1
Some studies on adsorbents used in textiles for odour adsorption

Textile material	Adsorbent class	Adsorbent precursor/type	Adsorbent application method	Tested odorous substance	Odour test method	Reference
Polyester nonwoven fabric	Activated carbon	Bamboo	Dredging of activated carbons, followed by binder spraying	NH_3	Detector tube method	Yen et al. (2012)
Cotton and polyester fabrics	Activated carbon	Coconut shells, Oil palm shells	Printing, coating	Onion odour	Olfactometry and electronic nose	Eza et al. (2014)
Cotton and bamboo fabrics	Activated carbon	İ	Spray coating of neat and microencapsulated activated carbon	İ	Subjective evaluation after wear tests	Pragadheeswari and Sangeetha (2017)
Cotton fabric	Activated carbon	Coconut shells	Printing	İ	Subjective evaluation after wear tests	Mohan and Dhurai (2020)
Cotton knitted fabric	Biochar	Cotton, cotton/polyester, and acrylic textile wastes	Printing	Isovaleric acid	GC-MS	Cay et al. (2020)
Polyester knitted fabrics	Charcoal	Bamboo	Incorporation during fibre spinning	NH_3	Detectro tube method	Lou et al. (2012)
Polyvinyl alcohol fibres	Charcoal	Bamboo	Incorporation during fibre spinning	NH_3	Detector tube method	Lin, An, and Hsu (2007)
-	Cyclodextrin	α-CD, β-CD, γ-CD, Meβ-CD, HPβ-CD, G2β-CD, MCTβ-CD	İ	Aging odour (unsaturated aldehydes)	GC-FID	Hara et al. (2002)
Cotton, polyester and wool fabrics	Cyclodextrin	MCT-β-CD	Impregnation	Isovaleric acid	Olfactometry	Hammer, Berner-Dannenmann and Hoefer (2012)

Cotton, nylon, and polyester fabrics	Cyclodextrin	β-CD		1-octen-3-one, octanal and isovaleric acid	GC-MS and subjective evaluation	Chung et al. (2013)
Cotton fabrics	Cyclodextrin	MCT-β-CD	Impregnation	Smoke odour	Electronic nose	Setthayanond et al. (2017)
Polyester fabrics	Cyclodextrin	β-CD	Impregnation	NH_3	Detector tube method	Voncina, Vivod and Chen (2009)
Cotton fabrics	Zeolite	NaY and AgNaY	Coating	NH_3	Detector tube method	Lim et al. (2006a)
Polypropylene nonwoven fabrics	Zeolite	Natural and Ag-H	Coating	Formaldehyde, benzene, and NH_3	Detector tube method	Lim et al. (2006b)
Polypropylene nonwoven fabrics	Zeolite and silica	Na-A, Na-A HDTMA, Na-P1, NaP1 HDTMA, Molecular sieve	Incorporation during melt blown process	Acetone, cyclohexane and NH_3	Gas adsorption measurement by gas analysers	Brochocka et al. (2019)

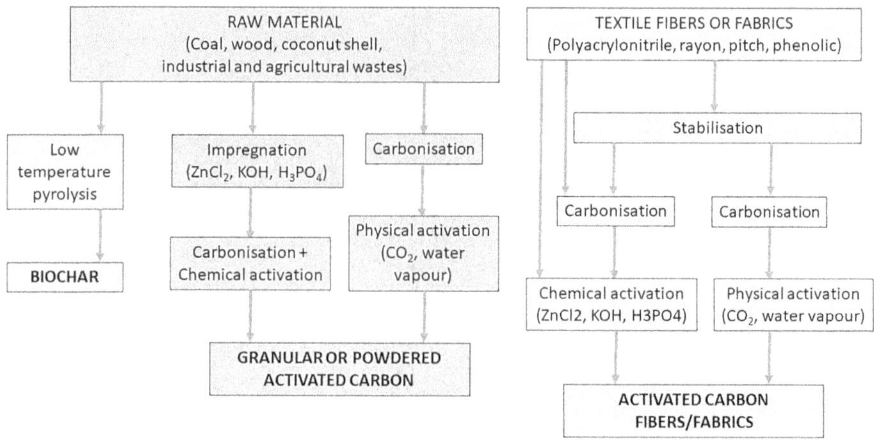

FIGURE 5.2 Schematic diagram of production processes of carbonaceous adsorbents.

odour adsorption performance increased as the amount of activated carbon on the fabric increased. It was determined that coated fabrics perform better than printed ones. Pragadheeswari and Sangeetha (2017) used activated carbon in powder form and microencapsulated activated carbon, which they spray-coated onto cotton and bamboo nonwoven fabrics. A good level of odour control was indicated through subjective evaluation after wear tests in shoes. Mohan and Dhurai (2020) applied printing paste that included activated carbon on both sides of a cotton fabric by screen-printing method. They developed a multi-layer structure with a water-repellent fabric and a cotton fabric on the outer layers, while the printed fabric remained as the middle layer. Anti-odour measurements were carried out by placement of the product under the shoe insole, and a good level of odour control was reported.

In addition to activated carbons, there has also been some work with materials called biochar, which are carbon-rich materials produced by low temperature pyrolysis but not activated. Cay et al. (2020) studied the printing of biochars derived from textile wastes onto cotton fabrics. In their study, isovaleric acid adsorption capability was also investigated. They concluded that biochars produced from cotton/polyester-blended textile wastes have a potential for use in odour masking, because they adsorbed isovaleric acid but were not totally desorbed. Additionally, the biochar application showed to increase the water vapour permeability and drying rate of cotton fabrics.

Another application of powdered activated carbon is incorporation into fibre spinning solution or polymer master batch (Kandhavadivu et al. 2011). Thus, viscose, polyester, or polyamide fibres/yarns, including activated carbon particles, can be produced. The fabrics produced from these yarns are washable without diminishing effectiveness since the activated carbon particles are distributed inside the fibres (Greenyarn 2021). Bamboo-based activated carbons are generally used for this application and named in the market as bamboo charcoal yarns. Lou et al. (2012) investigated the properties of knitted fabrics, including bamboo charcoal polyester yarns, and reported that bamboo charcoal polyester samples exhibited higher deodorization rate (Ammonia (NH_3) has been used as odour substance) compared to neat polyester samples. Lin, An, and Hsu (2007) produced bamboo charcoal polyvinyl alcohol (PVA) fibres through mixing of bamboo-based activated carbon with PVA spinning solution. A deodorization test has been applied by using NH_3, and it indicated that bamboo charcoal fibres, including 3% and higher concentrations of bamboo-based activated carbon, exhibited 100% deodorization rate, whereas neat polyvinyl alcohol (PVA) fibre had the deodorization rate of 3%. Along with academic studies, bamboo charcoal fibres are also produced commercially and marketed for products such as underwear, socks, and masks (i.e., Simplifi Fabric – Bamboo Charcoal (Simplifi Fabric n.d.), Greenyarn – Technology in Nature (Greenyarn n.d.), Litan– Bamboo Charcoal Yarn (Libolon n.d.), etc).

Although it is possible to apply powdered activated carbon onto fibres/fabrics, due to its black colour, it creates problems in terms of design and fashion for the development of leisure clothes or sportswear. For this reason, the use of activated carbons for the textile industry has gained importance, especially in the technical and medical fields. In this context, we come across activated carbon fibres (ACFs). ACFs are highly porous structures with a surface area of nearly 2000–2500 m^2/g (Tripathi et al. 2018) in fibrous form that can be produced in different textile structures, such as knitted, woven, or nonwoven fabrics. The ability to produce flexible materials thanks to the fibrous structure adds versatility to products made from ACFs. Moreover, the fibrous structure provides both macroporosity between the yarns and fibres, and microporosity on the fibre surface (Chen 2017). Compared to granular or powder-activated carbons, ACFs have the advantage of high rates of adsorption, easy separation, uniform tailored porosity, and purity (Carrott, Riberio Carrott, and Correia 2018). Additionally, while a common type of activated carbon has a ladder-like pore structure, ACF was reported to have a large amount of micropores directly on the surface, which promotes adsorption efficiency (Lee et al. 2014; Tripathi et al. 2018).

Activated carbon fibres are mostly obtained from polyacrylonitrile, cellulose (rayon), phenolic polymers, polyimide, and pitch-based carbon fibres (Thiruvenkatachari et al. 2009). The production steps are stabilisation (i.e., oxidation for polyacrylonitrile fibres, pretreatment with flame retardants for cellulosic precursors, etc.), carbonisation in an inert gas atmosphere and activation (Figure 5.2). Physical activation involves treatment at high temperature (700–900°C) with carbon dioxide or water vapour, and chemical activation involves treatment with chemicals such as zinc chloride, potassium hydroxide, or phosphoric acid at high temperature (400–900°C) in inert gas (Hassan et al. 2020).

The low mechanical strength and black colour of ACFs have restricted their use in apparels; on the other hand, they are used in protective wearable products in the fields of medicine, healthcare, and manufacturing (Wang et al. 2020). From the point of view of body odour prevention, ACF is used as an intermediate layer of commercial odour-adsorbing wound dressings to adsorb malodours that arise from wounds. Wound dressings with an ACF layer showed enhanced performance for controlling malodours compared to non-odour adsorbent basic dressings (Lee et al. 2010; Thomas et al. 1998). ActisorbTM Silver 2020 (Acelity n.d.), Carbonet (Smith-Nephew n.d.), Pharmapad (Pharmaplast Healthcare, n.d.), and Clinisorb (Clinimed n.d.) are examples of some commercial wound dressings that include ACF.

5.3.2 Cyclodextrins

CDs are cyclic oligosaccharides produced during the enzymatic degradation of starch. They consist of 6 (α-CD), 7 (β-CD), 8 (γ-CD), or more glucopyranose units joined by α(1→4)-glycosidic bonds (Buschmann, Knittel, and Schollmeyer 2001; Del Valle 2004; Szejtli 1998). They have a torus shape with a cavity in its centre. Inside the cavity, ether oxygen or polar hydrogen groups are located, which makes the cavity hydrophobic, whereas the outer layer is hydrophilic due to the external hydroxyl groups (Bezerra et al. 2020). The hydrophobic cavity acts as host and can form noncovalent inclusion complexes with various compounds, such as aromatics, alcohols, fatty acids, and esters (Nahar 2020; Voncina and Vivod 2013). Therefore, CDs exhibit odour control through adsorption of unpleasant odours.

CD applications are commonly on aroma finishing of textiles for odour masking, and the studies conducted are mostly related to the attachment of CDs onto textile materials and pleasant odour retention and release performance through CD/fragrance inclusion complexes (i.e., Martel et al. 2002; Sarıışık, Okur, and Asma 2012; Sricharussin et al. 2009; Wang and Chen 2004). On the other hand, studies encountered for CDs to provide odour control by adsorbing malodours will be discussed here. Hara et al. (2002) investigated the capture of unpleasant aging odours, which are unsaturated aldehydes that present at increased levels in the middle ages and by the elderly, by different types of CDs without textile application. It was concluded that all CDs have the capability of reducing aging odour. Hammer, Berner-Dannenmann, and Hoefer (2012) applied β-CD finishing in their study in which they examined the odour retention and release properties of cotton, wool, and polyester fabrics. Isovaleric acid has been used as a malodour agent. The presence of β-CD enhanced the retention of odorous substances, especially for cotton and polyester samples. They reported a complete entraption of isovaleric acid within the cavities of β-CD for these samples. The low performance of β-CD-finished wool fabrics was indicated due to the incomplete and unfixed fabric finish. Moreover, sensory evaluation by olfactometry showed that β-CD-finished cotton and polyester samples released less odour compared to unfinished samples. In another study, Chung et al. (2013) applied β-CD onto cotton, nylon, and polyester fabrics. 1-octen-3-one, octanal, and isovaleric acid were used as malodour substances. The odour intensity of β-CD-finished fabrics was found to be lower than unfinished ones. On the other hand, it was also reported that odorous substance retention after washing was higher in β-CD-finished fabrics than unfinished fabrics. Setthayanond et al. (2017) investigated that MCT-β-CD-treated cotton textiles reduced third-hand smoke odour release intensity. Voncina, Vivod, and Chen (2009) grafted β-CD onto polyester fabrics and ammonia adsorption tests and showed that the adsorption effect of textiles increased with β-CD application.

These studies reveal that CDs have a high potential to be used in odour prevention. However, the problem of CDs as odour adsorbents is that they are not equally effective for all odour substances since different odour compounds have various sizes and shapes. Therefore, CDs can host specific odour molecules with certain sizes (Bashari et al. 2018). Recently, CD-based finishing chemicals have been developed for odour control commercially. Wacker is marketing parent CDs and chemically modified CD derivatives under the trademarks of Cavamax® and Cavasol®, respectively (Wacker 2021). Cyclofresh®, developed by Pulcra Chemicals (Pulcra Chemicals, n.d.) is a CD-finishing chemical. Another product is Cyclofresh® Plus, containing silver as well. A dual purpose, releasing stored fragrances and capturing organic molecules in sweat that cause unpleasant odour, was targeted with this product. CDs adsorb pleasant fragrances from washing detergents, softeners, spray-on perfumes, etc. The working mechanism of the finish is explained by the contact of the CDs with moisture. In this process, the fragrances kept in the CD are released, and at the same time, unpleasant odours caused by sweating are adsorbed and retained.

5.3.3 Other Adsorbents for Body Odour Control

Porous materials, such as zeolites and silica materials, also have potential for the production of textile materials that provide odour control. Zeolites are micro-porous crystalline materials composed of a three-dimensional network of SiO4 and AlO4 tetrahedra (Belviso 2020; Kianfar et al. 2020; Maghfirah et al. 2020). They are employed as adsorbents, ion-exchangers, catalysts, and detergent builders. Lim et al. (2006a) investigated the coating of different types of zeolites with different sizes onto cotton fabric. Ammonia deodorisation tests showed that zeolites with smaller particle size ensured more effective deodorization. In another study by Lim et al. (2006b), zeolite powder was applied to nonwoven textiles for indoor air quality control. Zeolite-coated polypropylene nonwoven fabrics were subjected to a deodorisation test, and it was found that the applied binder type for coating is important for odour adsorption performance. Söylemez, Türkmen, and Özen (2020) studied the zeolite coating of breathable polyethylene films, which are one of the important components of baby and adult incontinence diapers and sanitary napkins. Hydrogen sulphide (H_2S) adsorption studies showed that raw zeolite exhibited higher adsorption values compared to zeolite-coated films, due to the closure and clogging of pores of zeolite after the coating process. It was reported that at least 40% of zeolite should be used for coating for efficient H_2S odour removal. Brochocka et al. (2019) incorporated modified zeolite and silica materials during the melt-blown process to produce polypropylene nonwoven fabrics. Adsoption tests with acetone, cyclohexane, and ammonia have shown the volatile organic compound adsorption potential of zeolite- and silica-incorporated polypropylene nonwovens. As is seen, the studies on the use of zeolites for body odour control are limited in the literature. On the other hand, finishing chemicals containing zeolite for deodorization are commercially available. Active XL® and Lava® products of Sciessent (Sciessent n.d.), and ZEOflair® of Lawrence Industries (Lawrence Industries n.d.) are some examples of commercial applications.

Another candidate for the production of odour-adsorbing textiles are dendritic materials, which are artificial nano-sized highly branched materials. Due to abundant functional groups and hollow interior branches, dendritic materials act as host and are able to form inclusions or ion-pairs with various compounds (Dykes 2001; Frechet 2003; Seiler 2006). This host-guest property of the dendritic materials led to several potential uses in textile applications as dyeing auxiliary, adsorbent, flame retardant, water repellent, antibacterial finish, etc. (Akbari and Kozlowski 2019). Studies in the literature on dendritic materials in terms of odour control have been found to be the development of fragrance fabrics for odour masking (Akbari et al. 2016; Mounesan, Akbari and Brycki 2021; Shakoorjavan et al. 2016). Applications related to odour control by adsorption of unpleasant body odours with dendritic materials will be of interest. On the other hand, no study on the use of dendritic material applied textile materials for unpleasant odour adsorption has been found yet in the open literature.

5.4 CONCLUSION

This chapter reports body odour control by adsorption for textile materials. Academic studies and commercial applications showed that activated carbons and CDs have the highest application potential for the production of odour-adsorbent textile materials. These adsorbents can be applied into the fibres during fibre spinning or onto fabrics through finishing processes. Another method is to convert the textile material itself into an adsorbent, as is the case with activated carbon fibres, which have found commercial applications, especially in medical textile applications, such as anti-odour wound dressings. The black colour of activated carbon/carbon fibres, however, is problematic in design, especially in leisure clothes, and appropriate design studies are required in this regard. Despite all the work carried out, currently the odour-adsorbent textile materials have not reached the expected commercial potential. Therefore, it is believed that, through extending the latest developments in the textile field approaching the functionalisation of textile materials with odour adsorbents, especially in casual products such as sports and leisure clothes, underwear, socks, and technical textiles such as wound dressings and masks; industrial uptake of these innovative functional textile materials into textile markets could be obtained.

REFERENCES

Abdul-Bari, M.M., McQueen, R.H., Nguyen, H., Wismer, W.V., de la Mata, A.P., and Harynuk, J.J. 2018. Synthetic clothing and the problem with odor: Comparison of nylon and polyester fabrics. *Clothing and Textiles Research Journal*, 36(4): 251–266.

Acelity. n.d. "Actisorb™ Silver 220 Activated Charcoal Dressing with Silver". Accessed March 28, 2021. https://www.acelity.com/healthcare-professionals/global-product-catalog/catalog/actisorb-silver-220-dressing.

Akbari, S., Akbari, M., Kish, M.H., and Mazaheri, F.M. 2016. Preparation of long-lasting fragrant worsted fabrics using polypropylene-imine (PPI) dendrimer. *International Journal of Clothing Science and Technology*, 28(5): 699–711.

Akbari, S., and Kozlowski, R.M. 2019. A review of application of amine-terminated dendritic materials in textile engineering. *The Journal of the Textile Institute*, 110(3): 460–467.

Al-Ghouti, M.A., and Da'ana, D.A. 2020. Guidelines for the use and interpretation of adsorption isotherm models: A review. *Journal of Hazardous Materials*, 393: 122383.

Ayawei, N., Ebelegi, A.N., and Wankasi, D. 2017. Modelling and interpretation of adsorption isotherms. *Journal of Chemistry*, 2017: 3039817.

Bansal, R.C., and M. Goyal. 2005. *Activated Carbon Adsorption*. CRC Press.

Bashari, A., Shakeri, M., Shirvan, A.R., and Najafabadi, S.A.N. 2018. Functional finishing of textiles via nanomaterials. In S. Islam, and B.S. Butola (ed.), *Nanomaterials in the Wet Processing of Textiles* (pp. 1–70). John Wiley & Sons.

Belviso, C. 2020. Zeolite for potential toxic metal uptake from contaminated soil: A brief review. *Processes*, 8: 820.

Ben-Mansour, R., Habib, M.A., Bamidele, O.E., Basha, M., Qasem, N.A.A., Peedikakkal, A., Laoui, T., and Ali, M. 2016. Carbon capture by physical adsorption: Materials, experimental investigations and numerical modeling and simulations – A review. *Applied Energy*, 161: 225–255.

Bezerra, F.M., Lis, M.J., Firmino, H.B., da Silva, J.G.D., Valle, R.C.S.C., Valle, J.A.B., Scacchetti, F.A.P., and Tessaro, A.L. 2020. The role of β-cyclodextrin in the textile industry-review. *Molecules*, 25: 3624.

Brochocka, A., Nowak, A., Panek, R., and Franus, W. 2019. The effects of textural parameters of zeolite and silica materials on the protective and functional properties of polymeric nonwoven composites. *Applied Sciences*, 9: 515.

Buschmann, H.J., Knittel, D., and Schollmeyer, E. 2001. New textile applications of cyclodextrins. *Journal of Inclusion Phenomena and Macrocyclic Chemistry,* 40: 169–172.

Carrott, P.J.M., Ribeiro Carrott, M.M.I., and Correia, P.F.M.M. 2018. Evolution of porosity of activated carbon fibres prepared from pre-oxidized acrylic fibres. *Microporous and Mesoporous Materials*, 264: 176–180.

Cay, A., Yanik, J., Akduman, C., Duman, G., and Ertas, H. 2020. Application of textile waste derived biochars onto cotton fabric for improved performance and functional properties. *Journal of Cleaner Production*, 251: 119664.

Chen, J.Y. 2017. *Introduction in Activated Carbon Fiber and Textiles*. J.Y. Chen (ed.), 3–20. Woodhead Publishing.

Chung, H., Hwang, N., Kim, J.Y., and S.Y. Shin. 2013. Adsorption and removal of volatile organic compounds from fabrics with β-Cyclodextrin finish. *Journal of the Korean Society of Clothing and Textiles*, 37(1): 113–123.

Clinimed. n.d. "Clinisorb – Wound care products". Accessed March 28, 2021. https://www.clinimed.co.uk/wound-care/products.

Dabrowski, A. 2001. Adsorption – from theory to practice. *Advances in Colloid and Interface Science*, 93(1-3): 135–224.

Del Valle, E.M. 2004. Cyclodextrins and their uses:a review. *Process Biochemistry*, 39: 1033–1046.

Dykes, G.M. 2001. Dendrimers: A review of their appeal andapplications. *Journal of Chemical Technology and Biotechnology*, 76: 903–918.

Eza, T.S.M., Ahmad, W.Y.W., Ahmad, M.R., Omar, K., and Ahmad, M.N. 2014. Effectiveness of activated carbon produced from coconut and oil palm shells as intiodour on textile fabrics. *Indian Journal of Fibre & Textile Research*, 39: 190–195.

Foo, K.Y., and Hameed, B.H. 2010. Insights into the modeling of adsorption isotherm systems. *Chemical Engineering Journal*, 156: 2–10.

Frechet, J.M.J. 2003. Dendrimers and other dendritic macromolecules: frombuilding blocks to functional assemblies in nanoscienceand nanotechnology. *Journal of Polymer Science: Part A: Polymer Chemistry*, 41: 3713–3725.

Greenyarn. n.d. "Technology in nature". Accessed March 22, 2021. https://www.greenyarn.com/technology.htm

Hammer, T.R., Berner-Dannenmann, N., and Hoefer, D. 2012. Quantitative and sensory evaluation of malodour retention of fibre types by use of artificial skin, sweat and radiolabelled isovaleric acid. *Flavour and Fragrance Journal*, 28: 238–244.

Hara, K., Mikuni, K., Hara, K., and Hashimoto, H. 2002. Effects of cyclodextrins on deodoration of "aging odor". *Journal of Inclusion Phenomena and Macrocyclic Chemistry*, 44: 241–245.

Hassan, M.F., Sabri, M.A., Fazal, H., Hafeez, A., Shezad, N., and Hussain, M. 2020. Recent trends in activated carbon fibers production from various precursors and applications-A comparative review. *Journal of Analytical and Applied Pyrolysis* 145: 104715.

Kandhavadivu, P., Vigneswaran, C., Ramachandran, T., and Geethamanohari, B. 2011. Development of polyester-based bamboo charcoal and lyocell-blended union fabrics for healthcare and hygienic textiles. *Journal of Industrial Textiles*, 41(2): 142–159.

Kianfar, E., Hajimirzaee, S., Mousavian, S., and Mehr, A.S. 2020. Zeolite-based catalysts for methanol to gasoline process: A review. *Microchemical Journal*, 156: 104822.

Klaykruayat, B., Siralertmukul, K., and Srikulkit, K. 2010. Chemical modification of chitosan with cationic hyperbranched dendritic polyamidoamine and its antimicrobial activity on cotton fabric. *Carbohydrate Polymers*, 80(1): 197–207.

Laing, R.M. 2019. Natural fibres in next-to-skin textiles: Current perspectives on human body odour. *SN Applied Sciences*, 1: 1329.

Lawrence Industries. n.d. "ZEOflair Speciality Zeolites". Accessed May 10, 2021. https://www.l-i.co.uk/products/zeoflair-voc-odor-trapping-zeolite

Le Blan, T., and Vatinel, A. 2018. Odor reduction thanks to textile materials. Paper presented at the 6th Reset Seminar on "New materials and new applications".

Lee, G., Anand, S.C., Rajendran, S., and Walker, I. 2010. Development of odour (volatile molecule) adsorbent materials for healthcare. In S.C. Anand, J.F. Kennedy, M. Miraftab, and S. Rajendran (eds.), *Medical and Healthcare Textiles* (pp. 206–214). Woodhead Publishing.

Lee, T., Ooi, C.H., Othman, R., and Yeoh, F.Y. 2014. Activated carbon fiber – The hybrid of carbon fiber and activated carbon. *Reviews on Advanced Materials Science*, 36: 118–136.

LeMinh, N., Sivret, E.C., Shammay, A., and Stuetz, R.M. 2018. Factors affecting the adsorption of gaseous environmental odors by activated carbon: A critical review. *Critical Reviews in Environmental Science and Technology*, 48(4): 341–375.

Libolon. n.d. "LitanBamboo Charcoal Yarn". Accessed March 25, 2021. https://www.http://www.libolon.com/product277

Lim, H.M., Jung, J.S., Kim, B.Y., and Lee, S.H. 2006a. Application of zeolites on cellulose fiber. *Key Engineering Materials*, 317-318: 777–780.

Lim, H.M., Jung, J.S., Kim, D.S., Lee, D.J., Lee, S.H., and Kim, W.N. 2006b. Modification of natural zeolite powder and its application to interior non-woven textile for indoor air quality control. *Materials Science Forum*, 510-511: 934–937.

Lin, C.A., An, T.C., and Hsu, Y.H. 2007. Study on the far infrared ray emission property and adsorption performance of bamboo charcoal/polyvinyl alcohol fiber. *Polymer-Plastics Technology and Engineering*, 46(11): 1073–1078.

Lou, C.W., Lu, C.T., Lin, C.W., Chen, A.P., Jang, S.B., and Lin, J.H. 2012. Manufacturing technique and deodorization effectiveness against ammonia gas of bamboo charcoal/spandex complex knitted fabrics. *Journal of Engineered Fibers and Fabrics*, 7(2): 109–117.

Maghfirah, A., Ilmi, M.M., Fajar, A.T.N., and Kadja, G.T.M. 2020. A review on the green synthesis of hierarchically porous zeolite. *Materials Today Chemistry*, 17:100348.

Martel, B., Morcellet, M., Ruffin, D., Vinet, F., and Weltrowski, M. 2002. Capture and controlled release of fragrances by CD finished textiles. *Journal of Inclusion Phenomena and Macrocyclic Chemistry*, 44: 439–442.

McQueen, R.H., Laing, R.M., Brooks, H.J.L., and Niven, B.E. 2007. Odor intensity in apparel fabrics and the link with bacterial populations. *Textile Research Journal*, 77(7): 449–456.

Mohan. A.V., and Dhurai, B. 2020. Development of anti odor and air filtering work wear using activated charcoal. *International Research Journal of Engineering and Technology*, 7: 34–40.

Mounesan, M., Akbari, S., and Brycki, B.E. 2021. Extended-release essential oils from poly (acrylonitrile) electrospun mats with dendritic materials. Industrial Crops & Products, 160: 113094.

Nahar, K. 2020. A review on anti odor finishing of textiles, its application method and commercialization. *ACTA TechnicaCorviniensis – Bulletin of Engineering*, XIII: 79–84.

Pharmaplast Healthcare. n.d. “Pharmapad- Wound dressing containing Activated Charcoal layer”. Accessed March 28, 2021. https://www.pharmaplast-online.net/product-category/wound-care/deodorizing-wound-dressings/

Pragadheeswari, R., and Sangeetha, K. 2017. Functional properties of activated carbon treated textile material. *International Journal for Scientific Research & Development*, 5: 1766–1768.

Pulcra Chemicals. n.d. “The world of Cyclofresh® - A world of freshness all day long”. Accessed March 25, 2021. https://www.cyclofresh-pulcra.com/#technology.

Richter, T.M., Bremer, P.J., Silcock, P., and Laing, R.M. 2018. Textile binding and release of body odor compounds measured by proton transfer reaction – mass spectrometry, *Textile Research Journal*, 88(22): 2559–2567.

Ryu, Z., Zheng, J., Wang, M., and Zhang, B. 1999. Characterization of pore size distributions on carbonaceous adsorbents by DFT. *Carbon*, 37(8): 1257–1264.

Saha, D., and Grappe, H.A. 2017. Adsorption properties of activated carbon fibers. In J.Y. Chen (ed.), A*ctivated Carbon Fiber and Textiles* (pp. 143–165). Woodhead Publishing.

Sarıışık, M., Okur, S., and Asma, Ş. 2012. Odor adsorption kinetics on modified textile materials using quartz microbalance technique. *ActaPhysicaPolonica Series a*, 121(1):243–246.

Sciessent, n.d. “Odor control solutions”. Accessed May 10, 2021. https://www.sciessent.com/solutions/odor-control/

Seiler, M. 2006. Hyperbranched polymers: Phase behavior and new applications in the field of chemical engineering. *Fluid Phase Equilibria*, 241: 155–174.

Setthayanond, J., Sodsangchan, C., Suwanruji, P., Tooptompong, P., and Avinc, O. 2017. Influence of MCT-β-cyclodextrin treatment on strength, reactive dyeing and third-hand cigarette smoke odor release properties of cotton fabric. *Cellulose*, 24: 5233–5250.

Shakoorjavan, S., Akbari, S., Kish, M.H., and Akbari, M. 2016. Correlation of sensory analysis with a virtual sensor array data for odour diagnosis of fragrant fabrics. *Measurement*, 90: 396–403.

Shishoo, R. 2015. *Introduction to textiles in sport, in Textiles for sportswear*. R. Shishoo (ed.), (pp. 3–16). Woodhead Publishing Limited.

Simlifi Fabric. n.d. “Bamboo Charcoal”. Accessed March 20, 2021. https://www.simplififabric.com/pages/bamboo-charcoal.

Sing, K.S.W., Everett, D.H., Haul, R.A.W., Moscou, L., Pierotti, R.A., Rouquerol, J., and Siemieniewska, T. 1985. Reporting physisorption data for gas/solid systems with special reference to the determination of surface area and porosity. *Pure & Applied Chemistry*, 57(4): 603–619.

Smith-Nephew. n.d. “Carbonet”. Accessed March 28, 2021. https://www.smith-nephew.com/australia/healthcare/products/product-types/exudate-management-dressings/low-adherent--absorbent-dressings/carbonet-/

Söylemez, C., Türkmen, İ., and Özen, İ. 2020. Achieving dual-functionality by surface coating of zeolite with stearic acid: Combining breathability and odor control properties in polyethylene/zeolite composite films. *Macromolecular Research*, 28(12): 1149–1159.

Sricharussin, W., Sopajaree, C., Maneerung, T., and Sangsuriye, N. 2009. Modification of cotton fabrics with β-cyclodextrin derivative for aroma finishing. *Journal of the Textile Institute*, 100(8): 682–687.

Sudaryanto, Y., Hartono, S.B., Irawaty, W., Hindarso, H., and Ismadji, S. 2006. High surface area activated carbon prepared from cassava peel by chemical activation. *BioresourceTechnology*, 97: 734–739.

Szejtli, J. 1998. Introduction and general overview of cyclodextrin chemistry. *ChemicalReviews*, 98(5): 1743–1754.

Thiruvenkatachari, R., Su, S., An, H., and Yu, X.X. 2009. Post combustion CO_2 capture by carbon fibre monolithic adsorbents. *Progress in Energy and Combustion Science*, 35: 438–455.

Thomas, S., Fisher, B., Fram, P.J., and Waring, M.J. 1998. Odour-absorbing dressings. *Journal of Wound Care*, 7(5): 246–250.

Timur, S., Kantarli, I.C., Onenc, S., and Yanik, J. 2010. Characterization and application of activated carbon produced from oak cups pulp, *Journal of Analytical and Applied Pyrolysis*, 89: 129–136.

Tripathi, H.K., Singh, V.V., Sathe, M., Thakare, V.B., and Singh, B. 2018. Activated carbon fabric: An adsorbent material for chemical protective clothing. *Defence Science Journal*, 68(1): 83–90.

Voncina, B., and V. Vivod. 2013. Cyclodextrins in textile finishing. In M. Gunay (ed.), *Eco-friendly Textile Dyeing and Finishing* (pp. 53–75). Intech Open.

Voncina, B., Vivod. V., and Chen, W.T. 2009. Surface modification of PET fibers with the use of β-cyclodextrin. *Journal of Applied Polymer Science*, 113: 3891–3895.

Wacker Chemie A.G. n.d. Cavamax® and Cavasol® Cyclodextrins – The Innovator's Molecule. Introduction paper, *Wacker*. https://www.wacker.com/h/medias/6223-EN.pdf (accessed March 25, 2021).

Wang, C.-X., and Chen, S.L. 2004. Surface modification of cotton fabrics with β-cyclodextrin to impart host-guest effect for depositing fragrance. *AATCC Review*, 4(5): 25–28.

Wang, J., Lu, X., Wang, J., and Wang, X. 2019. Quantitative and sensory evaluation of odor retention on polyester/wool blends. *Textile Research Journal*, 89(1): 2729–2738.

Wang, J., Tang, B., Bai, W., Lu, X., Liu, Y., and Wang, X. 2020. Deodorizing for fiber and fabric: Adsorption, catalysis, source control and masking. *Advances in Colloid and Interface Science*, 283:102243.

Webb, P.A. 2003. Introduction to chemical adsorption analytical techniques and their applications to catalysis. MIC Technical Publications. https://www.micromeritics.com/Repository/Files/intro_to_chemical_adsorption.pdf (accessed March 22, 2021).

Yen, M.S., Chen, C.W., Hsiao, C.N., and M.C. Kuo. 2012. Medical effects of poly-ethylene terephthalate (PET) non-woven fabrics treated with bamboo activated charcoal. *African Journal of Biotechnology*, 11: 11116–11125.

Zhu, L., Shen, D., and Luo, K.H. 2020. A critical review on VOCs adsorption by different porous materials: Species, mechanisms and modification methods. *Journal of Hazardous Materials*, 389: 122102.

6 Odour Control or Inhibition Using Antimicrobial Finishing

Rosie Broadhead, Laure Craeye, and Dr. Chris Callewaert
Ghent University, Center for Microbial Ecology and Technology, Coupure Links, Ghent, Belgium

CONTENTS

DOI: 10.1201/9781003141426-6

6.1 INTRODUCTION: BACKGROUND AND DRIVING FORCES

The clothing and fashion industry is a multi-billion market with a fast pace of innovation. From 1990 to 2000, laundering habits have shifted from chemical-based detergents and hot washes towards enzyme-based detergents and cold washes, driven by environmental and economic reasons (Laitala and Mollan Jensen 2010). The side effect of the lower washing temperatures, enzyme-based cleaning technology, decreased water consumption, and absence of bleaching agents is reduced microbial and volatile removal in the laundry process (Callewaert et al. 2015). Certain textile fibers are more prone to odour development, as is the case for polyester (Callewaert et al. 2015). Clothing textiles absorb skin sweat and microbes, which can lead to increased odour generation. Besides, textile microorganisms can cause staining, fabric deterioration, and even physical irritation, such as skin allergies or infections (Szostak-Kotowa 2004). In recent decades, a wide range of textile finishes, antimicrobial techniques, and nanoparticle applications have been developed to answer this need.

6.1.1 The Skin Microbiome

The skin's surface is the first line of defense for the body, and 'a sensory and excretory organ' (Jablonski 2013). Not only this, but the skin is home to hundreds of millions of microorganisms that live and feed on the skin's surface secretions (Jablonski 2013). These millions of microorganisms, such as bacteria, fungi, and viruses, form the so-called skin microbiome (Byrd, Belkaid, and Segre 2018; Schommer and Gallo 2013). It is known that these organisms are more numerous in comparison to the body's own cells: there are approximately 30 trillion body cells in comparison to 39–100 trillions microorganisms. As the skin covers a surface area of approximately two square metres, the microbial density is estimated to be 10^6 per square centimetre (Elias 2007; Lyman 2020; Scharschmidt and Fischbach 2013).

In general, at least 19 phyla are characterized within the microbiome, of which four phyla dominate the skin: Actinobacteria (51.8%), Proteobacteria (16.5%), Firmicutes (24.4%), and Bacteroidetes (6.3%). Furthermore, three genera represent 60% of the skin microbiome: *Staphylococcus* spp., *Corynebacterium* spp., and *Cutibacterium* spp. (Byrd, Belkaid, and Segre 2018; Lyman 2020; Scharschmidt, and Fischbach 2013; Schommer and Gallo 2013).

6.1.2 Odour Formation in Textiles

Clothing textiles are important contributors to odour development. Bacteria living on the textiles can convert sweat secretions, sebum, skin debris, and other external molecules into volatile compounds. Shelley and Hurley (1953) stated that odour intensity is potentially more intense in clothes compared to the axillary skin (Shelley and Hurley 1953). Dravnieks et al. (1968) described axillary skin odour as primary odour, and axillary clothing odour as secondary odour (Dravnieks et al. 1968). Bacteria attaching and living on and in the fibers use dirt or sebum compounds as substrate and generate volatiles as by-products. That odour can be vastly different from the odour from the axillary skin. There are two main reasons for this: the build-up of sweat secretions in the textiles, and the microbiome converting these molecules into malodour (Van Herreweghen et al. 2020).

Over time, the metabolite load can increase in the fibers, as more sweat secretions are transferred to the clothes and are insufficiently removed by the laundering process. Specifically, hydrophobic molecules, such as apocrine and sebaceous secretions, can adhere and bind to clothes, which leads to a build-up of metabolite precursors (Van Herreweghen et al. 2020). This might be an important explanation for permastink clothes. Not only a load of sweat transferred but also the type of sweat is of influence. It was shown that a mutation in the human ABCC11 gene, which is prominent among East-Asian people, impacts apocrine sweat secretions (Martin et al. 2010) and, as such, malodour build-up in clothes. Additionally, gender, hygiene habits, diet, and body mass index can have an impact on odour build-up in clothes.

The microorganisms living in clothing textiles similarly have a big influence on odour formation and have been a major target for odour control. Bacteria are transferred in high numbers from armpits to clothing fibers due to their close contact. However, axillary and textile microbiomes of one person are not the same and differ in composition. The odour-causing bacteria in armpits, *Corynebacterium* spp., could not be isolated from worn textiles (Callewaert et al. 2014; McQueen et al. 2014). Particular low abundant skin microorganisms can selectively enrich synthetic textiles (Teufel et al. 2010). Rather, the textile microbiome is composed of *Staphylococcus*, *Micrococcus*, *Bacillus*, Enterobacteriaceae, and *Acinetobacter*, among others (Callewaert et al. 2014). Selective enrichment of particular odour-associated taxa on different fiber types is an explanation of why some fibers, like polyester, smell less pleasant. *Micrococcus* was enriched on malodorous synthetic clothes and is known to have the enzymatic potential to cause malodour from apocrine sweat (James et al. 2013; Teufel et al. 2010). *Moraxella* and *Pseudomonas* were similarly found on washed laundry and associated with malodour formation (Goto et al. 2016). Unlike the microbial composition, the microbial load does not seem to be a determining factor in odour formation in clothes (McQueen et al. 2014). Nonetheless, the textile industry is making use of a variety of anti-odour techniques and antimicrobial agents, as described below.

6.2 ANTIMICROBIAL AGENTS IN THE TEXTILE INDUSTRY

The ability to retain moisture and the large surface area makes textiles contributory to microorganisms' growth. This microorganism susceptibility can cause a range of undesirable effects on the textile itself and for the user (Gao and Cranston 2008; Morais, Guedes, and Lopes 2016). Such effects can include a change in the mechanical strength of the textile, discolouration, a tendency toward user contamination, and most commonly the development of unpleasant odour (Shahidi and Wiener 2012). Over the last few years, there has been a developing public awareness of these bacterial effects on textiles and health. More specifically, there is a rising concern about the microbial propagation in textiles used for health care in hospitals. Besides this, with the focus in our society changing to more active lifestyles, the demand for anti-odour sportswear has been increasing (Gao and

Cranston 2008). Therefore, research and development in this field have been intensified in order to minimize or even eradicate microbes' growth, and antimicrobial textiles have been developed (Morais, Guedes, and Lopes 2016).

Antimicrobial textiles with odour-resistance properties are becoming a new requirement in the market. (Buschmann, Dehabadi, and Wiegand 2015). These textiles contain antimicrobial and/or antifungal compounds. An antimicrobial chemical agent is classified according to the mode of action against the cells' function. If it initiates the inhibition of cell growth, the microbial agent is biostatic, but if it can kill microorganisms, its effect is called biocidal. However, antimicrobial agents often have both characteristics (Gao and Cranston 2008; Rahman, Ahsan, and Islam 2010; Zille et al. 2014). Fabrics provided with such biostatics or biocides can neutralise odours and deodorise the wearer, creating antiseptic environments or even enable textiles to clean themselves (Quinn 2010).

There are two *modus operandi* that can be applied within antimicrobial textiles. The first being the prevention of sweat decomposition by killing the bacteria on the skin. For this, several antimicrobial agents have already been used: nanosilver particles, triclosan, quaternary ammonium compounds, ammonium salts, chlorine derivatives of phenols, drugs, etc. The main synthetic nanoparticle materials and finishes used in the textile industry are discussed in more detail below (Buschmann, Dehabadi, and Wiegand 2015; Höfer 2006). Besides the use of antimicrobial compounds, a second strategy is to chelate chemical, odorous substances originating from bacterial decomposition of sweat. Examples of chelating molecules are cyclodextrins or calixarenes. An advantage of this technique in comparison to the use of antimicrobials is that it does not affect the skin microbiome. With this technique, however, only odour is combated (Buschmann, Dehabadi, and Wiegand 2015).

There are different possible ways to apply antimicrobial compounds to textiles. First, textile fibers can be impregnated with a solution, suspension, or emulsion of the antimicrobial product. Secondly, an insoluble suspension containing the antimicrobial compound can be made. Afterwards, this solution can be applied onto the textile. Thirdly, it is possible to apply the product to the textile through chemical bonding. Finally, during the spinning process, fibers can be immersed in the antimicrobial product during the production process (Coman et al. 2010).

6.2.1 Synthetic Antibacterial Finishes and Nanoparticles

In the textile industry, biocides are the most used antimicrobial agent and include products such as metallic nanoparticles and their salts, quaternary ammonium compounds (QACs), triclosan, silver, etc. Nanoparticles in particular are often used in textile finishing because of the ease of incorporation in the fibers (Zille et al. 2014). The chemical properties and structure of these components determine the possible modes of action, which is referred to as the different ways of killing the microorganisms. Examples of such modes of action include damage to the cell wall or cell membrane, and inhibition of the synthesis of these structures, leading to cell leakage and therefore cell death. Inhibition of DNA/RNA, protein synthesis, or specific metabolic processes within the cell are other possibilities (Morais, Guedes, and Lopes 2016).

Different antimicrobials are available to be used as textile finishes, but several requirements are set: nontoxicity and more specific, not giving rise to irritation or allergy on the skin, efficient against microorganisms, suitable for textile processing, durable during laundering, and no impact on the quality or appearance of the textile, etc. (Morais, Guedes, and Lopes 2016).

6.2.1.1 Metals and Metallic Salts

Until now, metals such as silver, copper, zinc, and cobalt are most used in the textile industry on fabrics like cotton, wool, polyester, and nylon. Not only metals but also the equivalent oxides or salts are used as antimicrobial finishes. Metals are known to be extremely toxic in very low concentrations, both in free or bounded state. Another applicability of metals is usage as nanoparticles, which have a higher surface, resulting in stronger antimicrobials. Besides that, nanoparticles are easier to embed in textile fibers (Morais, Guedes, and Lopes 2016).

6.2.1.1.1 Silver

Of all the metals, silver is the best and most used one in the textile industry (Shahidi and Wiener 2012; Zille et al. 2014). It has broad-spectrum antibacterial properties against gram-positive and negative bacteria such as *Pseudomonas aeruginosa, Staphylococcus aureus, Staphylococcus epidermidis, Escherichia coli,* and *Klebsiella pneumonia.* Even antimicrobial activity against *Staphylococcus aureus* methicillin-resistant strains is reported (MRSA) (Hasan, Crawford, and Ivanova 2013). In addition, silver-based antimicrobials are also functional against viruses or even against eukaryotic microorganisms (Zille et al. 2014). Silver ions, for instance, bind disulfide or sulfhydryl groups of proteins present in the bacterial cell wall, therefore disturbing cellular processes leading to death (Shahidi and Wiener 2012). Other modes of action include interaction with the cell membrane, resulting in loss of permeability and subsequent intracellular reactions, such as DNA condensation, or stimulation of reactive oxygen species (ROS) production, leading to lipid and DNA damage (Shahidi and Wiener 2012; Zille et al. 2014). In the case of silver nanoparticles, microbial toxicity can be controlled by the size, shape, and crystalline structure of the nanoparticles. It is known that smaller particles with a size of one to ten nanometer, a triangular shape and a bigger surface area have a higher impact on cells in comparison to bigger particles (Zille et al. 2014).

Silver or AgNPs encapsulate or coated textiles are widely used in the textile industry as they are known to significantly reduce gram-positive bacterium *Staphylococcus aureus and Klebsiella pneumoniae* on textiles (Buschmann, Dehabadi, and Wiegand 2015; Dhiman and Chakraborty 2015; Gowramma et al. 2015). These organisms are known pathogens, but they are not only responsible for developing odour in textiles through the metabolism of human sweat (McQueen 2011). The effect of Ag-treated textiles on the skin microbiome has not been discussed in literature yet. We have done a study at UC San Diego and found a low but, surprisingly, unfavorable effect on the skin microbiome, with higher bacterial load and more odour-associated taxa on skin when wearing Ag-treated textiles (unpublished data).

6.2.1.1.2 Copper

Copper used as an antimicrobial is comparable to silver as it also can bind disulfide or sulfhydryl groups of proteins in the cell, resulting in cell death (Shahidi and Wiener 2012). Products like copper oxide (CuO) are cheaper than silver and have antimicrobial activity against gram-positive and negative microorganisms. However, higher concentrations of CuO are needed to achieve the same bactericidal effect as silver-based products. Moreover, the synthesis of this antimicrobial is more challenging (Zille et al. 2014). Besides CuO, copper ions or copper salts, such as copper sulfate ($CuSO_4$), are used in antimicrobial finishes (Shahidi and Wiener 2012).

Although silver nanoparticles and salts are the most widely used metal for antibacterial textile application (Morais, Guedes, and Lopes 2016), CuO, ZnO, and TiO nanoparticles have been developed for textile applications, often using the sol-gel method. The antimicrobial effect of ions from CuO nanoparticles differs to the effect of ions from copper salts, at the same concentrations (Gunawan et al. 2011). This study highlights the complex behavior of antimicrobial metal nanoparticles as both the nanoparticle itself or the ions can inform the biocidal action (Dastjerdi and Montazer 2010; Palza 2015).

6.2.1.1.3 Zinc

Zinc is a trace element that is vital for the immune system, sensory functions, and metabolism as well as skin regeneration and protection (O'Dell 1981; Schubert et al. 2015). Moreover, zinc ions, zinc oxides, or zinc salts, such as zinc acetate, zinc chloride, or zinc sulfate, are also used as antimicrobial finishes (Zille et al. 2014). Zinc oxide nanoparticles do have antibacterial activity against gram-positive and negative microorganisms. Bacterial growth is inhibited after the nanoparticle passes through the cellular envelope and disorganizes the cell membrane. Moreover, it can generate hydrogen peroxide molecules, which damage the cell. These zinc nanoparticles have an equivalent performance as the silver NP's or copper oxide. Zinc nanoparticles have a size of about 30–40 nanometers, which is inversely proportional to the particles' antibacterial activity (Morais, Guedes, and Lopes 2016; Zille et al. 2014).

A zinc nanoparticle-chitosan composite was developed in a study. The composite was coated onto cotton fabrics in order to fix the antimicrobial properties of ZnO and chitosan and improve the chemical bonding of chitosan to textile substrates. The ZnO nanoparticle-chitosan composite presented a better antibacterial activity at a lower molecular weight, which was likely due to the improved movement of the polymeric chains in solution (Farouk et al. 2012).

6.2.1.1.4 Others

There are a lot of different metals that can be used as antimicrobials. Besides the most used ones being silver, copper, and zinc, also titanium (oxide), magnesium, gold, tin, antimony, zirconium, and nickel are reported as antimicrobials. However, titanium, tin, and antimony are not very efficient. Moreover, nickel is less efficient than silver (Morais, Guedes, and Lopes 2016; Shahidi and Wiener 2012; Zille et al. 2014).

6.2.1.2 Triclosan

Triclosan (2,4,4'-trichloro-2'hydroxydiphenyl ether) is an odourless, synthetic, chlorine-containing derivative of phenol whereof its antimicrobial activity is concentration and formulation dependent. Triclosan is a broad-spectrum antimicrobial effective against gram-positive and negative organisms and is known to reduce MRSA (Bhargava and Leonard 1996; Morais, Guedes, and Lopes 2016; Shahidi and Wiener 2012). Triclosan has two modes of action: affecting the cell membrane's integrity by obstructing lipid biosynthesis and/or inhibiting RNA and protein synthesis within the cell. Due to its strong antimicrobial potency, triclosan is often used in the textile industry on polyester, nylon, polypropylene, cellulose acetate, or acrylic textiles (Bhargava and Leonard 1996; Morais, Guedes, and Lopes 2016). In recent years, however, it has been associated with harmful side effects on human skin; therefore, the use of triclosan in textiles is expected to decline (Ruszkiewicz et al. 2017).

To evaluate the odour-inhibiting properties of Tricosan, Mao and Murphy (2001) carried out a wear trial on treated textiles using Tinosan AM 100, a triclosan-based antimicrobial (Mao and Murphy 2001). During the study, the participants were asked to evaluate the odour emanating from a treated and untreated fabric worn under each armpit. It was found that 90% of the participants had a preference for the treated textiles, which had a 'fresher' odour over the untreated. After analysis it was found to have a 2–3 log decrease in *S. aureus* and *Klebsiella pneumoniae* strains after an inoculation period of 24 hours at 37°C from the triclosan treated textiles (Mao and Murphy 2001). Therefore, the study concluded the linkage between antimicrobial-treated textiles and odour intensity.

6.2.1.3 Quaternary Ammonium Compounds (QACs)

In 1930, quaternary ammonium compounds (QACs), a group of 191 components, were discovered to have antimicrobial activity (Morais, Guedes, and Lopes 2016; Tischer et al. 2012). Examples of QACs are hydrophobic, linear alkyl ammonium compounds, and their hydrophilic counterparts. The length of the alkyl chain, but also the number of cationic ammonium groups and the presence of perfluorinated groups determine the degree of antimicrobial activity (Morais, Guedes, and Lopes 2016). Other examples of QACs are cetylpyridinium chloride (CPC), cetyltrimethylammonium bromide (CTAB), benzyldimethylhexadecylammonium chloride (BDHAC), etc. (Tischer et al. 2012; Zhu and Sun 2004).

QACs are antimicrobials against gram-positive and negative bacteria, fungi, moulds, and some types of viruses (Messaoud et al. 2014; Morais, Guedes, and Lopes 2016; Shahidi and Wiener 2012; Tischer et al. 2012). However, gram-positive bacteria are more sensitive in comparison to gram-negative bacteria because the latter contains an extra outer cell membrane. Moreover, antibiotic-resistant staphylococci, such as the well-known MRSA, are not sensitive towards QACs as efflux pumps remove the QACs out of the cells (Tischer et al. 2012). QACs do have different modes of action. The first important target of some QACs is the cell membrane. Quaternary ammonium compounds are cationic and thus are able to bind the negatively charged cell membrane, leading to disorganization and eventually to cell leakage. In addition, intracellular degradation of nucleic acids and proteins is also possible (Morais, Guedes, and Lopes 2016; Shahidi and Wiener 2012; Tischer et al. 2012). QACs are used as antimicrobials on cotton, polyester, nylon, and wool (Morais, Guedes, and Lopes 2016).

QACs have a broad spectrum effect on bacteria, yeast, molds, algae, and viruses (Hegstad et al. 2010). Therefore, the QAC's anti-odour properties on textile are due to its ability to inhibit or kill the odour causing or pathogenic bacteria found in the axillary such as *Staphylococcus* aureus or *Corynebacterium* species (Taylor et al. 2003). In a study on the effectiveness of QACs, quaternary ammonium moieties was modified onto a poly(D,Llactide) (PDLLA) fibrous surface. The antibacterial efficiency with the addition of QACs was about 99.999% against gram-positive (*Staphylococcus* aureus) and gram-negative (*Escherichia coli*) bacteria (Yao et al. 2010).

Despite the effectiveness of QACs against gram-positive and negative bacteria, the lack of physical bonding of the compounds can lead to leaching from the textiles (Morais, Guedes, and Lopes 2016). Leaching is an environmental concern for many of the synthetic or metallic compounds, which is described in the detail below.

6.2.2 Alternative Antimicrobials in the Textiles Industry

6.2.2.1 Environmental and Toxicity Concerns

The antimicrobials used today for textile finishing bring up several concerns. First, antibiotic resistance is a global problem that humanity faces today (Morais, Guedes, and Lopes 2016; Shahidi and Wiener 2012). Gaining resistance is an evolutionary process that occurs within microbial cells. Microorganisms acquire genes, leading to antibiotic resistance mechanisms, such as the development of efflux pumps, for instance, to avoid antimicrobials entering the cell. As such, MRSA is not sensitive to quaternary ammonium compounds (Tischer et al. 2012). Moreover, resistance is a trait that can be interchanged between cells. Besides resistance against QACs, a lot of resistance against silver particles is reported, too. This means that antimicrobials are losing their efficiency in killing microorganisms, and therefore, in the context of antimicrobial finishing, lose their ability in reducing odour (Morais, Guedes, and Lopes 2016).

Antimicrobials can have negative health effects as well (Uddin 2014). Biocidal materials can have a detrimental effect on the skin microbiome and potentially worsen body malodour in the long term (Callewaert et al. 2014). These synthetic finishes tend to provide a short-term solution to prevent malodour in textiles and do not tackle the root of the problem (Quinn 2010).

There are also concerns about the permeability, neurotoxic effects, and potential toxicity, especially in the early development of the used antimicrobials (Ruszkiewicz et al. 2017). In this context, especially triclosan is of concern as it is known to be an endocrine disruptor and is able to move through human tissues. Trace elements of triclosan can be detected in organisms, and while it is known that triclosan causes oxidative stress, apoptosis, and inflammation in cells, it remains to be determined what the continuous, long-term, and low concentration exposure on (human) cells can cause. Regarding metallic compounds, it is already known that there is a risk of neurotoxicity (Morais, Guedes, and Lopes 2016).

Next to health effects or resistance problems, the environmental impact of antimicrobials should also be considered. Sustainability is an important topic driving us to think about ecological and environmentally friendly alternatives if products bring up implications for the environment. In the case of silver, triclosan, and quaternary ammonium compounds, there is a danger of wash-off during laundering of the textiles (Morais, Guedes, and Lopes 2016; Shahidi and Wiener 2012; Zille et al. 2014). Triclosan is even discouraged to be used in textiles or clothing that should be laundered regularly. The usage of antimicrobials is especially a problem when these cannot be removed during wastewater treatment. As such, antimicrobials can be toxic and harmful to aquatic organisms (Uddin 2014).

The environmental risks associated with silver nanoparticles compared with AG ions have been studied in the environment. Many hazard assessments for NPs, including AgNPs, have been conducted, specifically toxicity against aquatic organisms. As AgNPs are among some of the most soluble nanoparticles, there has been much discussion on the relative toxicity of AgNPs versus Ag ions. Nanotoxicology research, however, has found in 93.8% of studies conducted that AgNPs were less toxic than Ag ions (Dzhardimalieva et al. 2021) Yet, considering AgNPs often dissolve to silver ions, the environmental risks related with antimicrobial silver product leaching need to be weighed against the benefits of use (Mitrano et al. 2014; Reed et al. 2016).

Removal of antimicrobials can be done via sedimentation or biodegradation. Silver and triclosan can be removed efficiently; however, if triclosan is degraded under anaerobic conditions, more toxic byproducts are formed. Contrarily, quaternary ammonium compounds are not metabolized or degraded very well. However, due to the positive charge, absorption to negatively charged particles, such as sludge, soil, or sediments, is possible (Morais, Guedes, and Lopes 2016).

Excess silver nanoparticles that are not released during use or laundering will continue to leach from the textiles when disposed of in landfills (Reed et al. 2016). In a study on the environmental impact of silver and nano silver used in textiles, it was concluded that impact and performance should be considered. The study demonstrated that even low silver-containing fabrics can maintain a high antimicrobial efficacy before and after washing (Reed et al. 2016). However, not only the antimicrobials themselves but also the textile treatment process has an impact on the environment. It is clear that it's time to shift towards 'green technologies' and 'green antimicrobial agents' that are both renewable and sustainable (Morais, Guedes, and Lopes 2016).

6.2.2.2 New Directions

Some proposals to reduce the impact on the environment have been made. An example of this is the application of finishes that are delayed in the release of active ingredients, which would also increase the performance of antimicrobial textiles. Besides this, the usage of binders or cross-linking to apply the antimicrobials onto textiles makes the antimicrobials more resistant to the laundering process.

To tackle the problem of antibiotic resistance, in particular, odour management control should rather focus on selective inhibition of odour-associated bacteria instead of having broad-spectrum effects. Recent investigations have shown that microbial load is not necessarily the problem in odour formation in textiles; it is rather the microbial composition (McQueen et al. 2014; Van Herreweghen et al. 2020). There is, however, much more research needed to identify the odour-associated bacteria and find novel ways to selectively inhibit them.

Another possibility for both the resistance and environmental issues is the use of 'green antimicrobial agents', for instance, natural antimicrobials, such as chitosan, alginate, or starch or antimicrobials extracted from plants or herbs (Zille et al. 2014). Examples of the latter are terpenoids, flavonoids, tannins, etc. These natural antimicrobials are environmentally friendly and have several advantages, such as safety, easy availability, and nontoxicity to the skin, which is described in more detail below. Moreover, no antimicrobial resistance is known for these natural compounds (Morais, Guedes, and Lopes 2016). However, the search for new antimicrobials is continuing and other promising molecules, such as antimicrobial peptides (AMPs), are discovered. These are little peptides with an amphipathic structure, built of 12–50 amino acids, having a broad-spectrum microbial activity. AMPs affect RNA and DNA synthesis, but also induce membrane damage or even loss of ATP molecules within cells (Morais, Guedes, and Lopes 2016; Tischer et al. 2012).

6.2.3 Natural Antibacterial Finishes

The widespread use of antibiotic-coated surfaces has been linked to the emergence of several multi-drug resistant strains of infectious diseases. Not only this but the polluting effects of the textile industry starts from the manufacturing of the fiber itself. The cost of synthetic antimicrobial finishes to the environment is becoming more obvious, and at the same time, natural dyes and finishes in the textile are gaining significant momentum (Han and Lee 2009). This line of interest is motivated by the strict environmental standards imposed by many countries due to the overuse of chemicals, which can cause toxicity to the body. This direction of research has emerged in the field of textiles and apparel technology through utilising natural materials, on account of their compatibility with deodorising properties (Han and Lee 2009). This research into natural or naturally derived finishes is an important step from producing toxic-free manufacturing. When manufacturing clothing and textiles, the ecology of production methods, waste disposal, and the effects of the materials on the human skin are to all be considered (Paulose 2019). This increased focus on natural antibacterial finishes in the textile industry has led to some new innovations and reestablishing the old. Below, the main naturally antibacterial fibers and nanotechnology used for odour control and skin disease prevention are described.

6.2.3.1 Bamboo

In China, bamboo fiber is known as 'Air Vitamin' or 'long-lived element'. It contains anions that are helpful in purifying blood, calming the nervous system, relieving allergy symptoms, and according to Munjal and Kashyap (2015), this is beneficial to the health of the human body and skin. Bamboo is 100% natural, containing cellulosic fiber, hemicellulose, ash, and lignin (Li et al. 2014). It is one of the fastest-growing plants, which grows effectively without pesticides, fertilizers, or herbicides (Paulose 2019). The main manufacturing methods currently being operated in the production of sustainable bamboo textile are chemically based and mechanically based processes (Waite 2010). Yet, the fiber is completely biodegradable and is a solution for many of the environmental problems in the textile industry, as its manufacture from raw materials to decomposition does not cause any harmful effects to the environment (Paulose 2019).

Jothilinkam and Ramachandran (2017) state that bamboo fiber has a unique antibacterial bio-agent known as 'Bamboo Kun'. However, 'Kun' directly translates to the hydroxyl functional group (–OH), yet the exact chemical compound and origin of this antibacterial property was not clear until more recently (Afrin et al. 2012; (Jothilinkam and Ramachandran 2018). Like cellulose and hemicellulose, bamboo is a complex material consisting of hydrophobic lignin and hydrophilic carbohydrates, which are chemically bound (Afrin et al. 2012; Koshijima and Watanabe 2003). It was concluded in a 2011 study on the origin of the antibacterial function of bamboo that the antibacterial compound of bamboo is located in lignin (Afrin et al. 2012). Lignin contains aromatic and phenolic functional groups, which are thought to be responsible for bamboo's antimicrobial activity.

However, a 2020 study compared 12 commercial bamboo textiles for their antibacterial activity against S. aureus and K. pneumoniae. Significant antibacterial activity was found in the bamboo viscose textiles only, suggesting the antibacterial properties by bamboo viscose may not be a property from the bamboo, but the chemical processing and treatments during manufacture (Rocky and Thompson 2020).

An additional function of bamboo fiber includes its high absorbency properties, which wick sweat away from the human body very efficiently. As such, bamboo clothing is growing in the textile market with a claim for its antimicrobial properties and eco-friendly manufacturing systems. The absorbent properties of bamboo are said to be due to the microholes in the structure of bamboo caused by the cross-section of bamboo fibers, which is being utilised in the clothing and textiles industry (Shen et al. 2004).

6.2.3.2 Chitosan

Chitosan ($(C_6H_{11}O_4N)_n$) is a linear polysaccharide of β-1,4-glycosidic bounded D-glucosamine and N-acetyl-D-glucosamine units. It originates from deacetylated chitin, which in turn can be extracted from the cell walls of several fungi or from crustaceans like crabs and shrimps. Chitosan has antimicrobial activity because of the cationic properties of its primary amino groups under an acidic pH. As such, chitosan can interact with the negatively charged microbial cell membrane of a range of microorganisms (Morais, Guedes, and Lopes 2016; Muñoz-Bonilla et al. 2019). Moreover, the structural characteristics of chitosan, such as molecular weight, degree of polymerization, and deacetylation, actively affect its antimicrobial activity (Muñoz-Bonilla et al. 2019). Environmental conditions, such as temperature, pH, or ionic strength, also have an impact on chitosan's antimicrobial properties (Muñoz-Bonilla, Cerrada, and Fernández-García 2013). There are methods for improving the antimicrobial properties of chitosan, such as including a permanent positive charge through chemical modification of the chain structure.

Antimicrobial organic compounds, including essential oils and metallic nanoparticles, have also been combined with chitosan (Muñoz-Bonilla et al. 2019). These adjustments to the material propose to advance chitosan's solubility and antimicrobial activity without altering its biodegradability and biosafety properties (Rabea et al. 2003).

Chitosan's film-forming abilities and mechanical strength make it an ideal material for food packaging and biomedical devices. However, due to its abundance and desirable properties, such as biodegradability, there has been a focus on the use of chitosan in textiles for its unique antimicrobial bio-based properties (Muñoz-Bonilla et al. 2019). Nowadays, several chitosan-based products are already commercially available. When used, chitosan is mainly applied on textiles fabrications, such as cotton, polyester, and wool (Morais, Guedes, and Lopes 2016).

A practical example of the usage of chitosan in industry is that of a San Francisco-based footwear brand that launched its first line of apparel in 2020 utilising chitosan properties in textiles. The natural biopolymer is extracted from crustacean shells that the company has sourced from the fishing industry (Ahmed and Ikram 2017). The anti-odour properties, derived from its presence in crustacea help inhibit bacteria on the textile surface. The use of chitosan's bio-based antimicrobial properties is intended to reduce the need for washing, thereby helping to lower the garment's overall carbon footprint.

6.2.3.3 Essential Oils Nanoparticles

Due to the large surface area and high surface energy of nanoparticles, they are very efficient for usage in nano finishing of textiles. These benefits can mean improved durability of overall textile function and greater affinity to the fabrics (Sayed 2017; Shen et al. 2004). As such, the nanoencapsulation or finishing of essential oils into textiles has attracted more attention in recent years as a biobased alternative. Essential oils have anti-odour or antimicrobial properties and do not produce major side effects, as is the case with synthetic active agents. Essential oils, which are mainly composed of terpenes, are made up of a complex mixture of volatile compounds and are often found in aromatic plants (Dhifi et al. 2016). In the natural world, these 'essential' oils play a role in the protection of the plants because of the antibacterial, antiviral, antifungal, and insecticide properties (Shen et al. 2004; Sayed 2017). Since ancient times, plant extracts from peppermint, lemon, clove, and cinnamon have been used because of their antibacterial, inflammatory, and antioxidant activity on the skin. Due to these properties, these plant extracts can be used in wound healing products (Prabuseenivasan, Jayakumar, and Ignacimuthu 2006; Unalan et al. 2019). Other examples of natural oils used for their antibacterial properties are neem oil, castor oil, karanja oil, and citronella oil. Advantages would be the efficacy and absence of any adverse effects, so, consequently, there is an increased focus on applications that make use of these oils (Swamy, Akhtar, and Sinniah 2016). Additionally, pomegranate, orange, and lemon peel are presently the subjects of study for possible environmentally friendly performance textiles (Uddin and Farooq 2013).

An investigation into coated fabric shows antibacterial properties against gram-negative and gram-positive bacteria, indicating that this technique can be used in the textile industry as an antimicrobial finish of medical clothes and sports and leisurewear to inhibit odour in apparel (Mahesh, Reddy, and Kumar 2011). An example of the usage of essential oils in the textile industry for antimicrobial finishing is the use of peppermint oil by a European-based material science and apparel brand. The company utilises the anti-odour and antimicrobial properties of peppermint, by finishing the textile using padding technology (Parham et al. 2020). The incentive is to provide an alternative to toxic antimicrobials and to reduce the need to wash clothing due to the anti-odour activity of the finished textile.

6.2.3.4 Hydrogen Peroxide

Hydrogen peroxide (H_2O_2) is also capable of inhibiting invasive bacteria on textiles. As this compound is produced by human cells, hydrogen peroxide is also classified as natural antimicrobial. The antimicrobial activities of hydrogen peroxide can be used to inhibit bacteria, mold, fungi, and viruses, and it can also be found in honey, acting as a preservative (Uddin 2014). Additionally, certain concentrations of H_2O_2 are often used to remove staining on textiles, remove odours, and improve textile deterioration (Ferdush et al. 2019). Such finishes can accomplish antimicrobial effects in various textile products, which include the apparel, upholstery, and automotive industries (Uddin 2014; Quick-Med Technologies n.d.; Unalan et al. 2019). Its main drawback is the instability of the molecule, and thus the need to stabilize the molecule, or use it as fast as possible.

6.3 EMERGING TECHNOLOGIES

Skin microbiome research has become a hot field in skincare in the last few years, namely through means of bacteria therapy for malodour control. This research is also being acknowledged by the textile industry and new technologies are being developed continuously. An example of this is the biotechnical industry, which is gaining more importance and, as such, will also contribute to the future of the fashion industry for the production and development of odour-controlled textiles. Biotechnological and engineering methods will help to expand the functions, properties and potential of textiles surfaces along major frontiers (Jablonski 2013). Thereby, functions in order to treat diseases and injuries will be developed. Along the biotechnical advances, future technologies include bioinspired nanostructures, graphene, and microbiome-smart clothing, which are described below.

6.3.1 Microencapsulation

Traditionally known for their ability to absorb moisture, innovation in textile fiber technology reveals that they can release substances, too. Transdermal drug delivery is a method used in the healthcare sector in which medication can be administered directly through the skin via a textiles surface (Ratner et al. 2012). Through this kind of technology, it has enabled fiber and textile manufacturers to develop new concepts to the market for health, wellness, and odour control. Beneficial molecules and even microbes can be microencapsulated and attached to the textiles for sustained release. Vitamins, probiotics, antimicrobial ingredients, enzymes, fragrances, and others are common examples of products that utilise this technology.

6.3.2 Incorporating Beneficial Bacteria in Textiles

Bacterial and fungal species are known for the ability to decompose organic matter (sweat, blood, saliva, organic waste, etc.) with the use of specific enzymes. Species such as *Bacillus, Lactobacillus, Enterobacter, Streptococcus, Nitrosomonas, Nitrobacter, Pseudomonas, Alcaligens* and *Klebsiella spp.* are already used in odour control. A specific application of this are carpets for instance. After spilling organic matter on the carpet, dormant microorganisms are activated and degradation of the spill by enzymatic actions will prevent odour development. Other products in which bacteria are used to counteract odour are pillows, mattresses, towels, footwear, automotive products, body protective gear, etc. (Shrimali and Dedhia 2015). Bacterial usage in textiles for anti odour is a growing field of research.

Novel approaches are desired as a result of the increased demand for multipurpose and sustainable textile materials. Research into probiotic bacteria as an alternative for typical biocides and antimicrobials is being developed to reduce the current toxicity and resistance problems. The Food and Agriculture Organisation (FAO) and World Health Organization (WHO) have defined probiotics as living microorganisms that provide a positive health effect for the host (Călinoiu et al. 2019; Rokka and Rantamäki 2010). However, if probiotics are used, it is important to be sure that the bacteria stay alive on the substrate. The viability of the bacteria affects the efficiency of the product. To protect and guarantee bacterial viability, microencapsulation of the cells is often done in industrial applications. This principle is often used in the food industry for the production of probiotics and the use in dairy products (Rokka and Rantamäki 2010).

Over recent years, the use of microcapsules, or the microencapsulation of active ingredients for personal care and textiles products, has developed considerably. This process means active substances can be fixed either to the surface of the fiber or within the fiber itself depending on its composition. The aim of microencapsulation is to develop a coating protecting the core active ingredient from the environment. In fact, there are two groups of microparticles: those who are built of a core and surrounded by layers forming a protective shell or microparticles where the active ingredient is embedded in a matrix (Kanga 2006). Some chemical and fabric technology and companies have started to use microencapsulation to manufacture new fabric concepts. Current commercially available textile products containing probiotics mainly focus on their anti-allergic benefits. Examples include a chemical company based in Belgium, which has incorporated microbiota in fibers. The process encapsulates probiotic spores into a textile to significantly decrease the dust mite allergen. This technology is predominately used for incorporation into mattresses where the spores are released with friction. Their patented encapsulation process uses a polymer matrix to form shells around the active ingredients (Vandendaele and Temmerman 2010).

Similarly, a Swedish textile institute looked at the viability of probiotics incorporated into a textile to reduce infectious pathogens which can be acquired in hospitals. The institution experimented with an adapted sol–gel coating process to embed these probiotic beneficial spores on a woven polyester surface. The viability of bacteria was analysed along with the physical properties and characteristics of the coated fabric. The results demonstrate a successful encapsulation of the beneficial spores with a sufficient number of living organisms, before and after repeated washing cycles, as well as suitable tensile strength and abrasion resistance properties. The surface wettability remains an area of improvement in order to maintain adequate adhesion between substrate and coating (Dural-Erem, Biswas, and Nierstrasz 2018).

Another example is that of a Canadian-based biotechnology company, which uses spore-forming *Bacillus* spp. in cleaning agents to reduce odour on (textile) surfaces. The company focuses on developing bacterial and enzymatic strains, which are combined with surfactants to develop and manufacture cleaning products for surfaces and textiles.

Another patented technology pioneered by a biotech company based in Toronto utilises naturally occurring bacteria for odour management in footwear for athletics and leisure. Beneficial microbes found in soil are bonded to the surface of the fabric. This innovative process of applying live microorganisms to fibers, results in natural, nontoxic health benefits to consumers and the environment (Chow and Ciupa 2018; Vandendaele and Temmerman 2010).

There are many strains of probiotic bacteria that can be associated with different benefits inside the body and on its surface. Thus, the incorporation of probiotics into textiles has not been fully assessed; yet, they do suggest a promising technique to combat pathogenic bacteria on textile surfaces in the future (Niehaus 2016).

6.3.2.1 Microbiome-Smart Clothing

New research is exploring the impact of the textile microbiome of particular microbes on malodour development. Research is investigating so-called 'microbiome-smart' textiles, in which good microbes are added, their enzymatic potential, or the microbiome is steered towards non-odour-causing and healthy communities.

The microbiome plays a key role in body odour, and antibacterial fabric finishes on clothing have been designed in an attempt to reduce odour issues. In the past decades, innovation has mainly focused on broad-spectrum antimicrobial technologies. Yet, these are associated with many drawbacks. Research is now looking into the incorporation of healthy skin commensal into textiles (Broadhead et al, unpublished). The microorganisms are activated in contact with the moisture on the skin, allowing them to dominate other less beneficial bacteria. The probiotic clothing technology provides an alternative to antibacterial chemicals in clothing and cosmetics and a solution for natural odour management. Additionally, research is going into alternatives for the current antibacterial finishes. Such collaborative effort in science, technology, and fashion aims to use the recent advances in science into a novel approach in the textile industry.

6.3.2.2 Bio-Inspired Nanostructures

In recent years, there has been a new approach for realising bactericidal action through physical surface topography. Certain natural nanostructured surfaces have been identified to have the capacity to rupture the cell wall of the bacteria which is known as the 'contact killing mechanism' (Ivanova et al. 2012). This research is inspired by certain insects, which are inherently known for their bactericidal surfaces that are able to kill microbes on contact. This bactericidal effect is due to the sharp nanostructures, being 80–250 nm height and having a 50–250 nm nano-pillar and a 100–250 nm pitch, present on the insect's exterior surface (Tripathy et al. 2017). These sharp structures have the potential to pierce the cell wall of the bacteria, therefore rupturing its surface and killing the bacteria. This physical biocidal approach has become a new area of focus for tackling multi-antibiotic resistant bacteria (Lam et al. 2016; Tripathy et al. 2017).

Various techniques have been used to fabricate nanostructured surfaces on a variety of substrates, including silicone and polymer surfaces, which have been tested against different pathogenic bacteria (He et al. 2013; Fu et al. 2009; Seo et al. 2014; He et al. 2013; Fu et al. 2009; Seo et al. 2014). However, the scope of the bactericidal activity of such nanostructured surfaces depends on several criteria, including the shape of the structure, size, and density. While finding an optimised nanotopography technique, material and process remain a challenge, such developments have presented opportunities for further investigations (Pogodin et al. 2013; Tripathy et al. 2017). Yet, results have shown it to be a promising strategy in inhibiting pathogenic bacteria to solve antimicrobial resistance challenges in the future (Tripathy et al. 2017).

6.3.2.3 Graphene

Graphene's key components are carbon atoms organised in one-atom-thick planar sheets packed in a crystal lattice. Other graphene materials include graphene oxide and reduced graphene oxide. Besides its self-cleaning ability, flame retardancy, electrical conductivity, and UV blocking characteristics, graphene is known for its antimicrobial characteristics (Karimi et al. 2016). Graphene's antimicrobial features can be controlled by different parameters: its lateral size, number of carbon layers, shape, surface modifications, and agglomeration ability. Moreover, different modes of action are already suggested. First of all, graphene can act as so-called nano knives because of its sharp edges, which break up the cell membrane, giving rise to leakage of intracellular compounds. Secondly, graphene is able to induce oxidative stress, disturbing essential functions within the cell and causing cellular inactivation or even cell death. Thirdly, because of the thin structure of graphene, it's said to wrap around a bacterial cell, blocking nutrient transport towards the cell. Besides these three main mechanisms, other active modes, such as extraction of membrane lipids or interference in protein–protein interactions, are described as well (Zou et al. 2016). Yet, research regarding graphene and its antimicrobial properties is still at its infancy.

6.4 CONCLUSIONS

Laundering habits have changed since the 1990s towards more ecologically friendly conditions (lower temperature, less water use, and absence of chemical detergents). This is great but also led to reduced microbial and odour removal. As a result, we saw the rise of a wide array of antimicrobial finishes in clothes, which is not necessarily effective and is sometimes associated with unwanted side effects.

Recently, particular odour controlling agents have been encapsulated in nanoparticles to show a sustained effect.

The skin microbiome is shaped by its natural environment, and what is put on and next to the skin has an immediate impact. Designing microbiome-smart textiles can be a novel and alternative way to advance the functionality of clothing and to combat odour development in textiles. To make a shift in the effects of the antibacterial ingredients and toxic cosmetics that society has encouraged, the answers may be to look more closely at the skin's living ecosystem and natural skin biome. Our knowledge of microbes and their responsibility in odour formation in textiles is slowly but steadily growing. Research has shifted towards more natural and bio-inspired applications to control odour formation in textiles. As research digs deeper into the fundamental origins of malodour formation, novel odour-controlling techniques are developed to combat the malodour in clothes, enhance the skin feeling, and advance the overall well-being of the wearer.

REFERENCES

Afrin, T., Tsuzuki, T., Kanwar, R.K., and Wang, X. 2012. The origin of the antibacterial property of bamboo. *Journal of the Textile Institute*. doi: 10.1080/00405000.2011.614742.

Ahmed, S., and Ikram, S. 2017. *Chitosan: Derivatives, Composites and Applications*. John Wiley & Sons.

Bhargava, H.N., and Leonard, P.A. 1996. Triclosan: Applications and safety. *American Journal of Infection Control*, 24(3): 209–218.

Buschmann, H.-J., Dehabadi, V.A., and Wiegand, C. 2015. Medical, cosmetic and odour resistant finishes for textiles. *Functional Finishes for Textiles*. doi: 10.1533/9780857098450.1.303.

Byrd, A. L., Belkaid, Y., and Segre, J.A. 2018. The human skin microbiome. *Nature Reviews. Microbiology*, 16(3): 143–155.

Callewaert, C., De Maeseneire, E., Kerckhof, F.-M., Verliefde, A., Van de Wiele, T., and Boon, N. 2014. Microbial odor profile of polyester and cotton clothes after a fitness session. *Applied and Environmental Microbiology*, 80(21): 6611–6619.

Callewaert, C., Hutapea, P., Van de Wiele, T., and Boon, N. 2014. Deodorants and antiperspirants affect the axillary bacterial community. *Archives for Dermatological Research. Archiv Fur DermatologischeForschung*, 306(8): 701–710.

Callewaert, C., Van Nevel, S., Kerckhof, F.-M., Granitsiotis, M.S., and Boon, N. 2015. Bacterial exchange in household washing machines. *Frontiers in Microbiology*, 6(December): 1381.

Călinoiu, L.F., Ştefănescu, B.E., Pop, I.D., Muntean, L., and Vodnar, D.C. 2019. Chitosan coating applications in probiotic microencapsulation. *Coatings*, 9(3):194.

Chow, P., and Ciupa, C. 2018. Impregnated odour control products and methods of making the same. U.S. Patent No. 2018/0228164. Available: https://patentimages.storage.googleapis.com/99/34/c7/10af5e66a4a038/US20180228164A1.pdf.

Coman, D., Simona, O.B., and Vrînceanu, N. 2010. Biofunctionalization of textile materials by antimicrobial treatments: A critical overview. *Romanian Biotechnological Letters*. [cited 3 Jun 2021]. Available: http://www.rombio.eu/Archive/rbl1vol15/lucr%201%20Review%20Simona%20Oancea.pdf.

Dastjerdi, R., and Montazer, M. 2010. A review on the application of inorganic nano-structured materials in the modification of textiles: Focus on anti-microbial properties. *Colloids and Surfaces. B, Biointerfaces*, 79(1): 5–18.

Dhifi, W., Bellili, S., Jazi, S., Bahloul, N., and Mnif, W. 2016. Essential oils' chemical characterization and investigation of some biological activities: A critical review. *Medicines (Basel, Switzerland)*, 3(4). doi: 10.3390/medicines3040025.

Dhiman, G., and Chakraborty, J.N. 2015. Antimicrobial performance of cotton finished with triclosan, silver and chitosan. *Fashion and Textiles*, 2(1): 13.

Dravnieks, A., Krotoszynski, B.K., Lieb, W.E., and Jungermann, E. 1968. Influence of an antibacterial soap on various effluents from axillae. *Journal of the Society of Cosmetic Chemists*.

Dural-Erem, A., Biswas, T., and Nierstrasz, V. 2018. Incorporation of probiotics on textile surface by sol–gel coating. *Journal of Industrial Textiles*. doi: 10.1177/152808371 7750886.

Dzhardimalieva, G.I., Yadav, B.C., Lifintseva, T.V., and Uflyand, I.E. 2021. Polymer chemistry underpinning materials for triboelectric nanogenerators (TENGs): Recent trends. *European Polymer Journal*, 142(January): 110163.

Elias, P.M. 2007. The skin barrier as an innate immune element. *Seminars in Immunopathology*, 29(1): 3–14.

Farouk, A., Moussa, S., Ulbricht, M., and Textor, T. 2012. ZnO nanoparticles-chitosan composite as antibacterial finish for textiles. *International Journal of Carbohydrate Chemistry*. doi: 10.1155/2012/693629.

Ferdush, J., Nahar, K., Akter, T., Ferdoush, M.J., and Farhana Iqbal, S.M. 2019. Effect of hydrogen peroxide concentration on 100% cotton knit fabric bleaching. *European Scientific Journal ESJ*. doi: 10.19044/esj.2019.v15n33p254.

Fu, Y.Q., Colli, A., Fasoli, A., Luo, J.K., Flewitt, A.J., Ferrari, A.C., and Milne, W.I. 2009. Deep reactive ion etching as a tool for nanostructure fabrication. *Journal of Vacuum Science & Technology B: Microelectronics and Nanometer Structures*. doi: 10.1116/1.3065991.

Gao, Y., and Cranston, R. 2008. Recent advances in antimicrobial treatments of textiles. *Textile Research Journal*. doi: 10.1177/0040517507082332.

Goto, T., Hirakawa, H., Morita, Y., Tomida, J., Sato, J., Matsumura, Y., Mitani, A., et al. 2016. Complete genome sequence of moraxella osloensis strain KMC41, a producer of 4-Methyl-3-Hexenoic acid, a major malodor compound in laundry. *Genome Announcements*, 4(4). doi: 10.1128/genomeA.00705-16.

Gowramma, B., Keerthi, U., Rafi, M., and Muralidhara Rao, D. 2015. Biogenic silver nanoparticles production and characterization from native stain of corynebacterium species and its antimicrobial activity. *3 Biotech*. 10.1007/s13205-014-0210-4.

Gunawan, C., Teoh, W.Y., Marquis, C.P., and Amal, R. 2011. Cytotoxic origin of copper(II) oxide nanoparticles: Comparative studies with micron-sized particles, leachate, and metal salts. *ACS Nano*. 10.1021/nn2020248.

Han, M.-R., and Lee, J.-S. 2009. Natural dyeing of cotton fabrics with rumex Crispus L. root. *Journal of the Korean Society of Clothing and Textiles*. doi: 10.5850/jksct.2009.33.2.222.

Hasan, J., Crawford, R.J., and Ivanova, E.P. 2013. Antibacterial surfaces: The quest for a new generation of biomaterials. *Trends in Biotechnology*, 31(5): 295–304.

He, B., Yang, Y., Yuen, M.F., Chen, X.F., Lee, C.S., and Zhang, W.J. 2013. Vertical nanostructure arrays by plasma etching for applications in biology, energy, and electronics. *Nano Today*. doi: 10.1016/j.nantod.2013.04.008.

Hegstad, K., Langsrud, S., Lunestad, B.T., Scheie, A.A., Sunde, M., and Yazdankhah, S.P. 2010. Does the wide use of quaternary ammonium compounds enhance the selection and spread of antimicrobial resistance and thus threaten our health? *Microbial Drug Resistance*, 16(2): 91–104.

Höfer, D. 2006. Antimicrobial textiles, skin-borne flora and odour. *Current Problems in Dermatology*, 33: 67–77.

Ivanova, E.P., Hasan, J., Webb, H.K., Truong, V.K., Watson, G.S., Watson, J.A., Baulin,

V.A., et al. 2012. Natural bactericidal surfaces: Mechanical rupture of pseudomonas aeruginosa cells by Cicada wings. *Small*, 8(16): 2489–2494.

Jablonski, N.G. 2013. *Skin: A Natural History*. Univ of California Press.

James, A.G., Austin, C.J., Cox, D.S., Taylor, D., and Calvert, R. 2013. Microbiological and biochemical origins of human axillary odour. *FEMS Microbiology Ecology*, 83(3): 527–540.

Jothilinkam, M.D., and Ramachandran, T. 2018. A study of overall moisture management capability of plasma treated bamboo/cotton blended fabrics for medical applications. *International Research Journal of Pharmacy*, 8: 62–71. doi: 10.7897/2230-8407.0812252.

Kanga, V. 2006. Skin care delivery systems: An overview. In John J. Wille (ed.), *Skin Delivery Systems: Transdermal, Dermatological and Cosmetic Actives* (pp. 173–176). Blackwell Publishing.

Karimi, L., Yazdanshenas, M.E., Khajavi, R., Rashidi, A., and Mirjalili, M. 2016. Functional finishing of cotton fabrics using graphene oxide nanosheets decorated with titanium dioxide nanoparticles. *The Journal of The Textile Institute*. doi: 10.1080/00405000.2015.1093311.

Kashyap, M , and Radha, K. 2015. Bamboo fiber: An approach toward sustainable development. *International Journal of Science and Research (IJSR)*.

Koshijima, T., and Watanabe, T. 2003. Association between lignin and carbohydrates in wood and other plant tissues. *Springer Series in Wood Science*. doi: 10.1007/978-3-662-05191-7.

Laitala, K., and Mollan Jensen, H. 2010. Cleaning effect of household laundry detergents at low temperatures. *Tenside, Surfactants, Detergents*, 47(6): 413–420.

Lam, S.J., O'Brien-Simpson, N.M., Pantarat, N., Sulistio, A., Wong, E.H.H., Chen, Y.-Y., Lenzo, J.C., et al. 2016. Combating multidrug-resistant Gram-negative bacteria with structurally nanoengineered antimicrobial peptide polymers. *Nature Microbiology*, 1(11): 16162.

Li, Q., Song, J., Peng, S., Wang, J.P., Qu, G.Z., Sederoff, R.R., and Chiang, V.L. 2014. Plant biotechnology for lignocellulosic biofuel production. *Plant Biotechnology Journal*. doi: 10.1111/pbi.12273.

Lyman, M. 2020. *The Remarkable Life of the Skin: An Intimate Journey Across Our Largest Organ*. Atlantic Monthly Press.

Mao, J.W., and Murphy, L. 2001. Durable freshness for textiles. *AATCC Review*, 1: 28–31.

Mahesh, S., Reddy, A.H.M., and Kumar, G.V. 2011. Studies on antimicrobial textile finish using certain plant natural products. *International Conference on Advances in Biotechnology and Pharmaceutical Sciences*, 560059(9980819835): 253–258.

Martin, A., Saathoff, M., Kuhn, F., Max, H., Terstegen, L., and Natsch, A. 2010. A functional ABCC11 allele is essential in the biochemical formation of human axillary odor. *The Journal of Investigative Dermatology*, 130(2): 529–540.

McQueen, R.H. 2011. 17 - Odour control of medical textiles. In V.T. Bartels (ed.), *Handbook of Medical Textiles* (pp. 387–416). Woodhead Publishing.

McQueen, R.H., Harynuk, J.J., Wismer, W.V., Keelan, M., Xu, Y., and Paulina de la Mata, A. 2014. Axillary odour build-up in knit fabrics following multiple use cycles. *International Journal of Clothing Science and Technology*, 39(July): 88.

Messaoud, M., Chadeau, E., Chaudouët, P., Oulahal, N., and Langlet, M. 2014. Quaternary ammonium-based composite particles for antibacterial finishing of cotton-based textiles. *Journal of Materials Science & Technology*. doi: 10.1016/j.jmst.2013.09.012.

Mitrano, D.M., Ranville, J.F., Bednar, A., Kazor, K., Hering, A.S., and Higgins, C.P. 2014. Tracking dissolution of silver nanoparticles at environmentally relevant concentrations in laboratory, natural, and processed waters using single particle ICP-MS (spICP-MS). *Environmental Science: Nano*, 1(3): 248–259.

Morais, D.S., Guedes, R.M., and Lopes, M.A. 2016. Antimicrobial approaches for textiles: From research to market. *Materials*, 9(6). doi: 10.3390/ma9060498.

Munjal, K., and Kashyap, R. 2015. Bamboo fiber: An approach toward sustainable development. *Int. J. Sci. Res.*, 4: 1080–1083.

Muñoz-Bonilla, A., Echeverria, C., Sonseca, Á., Arrieta, M.P., and Fernández-García, M. 2019. Bio-based polymers with antimicrobial properties towards sustainable development. *Materials*, 12(4). doi: 10.3390/ma12040641.

Muñoz-Bonilla, A., Cerrada, M.L., and Fernández-García, M. 2013. CHAPTER 2. Antimicrobial activity of chitosan in food, agriculture and biomedicine. *Polymer Chemistry Series*. doi: 10.1039/9781782624998-00022.

Niehaus, K.-L. 2016. Viability and efficacy of probiotics printed on a textile material.

O'Dell, B.L. 1981. Metabolic functions of zinc — a new look. *Trace Element Metabolism in Man and Animals*. doi: 10.1007/978-3-642-68269-8_79.

Palza, H. 2015. Antimicrobial polymers with metal nanoparticles. *International Journal of Molecular Sciences*, 16(1): 2099–2116.

Parham, S., Kharazi, A.Z., Bakhsheshi-Rad, H.R., Nur, H., Ismail, A.F., Sharif, S., Krishna, S.R., and Berto, F. 2020. Antioxidant, antimicrobial and antiviral properties of herbal materials. *Antioxidants (Basel, Switzerland)*, 9(12). doi: 10.3390/antiox9121309.

Paulose, V. 2019. Analysis of anti-bacterial efficacy in bamboo and bamboo/cotton handloom fabrics finished with selected herbal extracts, no. November.

Pogodin, S., Hasan, J., Baulin, V.A., Webb H.K., Truong, V.K., Nguyen, The H.P., Boshkovikj, V., et al. 2013. Biophysical model of bacterial cell interactions with nanopatterned cicada wing surfaces. *Biophysical Journal*, 104(4): 835–840.

Prabuseenivasan, S., Jayakumar, M., and Ignacimuthu, S. 2006. In vitro antibacterial activity of some plant essential oils. *BMC Complementary and Alternative Medicine*, 6(November): 39.

Quick-Med Technologies: Quick-Med Technologies, Inc., Innovative technology brings scientific advance to everyday textiles (February 3, 2011), (Gainesville, Fla), https://www.globenewswire.com/fr/news-release/2011/02/03/439197/12481/en/Quick-Med-Announces-EPA-Registration-of-Stay-Fresh-R-Antimicrobial.html (cited 18 Feb 2021).

Quinn, B. 2010. *Textile Futures: Fashion, Design and Technology*. Berg Publishers.

Rabea, E.I., Badawy, M.E.-T., Stevens, C.V., Smagghe, G., and Steurbaut, W. 2003. Chitosan as antimicrobial agent: Applications and mode of action. *Biomacromolecules*, 4(6): 1457–1465.

Rahman, M.A., Ahsan, T., and Islam, S. 2010. Antibacterial and antifungal properties of the methanol extract from the stem of Argyreia Argentea. *Bangladesh Journal of Pharmacology*. doi: 10.3329/bjp.v5i1.4700.

Ratner, B.D., Hoffman, A.S., Schoen, F.J., and Lemons J.E. 2012. *Biomaterials Science: An Introduction to Materials in Medicine*. Academic Press.

Reed, R.B., Zaikova, T., Barber, A., Simonich, M., Lankone, R., Marco, M., Hristovski, K., Herckes, P., Passantino, L., Fairbrother, D.H., Tanguay, R., Ranville, J.F., Hutchison, J.E., and Westerhoff, P.K. 2016. Potential environmental impacts and antimicrobial efficacy of silver- and nanosilver-containing textiles. *Environmental Science & Technology*, 50(7): 4018–4026. doi: 10.1021/acs.est.5b06043.

Rocky, B.P., and Thompson, A.J. 2020. Investigation and comparison of antibacterial property of bamboo plants, natural bamboo fibers and commercial bamboo viscose textiles. *The Journal of The Textile Institute*. doi: 10.1080/00405000.2020.1807300.

Rokka, S., and Rantamäki, P. 2010. Protecting probiotic bacteria by microencapsulation: challenges for industrial applications. *European Food Research and Technology*. doi: 10.1007/s00217-010-1246-2.

Ruszkiewicz, J.A., Li, S., Rodriguez, M.B., and Aschner, M. 2017. Is triclosan a neurotoxic agent? *Journal of Toxicology and Environmental Health. Part B, Critical Reviews*, 20(2): 104–117.

Sayed, U. 2017. Application of essential oils for finishing of textile substrates. *Journal of Textile Engineering & Fashion Technology*. doi: 10.15406/jteft.2017.01.00009.

Scharschmidt, T.C., and Fischbach, M.A. 2013. What lives on our skin: Ecology, genomics and therapeutic opportunities of the skin microbiome. *Drug Discovery Today. Disease Mechanisms*, 10(3-4). doi: 10.1016/j.ddmec.2012.12.003.

Schommer, N.N., and Gallo, R.L. 2013. Structure and function of the human skin microbiome. *Trends in Microbiology*, 21(12): 660–668.

Schubert, C., Guttek, K., Reinhold, A., Grüngreiff, K., and Reinhold, D. 2015. The influence of the trace element zinc on the immune system. *LaboratoriumsMedizin*. doi: 10.1515/labmed-2015-0060.

Seo, J.-H., Park, J.H., Kim, S.-Il, Park, B.J., Ma, Z., Choi, J., and Ju, B.-K. 2014. Nanopatterning by laser interference lithography: Applications to optical devices. *Journal of Nanoscience and Nanotechnology*, 14(2): 1521–1532.

Shahidi, S., and Wiener, J. 2012. Antibacterial agents in textile industry. *Antimicrobial Agents*. doi: 10.5772/46246.

Shelley, W.B., and Hurley, H.J. Jr. 1953. The physiology of the human axillary apocrine Sweat Gland. *The Journal of Investigative Dermatology* 20 (4): 285–297.

Shen, Q., Liu, D.-S., Gao, Y., and Chen, Y. 2004. Surface properties of bamboo fiber and a comparison with cotton linter fibers. *Colloids and Surfaces. B, Biointerfaces*, 35(3-4): 193–195.

Shrimali, K., and Dedhia Dr E.M. 2015. Microencapsulation for textile finishing. *IOSR Journal of Polymer and Textile Engineering*, 2(2): 2348–181. doi:10.9790/019X-0220104.

Swamy, M.K., Akhtar, M.S., and Sinniah, U.R.. 2016. Antimicrobial properties of plant essential oils against human pathogens and their mode of action: An updated review. *Evidence-Based Complementary and Alternative Medicine: eCAM*, 2016(December): 3012462.

Szostak-Kotowa, J. 2004. Biodeterioration of textiles. In *International Biodeterioration and Biodegradation*. doi: 10.1016/S0964-8305(03)00090-8.

Taylor, D., Daulby, A., Grimshaw, S., James, G., Mercer, J., and Vaziri, S. 2003. Characterization of the microflora of the human axilla. *International Journal of Cosmetic Science*, 25(3): 137–145.

Teufel, L., Pipal, A., Schuster, K.C., Staudinger, T., and Redl, B. 2010. Material-dependent growth of human skin bacteria on textiles investigated using challenge tests and DNA genotyping. *Journal of Applied Microbiology*, 108(2): 450–461.

Tischer, M., Pradel, G., Ohlsen, K., and Holzgrabe, U. 2012. Quaternary ammonium salts and their antimicrobial potential: targets or nonspecific interactions? *ChemMedChem*. doi: 10.1002/cmdc.201100404.

Tripathy, A., Sen, P., Su, B., and Briscoe, W.H. 2017. Natural and bioinspired nanostructured bactericidal surfaces. *Advances in Colloid and Interface Science*, 248(October): 85–104.

Uddin, F. 2014. Environmental concerns in antimicrobial finishing of textiles. *International Journal of Textile Science*.

Uddin, F., and Farooq, M. 2013. Study of fruit waste in antimicrobial finishing of cellulose.

Unalan, I., Slavik, B., Buettner, A., Goldmann, W.H., Frank, G., and Boccaccini, A.R. 2019. Physical and antibacterial properties of peppermint essential oil loaded Poly (-Caprolactone) (PCL) electrospun fiber mats for wound healing. *Frontiers in Bioengineering and Biotechnology*, 7(November): 346.

Van Herreweghen, F., Amberg, C., Marques, R., and Callewaert, C. 2020. Biological and chemical processes that lead to textile malodour development. *Microorganisms*, 8(11). doi: 10.3390/microorganisms8111709.

Vandendaele, P., and Temmerman, R. 2010. Microcapsules containing microorganisms, issued 2010. https://patentimages.storage.googleapis.com/07/ef/50/19e105646001c0/US20120076864A1.pdf.

Waite, M. 2009. Sustainable textiles: The role of bamboo and a comparison of bamboo textile properties - Part I. *Journal of Textile and Apparel, Technology and Management*, 6. Available: http://ojs.cnr.ncsu.edu/index.php/JTATM/article/view/651.

Yao, C., Li, X.-S., Neoh, K.G., Shi, Z.-L., and Kang, E.T. 2010. Antibacterial poly(D,L-Lactide) (PDLLA) fibrous membranes modified with quaternary ammonium moieties. *Chinese Journal of Polymer Science*. doi: 10.1007/s10118-010-9094-x.

Zhu, P., and Sun, G. 2004. Antimicrobial finishing of wool fabrics using quaternary ammonium salts. *Journal of Applied Polymer Science*. doi: 10.1002/app.20563.

Zille, A., Almeida, L., Amorim, T., Carneiro, N., Esteves, M.F., Silva, C.J., and Souto, A.P. 2014. Application of nanotechnology in antimicrobial finishing of biomedical textiles. *Materials Research Express*. doi: 10.1088/2053-1591/1/3/032003.

Zou, X., Zhang, L., Wang, Z., and Luo, Y. 2016. Mechanisms of the antimicrobial activities of graphene materials. *Journal of the American Chemical Society*, 138(7): 2064–2077.

7 Herbal Antibacterial Agents as Odour Control Finish in Textiles

R. Rathinamoorthy
Departmentof Fashion Technology, PSG College of Technology, Coimbatore, India

G. Thilagavathi
Department of Textile Technology, PSG College of Technology, Coimbatore, India

CONTENTS

7.1 INTRODUCTION

The skin is the largest part of the human body and acts as a barrier to protect the body from environmental changes. Early research from the 1950s detailed the role of skin, surface microflora, and odour formation (Leyden et al. 1981; Shegadeh and Kligman 1963; Shelley et al. 1953; Strauss and Kligman 1956). Different parts of the body, such as the foot and axilla, are prone to generate odour due to the higher moist environment and development of microorganisms. Textile fabrics are worn very close to the skin and are capable of retaining odour molecules inside them. Hence, body odour development from the skin, along with retention behaviour of the clothing material, may cause the individual to have an unpleasant odour.

DOI: 10.1201/9781003141426-7

Apocrine sweat from the human axilla was identified as one of the main sources of odour generation by bacterial metabolism (Shelley et al. 1953). Though several deodorising cosmetic products exist in the market, control of body odour is still perceived to be an issue. This is because sweat secretion, odour formation, and odour intensity largely vary between individuals. When textiles are considered, cellulosic fibers are preferred as skin contact fabric due to their higher wear comfort, but, in the case of cellulosic fibers, the presence of carbohydrates acts as a nutrient for the growth of microorganisms and causes unpleasant odour, staining, and loss of mechanical strength in clothing (Raja et al. 2010). Odour formation in textiles also affects the health and hygiene of the wearer, along with their feeling of comfort. Hence, removal of odour from the textile is desirable.

In the case of body odour control, there are different methods adapted to prevent body odour. Sweat is the main element that helps in odour formation, so controlling sweat generation through different approaches will control odour formation. The second approach talks about the use of antibacterial products directly on the skin or in textiles to control the growth of odour-causing bacteria in body parts. The odour can also be controlled by the application of deodorants to the body or by masking it with fragrances either directly applied on the skin or on the textile (Evans 1968; Quatrale et al. 1981; Shegadeh and Kligman 1963). Despite several odour control methods, the use of antibacterial agents plays a vital role due to their versatile nature. Antibacterial agents can be either synthetic materials or natural substances. The basic reason behind the use of an antibacterial agent is to reduce or inhibit bacterial growth in textiles. Since the odour-forming substances are mainly produced by the biotransformation of bacterial strains, inhibiting the growth of microbes reduces the odour in the textile. Hence, most of the research reports measure the efficacy of the antibacterial property as a measure of the odour control ability of a particular antibacterial agent.

The survival ability of microorganism is mainly dependent on the stability of the outermost cell. The outer wall is mostly made of polysaccharides, which protect the microorganism from the environment. Antibacterial agents can be either bacteriostatic or bactericidal; the first type inhibits the growth of bacteria, and the second type kills the bacteria by affecting the cell wall or altering its permeability (Gao and Cranston 2008). The changes in the outer cell wall inhibit the cell membrane function, inhibit protein synthesis, and restrict other metabolic processes that are essential for the cell's survival (Rahman, Ahsan, and Islam 2010). Most of the common antibacterial agents that are used in textiles, such metal oxides, especially chemicals like triclosan, etc., are biocides (Gao and Cranston 2008). When it comes to textile application, there are a few additional requirements that are expected with antibacterial agents, like (i) affinity towards textile fibers and (ii) ability to withstand multiple washings during the lifetime. Further, as the fabrics are worn next to the skin, the antibacterial agent that is finished on the textile should not affect the resident skin microflora. The major focus is on the inhibition of bacterial growth on textile material, fabrics, or fibres. The different types of antibacterial agents are detailed in Figure 7.1.

There are several antibacterial agents used in the antibacterial treatment of textiles due to their various means of action. Most importantly, metals that have a positive surface charge are capable of killing bacterial strains with an electron negative surface (Kanazawa, Ikeda, and Endo 1993). Similarly, the components of

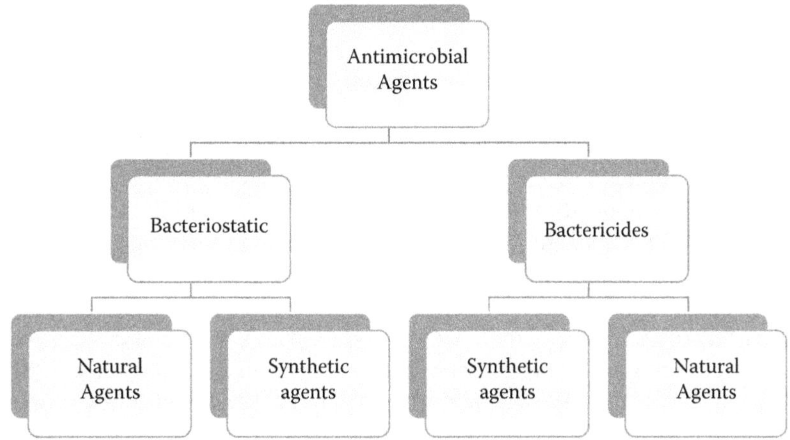

FIGURE 7.1 Classifications of antibacterial agents.

carboxylic acids can readily absorb odour molecules and reduce or degrade them through oxidation or reduction (Disuke and Kenichi 2001). The chapter aims to review the effectiveness of antibacterial agents or treatments on body odour reduction efficiency. The first part of this chapter details synthetic antibacterial agents and their application as odour control finishes in textiles; the potential results of previous research are analysed and reported. The second part of this chapter details application of herbal extracts and plant-based antibacterial agents, specifically *Terminalia Chenula* as odour control finishes.

7.2 SYNTHETIC ANTIBACTERIAL AGENTS AND ODOUR CONTROL ABILITY

Antibacterial material commonly used in textiles is based on silver, copper, zinc, and titanium metal salts or their oxide compounds. The salts of metals are commonly used due to their economic viability in various applications. The antibacterial effect of the metal oxides is mainly based on the reduction potential of the metals, the donor atom selectivity, or its capability in new species formation (Morais, Guedes, and Lopes 2016). The metals bind to donor ligands like O, N, and S by selective interaction and replace the original metals in biomolecules of bacteria and restrict the cell growth by cellular dysfunction (Palza 2015). Silver is one such material that is well known for its wide spectrum of antibacterial activity and commonly used in textile finishing as an antibacterial agent. The main mechanism of silver-based antibacterial finish acts in three stages. At first, after microbial growth, the silver ion adheres to the surface of the bacteria cells randomly and penetrates their structure. Second, it readily binds the electron donor groups like thiols (-SH) and restricts the enzyme development inside the cell structure. Last, silver ions restrict the replications of DNA and thus restrict the growth and odour development (Nahar 2020).

Similar to silver, various zinc salts and water-soluble zinc salts also have the potential to reduce the malodour formation from the body parts when they are used along with antiperspirants, natural androgen receptor expression inhibitors, and malodour carrier protein inhibitors. Out of several types of zinc salts reported by previous researchers (Ascione, Forestier, and Rollat-Corvol 2003; Li and Liu 2002), zinc oxide is one of the most common antibacterial agents used in body odour reduction. The odour molecule and zinc salt interaction were detailed by Kuhn et al. (2000). They reported that the interaction is a nucleophilic attack between the zinc salt and odour compound in the case of the aqueous medium. In an *in-vitro* study performed among 30 healthy volunteers, the application of zinc on their underarm for 13 consecutive days showed a significant reduction in the self-perceived axilla odour on the zinc salt-treated side. The bacterial extraction results also supported the responses of the subject with a statistically significant reduction in the growth of the odour-generating *Corynebacterium spp.* and *S. hominis* strains (Anon 2019).

Though several studies were performed on the antibacterial efficacy of metal oxide-treated textiles, the studies conducted to evaluate the odour reduction were very limited. McQueen et al. analysed the effectiveness of silver chloride in reducing axillary odour in textiles. They treated the textile with two different concentrations equating to 30 and 60 ppm of silver on fabric. After finishing, the fabrics were laundered for 30 washes, and researchers compared the odour reduction effectiveness through wear trial analysis. The worn textiles were evaluated for their odour reduction through sensory assessment. The results showed no significant difference in the odour intensity between the control sample and silver chloride finished samples. The results were also supported by bacterial count analysis. While analyzing the sliver chloride treated fabric, it had potential antibacterial activity *in-vitro* against a few odour causing bacterial strains. However, *in-vivo* studies did not show potential odour reduction or bacterial reduction in the silver chloride finished textiles (McQueen et al. 2013). Chen, Wang and Yeh (2010) used silver nanoparticle and titanium dioxide nanoparticle by mixing in waterborne polyurethane emulsion as a finish for polyester textiles. They evaluated the efficacy of odour reduction against the odour chemicals, namely ammonia gas, acetic acid, and trimethylamine. The results were positive, and they noted a significant reduction in odour intensity of nano-emulsion treated fabric. Further, they also analysed the effect of UV irradiation on the odour-control ability of the nano-emulsion. As UV irradiation increases the active sites of nanoparticles of metal oxides, it helps actively reduce the odour intensity in textiles. On comparing the effect of UV irradiation with odour reduction, a maximum of 90% odour elimination (for ammonia and acetic acid) was noted with the prepared nano-emulsion at 1.5 hours of UV irradiation. Whereas in the case of trimethylamine gas, a maximum reduction of 70% was noted (Chen, Wang and Yeh 2010).

Yin Xu (2012) used polyhexamethylene biguanide (PHMB) and zinc pyrithione (ZP) as antibacterial agents in polyester and cotton textiles to control odour formation. After the wear trial, the odour intensity of the worn textile was analysed and compared with the untreated textile sample. The results showed that PHMB showed a significant bacterial count reduction (CFU/mL) compared to the control sample and ZP after the wear trial. Though both ZP and PHMB reduced the bacterial count on textile, the maximum reduction was noted with PHMB. However, in the case of sensory panel analysis, the antibacterial treatment did not show any effect on odour reduction. Irrespective of the antibacterial treatment applied, the polyester fabric showed a higher odour intensity than the cotton fabric. Others reported the use of a Triclosan-based antibacterial agent for textile treatment. The *in-vivo* analysis among 20 participants reported a significant reduction in odour formation compared to the untreated samples (Mao and Murphy 2001). Walter et al. (2014) evaluated the application of synthetic antibacterial agents in textiles and their effect on skin microflora. The results reported that the use of an antibacterial agent may significantly alter the native skin microflora and cause an imbalance in healthy skin. The most common synthetic antibacterial agents and their mode of action against bacterial strains are provided in Table 7.1, along with the type of textiles (Morais, Guedes, and Lopes 2016).

Most of the other antibacterial studies performed on the textiles only measured the antibacterial effectiveness of the treatment *in-vitro* and related it to odour reduction indirectly. They did not directly measure the odour formation or reduction through a direct *in-vivo* or real-time analysis (Dhiman, and Chakraborty 2015; Frattarelli et al. 2018; Parthiban, Srikrishnan and Viju 2011; Shastri, Rupani, and Jain 2012; Wasif and Laga 2009). Hence, the effectiveness of the synthetic antibacterial treatment on textile odour control in a real-time situation is not clear. Researchers reported that the antibacterial agents may not react *in-vivo* in the same way as they do in the *in-vitro* analysis.

7.3 NATURAL ANTIBACTERIAL AGENTS IN ODOUR CONTROL

Though several research works were performed on odour control in textiles, the majority of the studies use synthetic antibacterial agents due to their improved performance. In the case of synthetic antibacterial agents, the interaction between the antibacterial agent and native skin microflora is unclear, and it may cause some negative effects to the wearers. For instance, metal salts can sometimes become skin irritants, and a few other chemicals like triclosan are suspected to be carcinogenic (Glaser 2004). Hence, a lot of studies desired to use natural herbal material in textiles to improve the environmental friendliness of the process and also to reduce the health hazards. Only a limited number of herb-based extracts were used in the case of body odour control studies. The limitation is due to durability issues and plant variety availability in different geographical regions of the globe.

Literature showed few traditional herbs can control body odour. Research developed a product for topical application for cosmetic or pharmaceutical purpose in the axilla region to control body odour. It is reported that people with typical strong body odour have a smaller amount of *Staphylococcus epidermidis*, a higher amount of *Anaerococcusoctavius,* and a slightly higher number in a certain type of *Corynebacterium* species. Hence, the study focused on developing probiotic activity on the skin. The researchers used various plant extracts, namely *Camellia spec., Hibiscus spec., Malva spec., Vitis spec., Daucus spec., Commiphora spec., Simmondsia spec., or Calendula spec.* or mixtures of those extracts to develop the product. The study developed formulation targets of only the odour causing bacteria compared to the skin-friendly and odour neural microorganism in the axilla region (Bockmuhl et al. 2007). Extracts of *Senna alata*

TABLE 7.1
Chemical structure and action modes of popular antibacterial agents (Reprinted with permission under Creative Commons licence)

S. No	Biocide	Chemical structures	Action mode	Textiles used
1.	Quaternary ammonium salts	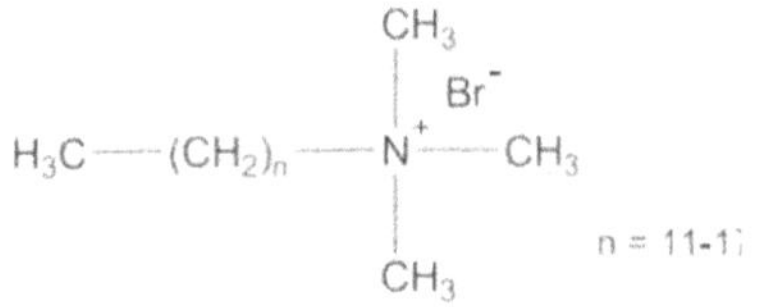 $H_3C-(CH_2)_n-N^+(CH_3)_3\ Br^-$, n = 11-17	Damages cell membranes. Denatures proteins. Inhibits DNA production, avoiding multiplication.	Cotton, Polyester, Nylon, Wool
2.	Triclosan	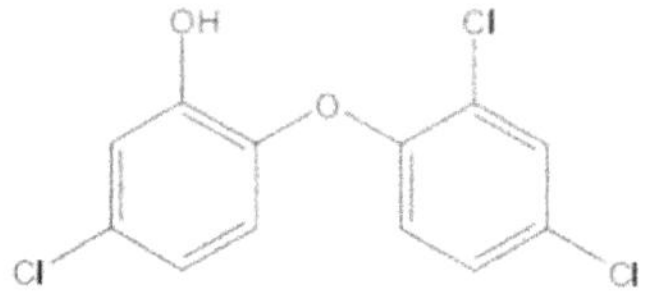	Blocks lipid biosynthesis, affecting the integrity of cell membranes.	Polyester, Nylon, Polypropylene, Cellulose acetate, Acrylic
3.	Metals and metallic salts	TiO_2 and ZnO	Generates reactive oxygen species damaging cellular proteins, lipids, and DNA.	Cotton, Wool, Polyester, Nylon
4.	Chitosan	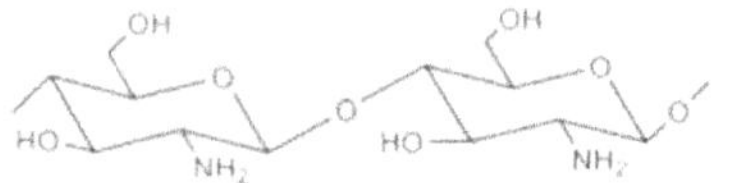	Low Mw: Inhibits synthesis of mRNA, preventing protein synthesis. High Mw: Causes leakage of intracellular substance or blocks the transport of essential solutes into the cell.	Cotton, Polyester, Wool
5.	PHMB	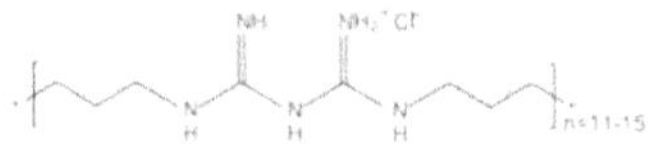 n = 11-15	Interacts with membrane phospholipids, resulting in its disruption and the lethal leakage of cytoplasmic material.	Cotton, Polyester, Nylon
6.	N-halamines	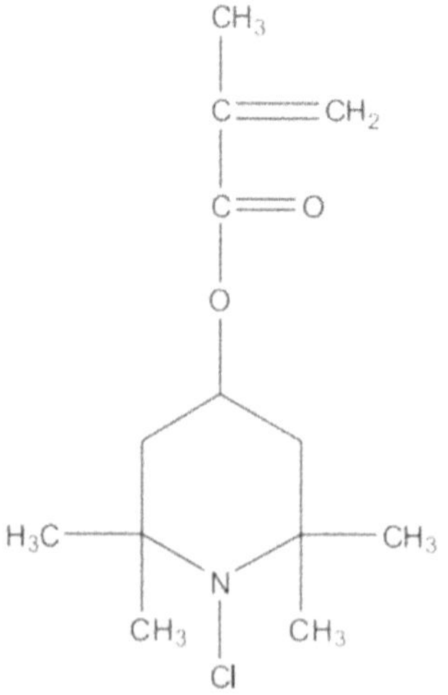	Precludes the cell enzymatic and metabolic processes, causing consequent microorganism destruction.	Cotton, Polyester, Nylon, Wool

were also identified as an effective traditional medicine of Thailand, that can neutralise body odour. The extract had higher potential against *S. epidermidis,* with moderate minimum inhibition concentration (MIC) and minimum bactericidal concentration (MBC) values. The study also reported the extract of *Garcinia mangostana* is very effective in inhibiting the growth of *S. epidermidis* with a MIC of 39 ppm. Hence, researchers reported it can control body odour formation if used as pharma and cosmetic material (Chomnawang et al. 2009).

In-vitro analysis of the *Harunganamadagascariensis* extract showed higher antibacterial activities against odour causing bacterial strains. The *Harunganamadagascariensis* is known for its higher biologically active components, like flavonoids, alkaloids, saponins, glycosides, and tannins. Due to this reason, the ethyl acetate extracts of *Harunganamadagascariensis* leaf had higher MIC and MBC values against the armpit odour causing bacterial strains. The important odour causing bacterial strains like *Corynebacterium xerosis* and S. *epidermidis* were inhibited at 200 and 250 ppm, respectively (Moulari et al. 2006; Okoli et al. 2002). Ueda, Tokimitsu, and Masatoshi (1990) analysed the effect of Licorice root extract for the development of a deodorant product that can control axillary odour. The extract was formulated in various forms like powder, stick, cream, lotion, and roll on. The presence of the active ingredient glycyrrhetic acid was noted as the reason for its higher inhibition against odour causing components. The study developed a product by combining the extract with tannic, sorbic, salicylic acid, and phenol, along odour-masking agents.

Terminalia Chebula retz fruit extract is one such potential antibacterial material identified by several researchers in the previous decade. The extracts of *Terminalia Chebula* showed inhibition to a wide spectrum of bacterial species. For instance, the extracts of *Terminalia Chebula* fruit showed potential antibacterial activity against *Bacillus substils*, *Staphylococcus aureus*, *staphylococcus epidermis*, *Escherichia coli*, *Staphylococcus flexineria,* and *Pseudomonas aeruginosa* (Manoj Kumar et al. 2009). The extracts also showed markable activity against *Aspergillus niger* and *candida Albicans*, the commonly found fungal strains (Sachin Kumar et al. 2011). The common active ingredient found in *Terminalia Chebula* is provided in Figure 7.2. Previous researchers reported the effectiveness of *Terminalia Chebula* extract by evaluating its medicinal properties like cardioprotective effect (Suchalatha and Shyamala Devi 2004), antifungal activities (Naqvi et al. 2010), inhibition of cancer cell growth (Saleem et al. 2002), immunomodulatory activity (Aher and Wahi 2011), molluscicidal activity (Sharma et al. 2011), wound healing (Choudhary 2011), gastrointestinal motility (Tamhane et al. 1997), anti-clastogenic effect (Wasim Raja et al. 2011), anti-aging activities (Manosroi et al. 2010), anti-hyperglycemic effect (Murali et al. 2004), antioxidant activity (Walia et al. 2011), hypolipidemic activity (Maruthappan and Sakthi Shree 2010), anti-lithiatic activity (Tayal et al. 2012), and anti-ulcerogenic activity (Sarabjit Kaur and Jaggi 2010).

7.3.1 *Terminalia Chebula* Extracts and Odour Control

Based on their multi-strain resistant nature, the application of *Terminalia Chebula* extracts as an odour control finish on different textile material was performed by Rathinamoorthy et al. (2011). Initial study results reported that, out of other herbals

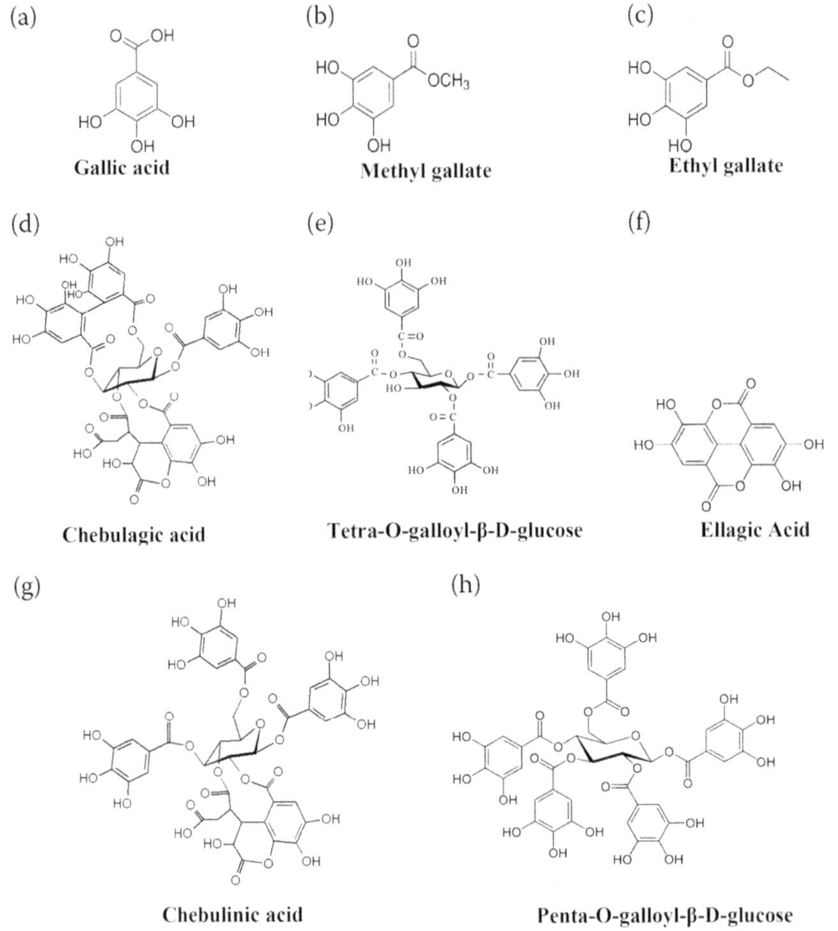

FIGURE 7.2 Structures of varius acids present in Terminalia Chebula extract. (a) Gallic acid, (b) Methyl gallate, (c) Ethyl gallate, (d) Chebulagic acid, (e) Tetra-O-galloyl-B-D-glucose, (f) Ellagic acid, (g) Chebulinic acid and (h) Penta-O-galloyl-B-D-glucose.

materials used in the experiments, the extracts of *Terminalia Chebula* provided superior antibacterial properties against odour causing bacterial strains like *Corynebacterium sp, Micrococcus sp, Bacillus sp,* and other than this, it showed good performance against common skin-born strains like *Staphylococcus aureus, Escherichia coli, Salmonella typhi, and Proteus vulgaris,* with a higher zone of inhibition value. Through a series of work, Rathinamoorthy and Thilagavathi (2020) analysed the effectiveness of *Terminalia Chebula* in odour reduction of different textile material. After the initial study, they optimized the extraction parameters for higher active ingredient yield through the extraction process. As solvent extraction is one of the most common and effective methods in the extraction process, organic solvent extraction is used to extract the components of *Terminalia Chebula.* Extraction temperature, time, and sonication time are the three parameters identified as major influencing parameters. The antibacterial activity of the extract against the most common odour causing bacterial strains like *Corynebacterium* sp (MTCC 8730), *Bacillus licheniformis* (MTCC 429), *Micrococcus luteus* (ATCC 49732) and *Corynebacterium acnes* (MTCC1951) are kept as responses. Using statistical methods, the process parameters were optimized and reported as 29 hours of extraction (immersion) time, 30 min of sonication process, and a temperature of 35°C provided a higher zone of inhibition against the selected odour-causing bacterial strains.

Further, to evaluate the active components responsible for the antibacterial activity, several analytical tests were performed on *Terminalia Chebula* extract. The results of phytochemical analysis represented various active biological components of *Terminalia Chebula,* as provided in Table 7.2. The methanol extract of the *Terminalia Chebula* showed the presence of phytochemicals, such as tannins, saponins, flavonoids, phenol, and glycosides. Out of all the components, a maximum contribution of 10.90% was noted from tannin, followed by 10.10% of phenolic components. Along with that, a significant portion of saponins (at 3.32%), and flavonoids (at 0.68%) were also identified as a part of the extract, using the quantitative evaluation method (Rathinamoorthy and Thilagavathi 2014). Flavonoids are nature's biological response modifiers as they efficiently restrict the body's reaction to the allergic substance, virus, and harmful bacterial strains (Middleton and Kandaswami 1994).

Similarly, the high-performance liquid chromatography (HPLC) analysis results reported prominent peaks with the retention time of 3.06, 3.32, 3.79, and 4.3 minutes as shown in Figure 7.3 It is a method used for the quantification of the component of a

TABLE 7.2

Phytochemical component analysis (From Rathinamoorthy and Thilagavathi (2014). (Reprinted with permission under Creative Commons licence from Morais, Guedes, and Lopes 2016)

Name of components	Methanol extract of *Terminalia Chebula*	Quantitative analysis results (%)
Flavonoids	+++	0.68
Alkaloids		
Dragondroff reagent	+	
Wagners reagent	-	
Mayers reagent	-	
Tannin	+++	10.90
Protein	++	
Carbohydrate – Fehlings method	+	
Saponin	++	3.32
Glycosides	++	
Phenols	+++	10.10
Thiols	-	
Steroids	+	
Triterpenoids	+	

+ - Presence, ++ Moderate presence, +++ High presence.

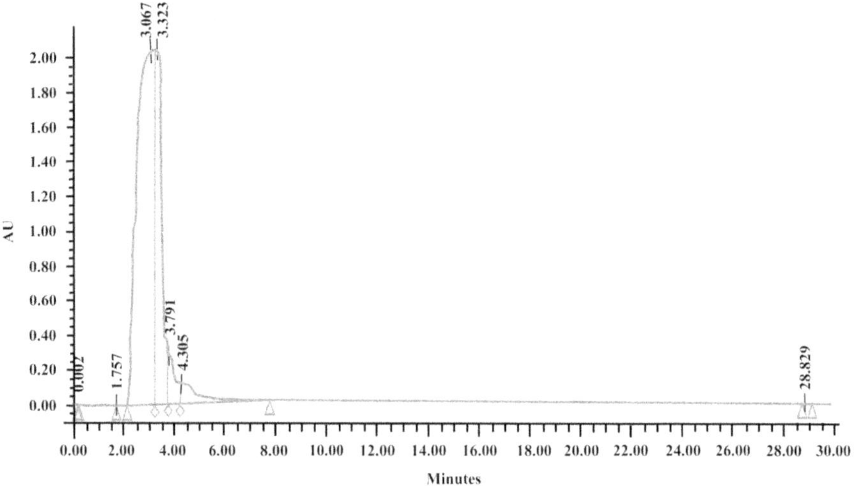

FIGURE 7.3 HPLC Spectrum of *Terminalia Chebula* extract. (From Rathinamoorthy and Thilagavathi (2014). Reprinted with permission under Creative Commons licence.)

mixture of compounds regarding UV/VIS light absorption. The retention time values of the graph help us identify the component by comparing a reference material. Comparing the standard retention times, out of the noted peaks, saponin, ascorbic acid, and gallic acid are represented by the peaks 3.32, 3.79, and 4.3 minutes, respectively (Naik et al. 2004). The same was further confirmed by other researchers, who reported the standard peak of saponin as 3.31 min in her study (Shrama et al. 2008). Hence, based on the analysis, the higher antibacterial activity of the *Terminalia Chebula* extract is mainly attributed to the presence of a higher number of phytochemical components, like tannin, saponin, alkaloids, phenols, and flavonoids. Tannin, saponin, and alkaloids are reported to have potential medicinal properties and are generally used in healing injuries and to stop bleeding from cuts. Similarly, the flavonoids present in *Terminalia Chebula* help increase the anti-allergic, anti-inflammatory, and anti-microbial properties of the extract (Figure 7.3).

The next major component reported is phenolic compounds. This is one of the most common constituents of plant extracts; however, the presence of phenolic component increases the anti-inflammatory, anti-atherogenic, antithrombotic, immune-modulating, analgesic activities and also acts as an antioxidant (Sun et al. 2009). Though several studies reported the antibacterial effectiveness of *Terminalia Chebula* extract, the application of the extract on textiles has been performed only in recent times. The major expectation of the textile finishing process is to embed the active substance onto the surface of individual fibers in the textile fabric. A study applied the extract of *Terminalia Chebula* on textile fabric and measured the presence of active substance through FTIR analysis. To ensure the presence of the active components in the fabric after treatment, FT-IR analysis was performed for both extract-treated and untreated fabric. The results are shown in Figure 7.4. The presence of more carboxyl group [C(=O)-OH] at the absorption region 3200 cm^{-1} to 3600 cm^{-1} and 1200 cm^{-1} to 1700 cm^{-1} of treated sample confirms the presence

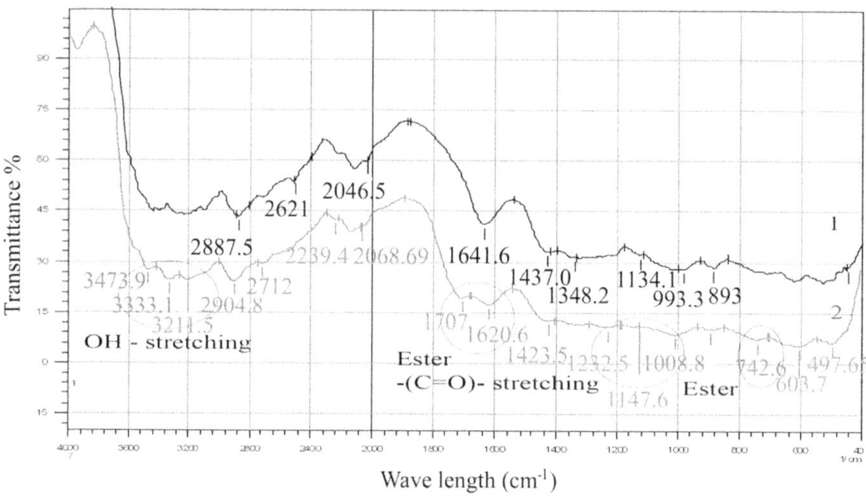

FIGURE 7.4 FTIR Spectra of *Terminalia Chebula* extract-treated textile (2) and control sample (1). (From Rathinamoorthy and Thilagavathi (2013). Reprinted with permission under Creative Commons licence.)

of -OH group stretching, which in turn represents the active substance like gallic and ascorbic acid in the extract. The presence of –(C=O)- group stretching is in the region of 1600 cm^{-1} to 1900 cm^{-1} in the treated sample. This stretching of 1760 cm^{-1} to 1670 cm^{-1} confirms the presence of the ester group. This might be due to the interaction of gallic acid and cellulose, which resulted in the ester as an end product on fabric. These studies confirm the presence of active components on the treated fabric (Rathinamoorthy and Thilagavathi 2013).

7.3.2 Effect of Herbal Antibacterial Treatment on Different Textile Fibers

Application of the extracts of *Terminalia Chebula* on different textile fibres like 100% cotton, viscose, linen, nylon, and 60/40 cotton/polyester and 100% polyester knitted fabrics were performed. All the materials were treated with methanol extracts of *Terminalia Chebula* and pad dried. The finished samples were affixed in the axilla region of the T-shirt and subjected to wear trial. After trial, the samples were collected and stored for 24 hours, and odour intensity of the textiles were analysed subjectively as per Swiss Norms (SN). They compared the odour intensity of the unfinished samples and finished samples and reported the effectiveness of *Terminalia Chebula* extract as an odour control substance. The control samples showed a higher odour intensity with synthetic textiles like polyester and nylon, followed by cotton polyester blend, and natural fibers with less odour intensity. The sequence from higher to lower is noted as polyester > nylon > cotton/polyester > linen > viscose > cotton (Rathinamoorthy et al. 2014).

The odour evaluation of the *Terminalia Chebula* treated fabric showed that the treatment significantly reduced the odour intensity of the textile fabrics after the wear trail. A maximum of 53% odour intensity reduced for polyester and polyester-blended fabric, followed by linen (45.5%), viscose (32%), cotton (18.3%), and nylon (13%) in the case of sedentary workers, as shown in Figure 7.5. The reduction rates are different in the case of nonsedentary workers, where linen showed a higher reduction of 35.1%, followed by polyester/cotton blend (32.9%), nylon (24%), cotton (17.3%), and polyester (9%). To confirm the odour reduction objectively, a bacterial count reduction analysis was performed after a wear trial for both herbal extract-treated and untreated fabrics. The results are presented in Figure 7.6, and it

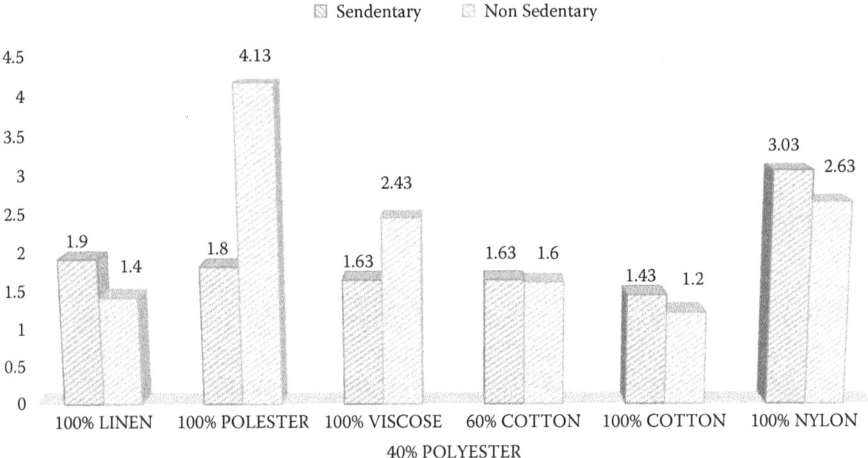

FIGURE 7.5 Odour control efficacy of *Terminalia Chebula* extract finished textiles. (From Rathinamoorthy et al. (2014). Reprinted with permission.)

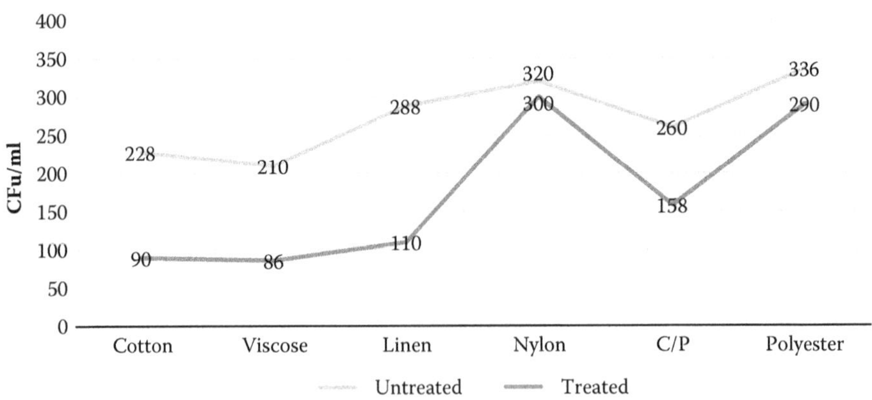

FIGURE 7.6 Bacterial populations on *Terminalia Chebula* extract treated fabrics (in CFU/mL). (From Rathinamoorthy et al. (2014). Reprinted with permission.)

shows that except for nylon, all the samples show a significant reduction in bacterial count (CFU/mL) after *Terminalia Chebula* treatment. As the odour formation is directly related to the bacterial population in the textile (McQueen et al. 2007), reduction in the bacterial count after herbal extract treatment was reported as the main reason for odour reduction.

To confirm the antibacterial activity, a separate agar diffusion test was performed on treated fabrics against the axilla-born odour causing strains like *Staphylococcus aureus* (gram-positive), *Cornybacterium, Cornybacterium 2, Bacillus licheniformis, Cornybacterium acnes, Pseudomonas Sp., Micrococci,* and *Klebsiella pneumonia.* The results were in support of the bacterial count analysis results, which were reported earlier. Though all the fabric showed good antibacterial activity against all the strains selected, nylon did not show any activity. The polyester showed a higher activity similar to cellulose-based fibers except for *Pseudomonas* Sp., and *Klebsiella.*

When the synthetic textiles were compared, a higher activity of polyester fabric compared to the nylon was noted, and this might be due to the methanolysis process, which modified the surface of the polyester and aided in the adhesion of herbal extract components (Rathinamoorthy et al. 2014). Nylon, due to its higher resistive nature against organic solvent, does not show any change. All the fabrics treated with herbal extract showed a significant reduction in odour formation after the wear trial except nylon. The results of the study showed the potential of *Terminalia Chebula* treatment against odour causing bacterial strains. As reported earlier, the higher active ingredients like tannin, saponin, gallic acid, phenols, and chebulic acids present in the extract were noted as the main reason for its supreme performance. The other researcher who analysed the extract of *Terminalia Chebula* also reported the presence of different carboxylic acids (Mahajan al. 2010).

7.3.3 Effect of Fabric Structure and Antibacterial Treatment

Effect of fabric structure on the antibacterial treatment and odour control ability of the antibacterial agent was also analysed to understand their interactions. In this research, 100% cotton knitted fabric with different structures was considered for odour analysis. They prepared two sets of samples of these three structures, out of which one set was treated with the methanol extract of *Terminalia Chebula*. Both the sets of samples were subjected to wear trials, and the odour formation in the fabric was evaluated subjectively and objectively. The results of the subjective odour analysis are provided in Figure 7.7.

The results showed that the odour intensity was higher in the case of rib and interlock structure compared to the single jersey fabric. This is mainly attributed to their structural thickness and mass per square meter variation. The single jersey structure possesses a very open structure compared to the other two selected structures. This increases the air transmission between the skin and atmosphere, so the axilla region remains drier than the other two fabrics. This is the reason for the lower odour in the single jersey structure. The *Terminalia Chebula* extract-treated fabric showed a significant reduction in odour for a similar wear time by the same participants. The results showed a maximum reduction of odour intensity (47%) with the rib structure. The reduction percentage is higher for rib fabric than the interlock and

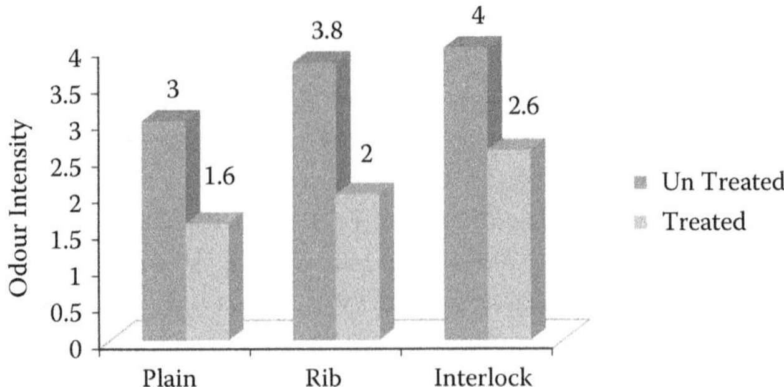

FIGURE 7.7 Average odour intensity rating of different knit structure with different thicknesses. (From Rathinamoorthy and Thilagavathi (2014). Reprinted with permission under Creative Commons licence.)

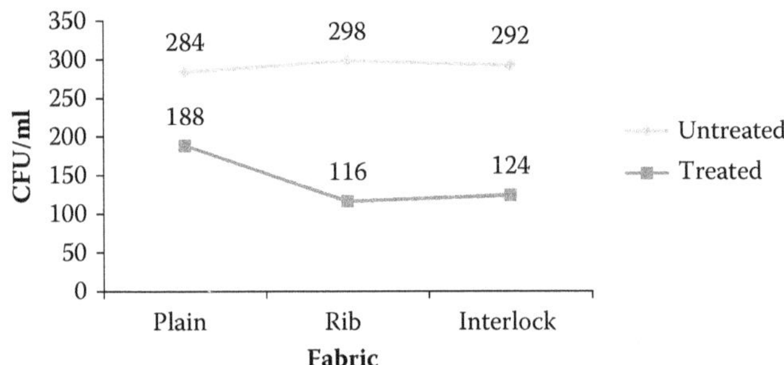

FIGURE 7.8 Bacterial populations on *Terminalia Chebula* treated knit structures (in CFU/mL). (From Rathinamoorthy and Thilagavathi (2014). Reprinted with permission under Creative Commons licence.)

single jersey structures. The higher thickness and grams per square meter of the rib structure consist of higher yarn density than the other two selected structures. This leads to higher extract absorption (pick up %) during the treatment process. This was evident from the lower odour formation of the rib fabric than other fabrics despite its higher thickness. For interlock fabric, a 35% odour reduction was noted, and a 46% reduction was noted for a single jersey fabric in terms of odour intensity. The research analysed the bacterial colony count in the worn textile to confirm the odour reduction objectively (Rathinamoorthy and Thilagavathi 2014).

After the wear trial, the isolated bacterial strains were incubated for 24 hours and then plated on nutrient agar and incubated for a day. The colony-forming units in the plates were counted and reported in Figure 7.8. The results showed a statistically significant reduction ($p < 0.05$) in the bacterial count on the *Terminalia Chebula* treated fabric than the untreated samples. After herbal extract treatment, the single jersey fabric showed a higher number of bacterial colonies (188 CFU/mL) than the rib and interlock structure (116 and 124 CFU/mL, respectively). The results were in contradiction with the untreated samples, where the single jersey fabric showed a lower bacterial count than other fabrics. The author detailed the reason as the structural variation of the fabric and thickness differences. The higher odour in the single jersey fabric than rib and interlock is mainly associated with its thickness and areal density. Due to its open structure and fewer yarns per unit area, it was able to absorb and retain a very limited amount of antibacterial agent in its structure. However, the case is reverse in interlock and rib as it has a higher thickness and mass per unit area. They confirmed the same by calculating the weight difference in all the samples, where the rib fabric showed a higher difference (addition) after herbal extract treatment. The results represented a stronger correlation between odour reduction and bacterial count reduction (Rathinamoorthy and Thilagavathi. 2014). All the fabrics were laundered five times as per standard and evaluated for their antibacterial property using an agar diffusion test. The results showed a significant reduction in antibacterial activity of the washed sample as provided in Table 7.3. From the result, it is evident that after five laundries, the rib fabric possessed a maximum zone of inhibition followed by interlock and a single jersey. This confirms the poor extract holding capacity of the single jersey fabric, which resulted in poor odour reduction after herbal extract treatment. The rib and interlock were able to retain the extract even after five laundries and so showed a better performance.

7.3.4 Objective Measurement of Odour Reduction – GC-MS Analysis

7.3.4.1 GC-MS Spectra of Cotton Fabric Worn Next to the Axilla

To analyse the effectiveness of the *Terminalia Chebula* extract against odour control properties of the textile objectively, gas chromatography–mass spectrometric (GC-MS) was analysis performed for both extract-treated and untreated samples after a wear trial. Rathinamoorthy and Thilagavathi (2016) identified and

TABLE 7.3
Durability evaluation of *Terminalia Chebula* extract finished knit structures (Rathinamoorthy 2016)

S.No.	Bacterial strains	*Terminalia Chebula* fruit methanol extract finished textile - zone of inhibition in mm after 5 washes			
		Control Fabric	Single Jersey	Rib (1X1)	Inter lock
1	*Staphylococcus aureus*	–	22	20	20
2	*Corynebacterium* Sp (*ATCC 3021*)	–	–	22	–
3	*Corynebacterium* Sp (*MTCC 8730*)	–	20	20	–
4	*B. licheniformis*	–	–	20	20
5	*M. luteus*	–	–	–	–
6	*Corynebacterium acnes*	–	–	22	-
7	*Pseudomonas* Sp.,	–	–	20	20
8	*Escherichia coli*	–	–	20	20

analysed the odour causing components from the worn textile materials. The results reported the presence of 5à-Androst-16-Ene-3-One, a type of steroid that results from the metabolism of *coryneform* bacteria. It is one of the distinctive steroids that are identified with male axilla. 16-androstenes and 5a-androstenol along with the above-mentioned steroid are the common steroids found in the apocrine sweat, causing axillary malodour (Gower et al. 1997). The literature also reported that 5à-Androst-16-Ene-3-One, along with 5a-androstenol and 5a-androstenone, were identified and reported as the metabolic activities of *Corynebacteria* and their subgroups. The urine-like smell characteristic of axillary odour is mainly due to the presence of steroids. However, no research completely detailed the bio-transformation of steroids by *Corynebacterial* genes due to their complex nature, unlike volatile fatty acids. The second most common way of odour generation is the conversion of skin lipids through degradation as volatile fatty acids. They reported the presence of more fatty acid groups in the isolated worn textile samples. At a retention time (RT) of 15.21, the methyl esters dodecanoic acids are identified. Similarly, the research also reported the fatty acids of lauric acids, diethyl esters of 1,2-benzenedicarboxylic acid at RT of 16.65, a derivative of phthalic acids, methyl esters of tetradecanoic acid at RT of 19.51, a saturated fatty acid of myristic acid, 3-methyl-hexanoic acid at RT of 27.49, a short-chain fatty acid, aminoisobutyric acid at RT of 32.52, a derivative of isobutric acid, 1,2-benzenedicarboxylic acid at RT of 33.12, a derivative of aromatic dicarboxylic acid called phthalic acid, Tetra-decanol at RT of 36.29, a derivative of myristyl alcohol and acetic acid at the RT of 36.91. These findings further confirm the bacteria as a source for odour generation in fabric (Rathinamoorthy and Thilagavathi 2016).

Another common odour causing component in the axilla is short- and medium-chain volatile fatty acids, principally originating from the hydrolysis of the triglyceride component of sebum and also from the bacterial lipase action on apocrine lipids. This biotransformation of triglycerides and fatty acids into short-chain volatile fatty acids leads to odour formation in the axilla. Among the several bacterial strains reported earlier, *Propionibacteria* and *Staphylococci* strains are noted to ferment glycerol from triacylglycerol hydrolysis and lactic acid. Other than this, *Staphylococci* strain converts the branched aliphatic acids like L-leucine into highly odourous short-chain volatile fatty acids like isovaleric acid (Leyden et al. 1981). In their results, they found several target fatty acids, particularly acetic, butyric, (or isobutyric), and isovaleric, in subjects, as reported in Table 7.4. (Rathinamoorthy and Thilagavathi 2016).

7.3.4.2 GC-MS Spectra of Terminalia Chebula *Finished Cotton Fabric Worn Next to the Axilla*

The same analysis was performed on *Terminalia chebula* treated cotton fabric after wear trail, and the results are presented in Table 7.5. The results showed that after wear trial the samples showed various components of *Terminalia Chebula* extract along with the sweat components, like volatile fatty acids and alcohols, including undecanol, tetradecanoic acid, 2-hydroxyethyl ester (CAS), and stigmast-5-En-3-Ol, (3á), also found at the retention times of 38.21, 25.35, and 35.18 min, respectively. Compared to the unfinished fabric, there was a significant amount of

TABLE 7.4

Odour causing components derived from axilla worn fabric (without treatment). (Rathinamoorthy and Thilagavathi 2016. Reprinted with permission)

S.-No	Retention Time (RT)	Compound Name	Molecular Weight	Molecular Formula	Type of Material	Property
1.	3.01	5à-Androst-16-Ene-3-One	301	$C_{20}H_{31}NO$	Steroid	(Odour-forming Compound)
2.	3.58	Boric Acid (H_3BO_3), Trimethyl Ester	104	$C_3H_9BO_3$	Ester	–
3.	10.85	2-Nonanol	144	$C_9H_{20}O$	Alcohol	(Odour-forming Compound)
4.	15.21	Dodecanoic Acid, Methyl Ester	214	$C_{13}H_{26}O_2$	Fatty Acids of Lauric Acid	(Odour-forming Compound)
5.	16.65	1,2-Benzenedicarboxylic Acid, Diethyl Ester	222	$C_{12}H_{14}O_4$	Diethyl Ester of Phthalic Acid	(Odour-forming Compound)
6.	19.51	Tetradecanoic Acid, Methyl Ester	242	$C_{15}H_{30}O$	Saturated Fatty Acid of Myristic Acid	(Odour-forming Compound)
7.	19.91	Fumaric Acid,	408	$C_{25}H_{44}O_4$	Unsaturated Dicarboxylic Acids	(Odour-forming Compound)
8.	23.13	4-Octadecenal	266	$C_{18}H_{34}O$	Alcohol	(Odour-forming Compound)
9.	23.59	Hexadecanoic Acid, Methyl Ester	270	$C_{17}H_{34}O_2$	Saturated Fatty Acids (Palmitic Acid)	(Odour-forming Compound)
10.	27.49	Hexanoic Acid, 3-Methyl-, Methyl Ester	144	$C_8H_{16}O_2$	Short Chain Fatty Acids	(Odour-forming Compound)
11.	29.51	1,1,3,3,5,5,7,7,9,9,11,11,13,13-Tetradecamethyl-H Eptasiloxane	504	$C_{14}H_{44}O_6S$	–	(Odour-forming Compound)
12.	30.63	9-Octadecenamide, (Z)	281	$C_{18}H_{35}NO$	Amide of Fatty Acid (Oleic Acid)	(Odour-forming Compound)
13.	32.52	Benzoic AcidMethyl Ester	431	$C_{24}H_{21}N_3\ O_3S$	Aromatic Carboxylic Acid	(Odour-forming Compound)
14.	32.52	Aminoisobutyric Acid	536	$C_{24}H_{39}\ F_3N_4O_6$	Amino Acid (Peptide)	(Odour-forming Compound)
15.	33.12	1,2-Benzenedicarboxylic Acid	390	$C_{24}H_{38}O_4$	Aromatic Dicarboxylic Acid (Phthalic Acid)	(Odour-forming Compound)
16.	35.61	Cholesterol	386	$C_{27}H_{46}O$	Modified Steroid	(Odour-forming Compound)
17.	36.29	Tetradecanol	214	$C_{14}H_{30}O$	(Myristyl Alcohol)	(Odour-forming Compound)
18.	36.91	Acetic Acid, Chloro-, Octadecyl Ester	346	$C_{20}H_{39}\ CLO_2$	–	(Odour-forming Compound)
19.	38.18	[2.2]Metacyclophane-4,12-Dialdehyde	264	$C_{18}H_{16}O2$	Aldehyde	(Odour-forming Compound)

TABLE 7.5

Odour causing components derived from axilla worn fabric after *Terminalia Chebula* treatment (Rathinamoorthy and Thilagavathi 2016. Reprinted with permission)

S.No	Retention Time	Compound Name	Molecular Weight	Molecular Formula	Type of Material	Properties expected
1.	3.10	3-Methyl-5-Chloro-N-Pentane-2-One	134	$C_6H_{11}CLO$	Straight-chain alkyl carboxylic acid	(Odour-forming compound)
2.	3.55	3-Methylhexan-2-One-6-D3	114	$C_7H_{11}D_3O$	Alkane	–
3.	5.88	1,3-Benzenedicarboxylic Acid	166	$C_8H_6O_4$	PlasticizerCompound	Anti-fouling, Antibacterial
		Benzaldehyde, 4-(Dimethylamino)	149	$C_9H_{11}NO$	Aldehyde	–
4.	6.84	2-Octanone	128	$C_8H_{15}DO$	Ketone	Plant source chemical, Flavoring agents
5.	7.28	H-Pyrazole-1-Propanoic Acid,A-[[(4-Oxo-2-Azetidinyl)Carbonyl]Amino]-, Methyl Ester	266	$C_{11}H_{14}N_4O_4$	Fatty Acid	Antibacterial, Anti-inflammatory, Anticancer, Fungicide
6.	10.61	N-Decane -2, 3-D2	142	$C_{10}H_{20}D_2$	Alkane Hydrocarbon	Antibacterial
7.	16.60	1,2-Benzenedicarboxylic Acid, Diethyl Ester	222	$C_{12}H_{14}O_4$	Organic Ester	Antibacterial and Antioxidant
8.	20.08	5,6,8,9-Tetramethoxy-2-Methylpepero(3,4,5-JK)-9,10-Dihydrophenanthracene	355	$C_{21}H_{25}NO_4$	Alkaloid	–
9.	23.55	Hexadecanoic Acid, Methyl Ester	270	$C_{17}H_{34}O_2$	Palmitic Acid / Ester Compound	Antioxidant, Hypocholesterolemic, Anti-androgenic, Hemolytic, Anti-inflammatory

	23.55	Pentadecanoic Acid, 14-Methyl-, Methyl Ester	270	$C_{17}H_{34}O_2$	Rare Fatty Acid In Nature; Saturated Fatty Acids of Lauric acid	Anti-microbial and Anti-inflammatory, Flavoring agents (Odour-forming compound)
10.	26.96	9-Octadecenoic Acid (Z)-, Methyl Ester	296	$C_{19}H_{36}O2$	Linoleic Acid	Anti-inflammatory, Nematicide, Insectifuge, Hypocholesterolemic, Cancer Preventive, Hepatoprotective, Anti-androgenic, Anticoronary
11.	29.38	(-)-Dehydroaromadendrene	204	$C_{15}H_{24}$	Hydrocarbons	–
		Butanedioic Acid, Bis (Trimethylsilyl) Ester (Cas)	262	$C_{10}H_{22}O_4SI_2$	Ester Compound	Antibacterial, Preservative
12.	31.04	Tetratetracontane	618	$C_{44}H_{90}$	–	–
13.	32.06	Acovulparine	437	$C_{23}H_{35}NO_7$	Alkaloid	–
14.	34.21	3,4,5,6,-Tetrahydrohippurate	183	$C_9H_{13}NO_3$	–	Dietary food metabolome
15.	35.18	Stigmast-5-En-3-Ol, (3á)-	414	$C_{29}H_{50}O$	Phytosterols (Plant Steroid)	–
16.	36.03	Cyclohexane, 1,1',1'',1'''-(1,6-Hexanediylidene)Tetrakis	414	$C_{30}H_{54}$	Sapogenins –Triterpenoid Saponins	–
17.	36.97	12-Methoxyibogamine-18-Carboxylic Acid Methyl Ester	368	$C_{22}H_{28}N_2O_3$	Alkaloids	–
	36.97	9,12,15-Octadecatrienoic Acid, Methyl Ester,	496	$C_{27}H_{52}OSI_2$	Fatty Acid Ester of Linolenic Acid	Antiinflammatory, Hypocholesterolemic, Cancer Preventive, Nematicide, Hepatoprotective, Antihistaminic, Antieczemic, Antiacne, Anticoronary,

(*Continued*)

TABLE 7.5 (Continued)

Odour causing components derived from axilla worn fabric after *Terminalia Chebula* treatment (Rathinamoorthy and Thilagavathi 2016. Reprinted with permission)

S.No	Retention Time	Compound Name	Molecular Weight	Molecular Formula	Type of Material	Properties expected
18.	38.21	Undecanol	172	$C_{11}H_{24}O$	Fatty Alcohol	Flavour, Perfumery, (Odour-forming compound)
19.	15.21	12-Phenyl-2,3,7,8-Tetramethoxy-5h-(1)-Benzopyrano[4,3-C] Isoquinoline	429	$C_{26}H_{23}NO_5$	Alkaloids	Anti-Hypertension, Antifungal
20	25.35	Tetradecanoic Acid, 2-Hydroxyethyl Ester	272	$C_{16}H_{32}O_3$	Fatty Acid	Antibacterial, Antifungal (Odour-forming compound)

reduction in odour forming components noted with *Terminalia Chebula* extract finished textiles. They reported the presence of extract component, even after wear trial that is responsible for the antibacterial, antioxidant, and anti-inflammatory properties. These findings again confirm that the presence of these components reduced the growth of bacterial in the axilla. Though the treatment showed a reduction in odour causing components in textile, the results of the GC-MS study also spotted 3-methyl-5-chloro-N-pentane-2-One at the retention time of 3.10, one of the odourous compounds. Similarly, other odour causing compounds like 1,2-benzenedicarboxylic acid (RT-16.60), 1,3-benzenedicarboxylic acid (RT-5.88), tetratetracontane (RT-31.04), hexadecanoic acid (RT-23.5), pentadecanoic acid (RT-25.33), and N-decane (RT-10.61) were also noticed in the *Terminalia Chebula* extract finished cotton fabric. These findings confirmed the presence of the active ingredient in the extract-finished textile and its potential in odour control or reduction during wear trial. The finished fabric also consists of phytochemical components, like flavonoids, alkaloids, saponins, and aldehydes, along with some carboxylic acid groups, as reported earlier (Rathinamoorthy and Thilagavathi 2016).

Other than this, the effect of surface modification of polyester fabric by alkali treatment on the odour- formation characteristics was evaluated by Rathinamoorthy and Thilagavathi (2017). They treated the polyester knitted textile with mild alkali (0–2%) and subjected it to odour analysis. After 8 hours, the samples were collected and stored for a day and analysed for their odour intensity. The results were supportive that the surface-modified fabric showed a lower odour intensity with an average intensity rating of 2.3 compared to the untreated polyester fabric (4.8). The research mentioned that the alkali treatment modifies the surface of the polyester and by creating surface short chains functional carboxylic acids and hydroxyl groups. This modification forms carboxyl and hydroxyl groups on the surface and thus it creates hydrophilic nature to the polyester. When the wearer uses the alkali-treated polyester, the polyester absorbs and transfers sweat through it and keeps the wearer's skin dry as much as possible. Hence the odour intensity of the polyester reduced significantly after the alkali treatment. From a technical viewpoint, while comparing the cotton and polyester, the lower odour intensity in cotton due to the absorption of an oily substance into the structure whereas, in polyester it is adsorption. However, due to the conversion of hydrophilic nature by surface modification process, the modified polyester showed a reduced odour intensity after wearing. They confirmed the odour reduction through isolation of bacterial strain from worn textile. The results showed a significant amount of bacterial reduction in alkali-treated textile (80 CFU/mL) from 340 CFU/mL in untreated textile. The findings revealed that the conversion of polyester fabric surface into hydrophilic material significantly change the oleophilic characteristics of polyester and so it reduces the odour development in the polyester textile (Rathinamoorthy and Thilagavathi 2017).

The results of synthetic antibacterial agents did not show much effect on the *in-vivo* odour reduction efficiency, the results of herbal extract, *Terminalia Chebula* extract, treatment showed a higher potential in this domain. Both the *in-vitro* and *in-vivo* analysis, along with subjective sensory analysis and objective measurements, confirmed the odour-reduction capacity of the herbal extract. The ecofriendly nature, nontoxic properties, and plant-based origin are the main advantages of these agents, which ultimately reduce chemical hazards during application, usage, and disposal of the apparel.

7.3.4 Summary

This chapter summarises the previous research works performed on odour control studies using antibacterial agents. The use of synthetic antibacterial agents on odour control through *in-vivo* studies did not show a potential reduction. However, the effects of synthetic antibacterial agents were analysed for a wide spectrum of bacterial strains. The use of *Terminalia Chebula* extract on odour control finishing of textile was reported. The effectiveness of *Terminalia Chebula* extract with respect to different fibers and fabric structures was also reported. Both the *in-vitro* and *in-vivo* analysis results showed a higher potential of the selected herb on body odour reduction. The same results were confirmed by subsequent gas chromatography analysis, in which the finished textiles showed a significant reduction in odour-causing substance than the untreated sample.

REFERENCES

Aher, V., and Wahi, A.K. 2011. Immunomodulatory Activity of Alcohol Extract of *Terminalia Chebula* Retz Combretaceae. *Tropical Journal of Pharmaceutical Research*, 10(5): 567–575.

Anon. 2019. Zinc oxide reduces body odor caused by bacteria and aids wound healing. Retrieved 6 May 2021 from https://medicalxpress.com/news/2019-04-zinc-oxide-body-odor-bacteria.html

Ascione, J.M., Forestier, S., and Rollat-Corvol, I. 2003. Deodorant composition comprising a water-Soluble Zinc Salt as Odor-Absorbing agent. US patent 6 632 421 B2. L'Oreal, Paris.

Bockmuhl, D., Hohne, H.M., and Jassoy, C. 2007. Substances with a probiotic action used in deodorants. US patent 0 190 004 A1. Paul & Paul.

Chen, C-C., Wang, C-C., and Yeh, J-T. 2010. Improvement of odor elimination and antibacterial activity of polyester fabrics finished with composite emulsions of nanometer titanium dioxide-silver particles-water-borne polyurethane. *Textile Research Journal*, 80: 291. doi: 10.1177/0040517508100626

Chomnawang, M.T., Surassmo, S., Wongsariya, K., and Bunyapraphatsara, N. 2009. Antibacterial activity of Thai medicinal plantsagainst methicillin-resistant Staphylococcus aureus. *Fitoterapia*, 80: 102–104.

Choudhary, G.P. 2011. Wound healing activity of the ethanolic extract of *TERMINALIA CHEBULA* retz. *International Journal of Pharma and Bio Sciences*, 2(1): 48–52.

Dhiman, G., and Chakraborty, J.N. 2015. Antibacterial performance of cotton finished with triclosan, silver and chitosan. *Fashion and Textiles*, 2: 13. doi: 10.1186/s40691-015-0040-y

Disuke, T., and Kenichi, H. 2001. Filter for Eliminating Formaldehyde, JP 2001340436.

Evans, N.M. 1968. The classification of aerobic diphtheroids from human skin. *British Journal of Dermatology*, 80: 81–83.

Frattarelli, D., Powers, L., Doshi, D., Vargo, K., Patel, B., Liboon, J., and Gallagher, M. 2018. Holistic management of textile odor using novel silver-polymeric complexes. *AATCC Journal of Research*, 5(4): 7–16. doi: 10.14504/ajr.5.4.2

Gao, Y., and Cranston, R. 2008. Recent advances in antibacterial treatments of textiles. *Textile Research Journal*, 78(1): 60–72. doi: 10.1177/0040517507082332.

Glaser, A. 2004. The ubiquitous triclosan: A common antibacterial agent exposed. *Pesticides and You: Beyond Pesticides/National Coalition Against the Misuse of Pesticides*, 24: 12–17.

Gower, D.B., Mallet, A.I., Watkins, W.J., Wallace, L.M., & Calame, J.-P. 1997. Capillary gas chromatography with chemical ionization negative ion mass spectrometry in the identification of odorous steroids formed in metabolic studies of the sulphates of androsterone, DHA and 5α-androst-16-en-3β-ol with human axillary bacterial isolates. *The Journal of Steroid Biochemistry and Molecular Biology*, 63, 81–89.

Kanazawa, A., Ikeda, T., and Endo, T. 1993. Polymeric phosphonium salts as a novel class of cationic biocides. II. Effects of counter anion and molecular weight on antibacterial activity of polymeric phosphonium salts. *Journal of Polymer Science Part A Polymer Chemistry* 31: 1441–1447.

Kaur, S., and Jaggi, R.K. (2010). Antinociceptive activity of chronic administration of different extracts of Terminalia bellericaRoxb, and *Terminalia Chebula* Retz. Fruits. *Indian Journal of Experimental Biology*, 48: 925–930

Kuhn, H., Müller, F., Peggau, J., and Zekorn, R. 2000. Mechanism of the odor-adsorption effect of zinc ricinoleate: A molecular dynamics computer simulation. *Journal of Surfactants and Detergents; Heidelberg*, 3(3): 335–343. doi: 10.1007/s11743-000-0137-9

Kumar, M., Agarwala, R.C., dey S., Rai, V.K., and Johnson, B. 2009. Antibacterial activity of aqueous extract of *Terminalia Chebula* Retz: On grampositive and gramnegative microorganisms. *International Journal of Current Pharmaceutical Research*, 1(1): 56–60.

Kumar, S., Choudhary, H.S., and Seniya, C. 2011. *In vitro* antibacterial study of aqueous and methanolic extracts of some selected medicinal plants. *Journal of the Society of Cosmetic Chemists*, 3(4): 854–860.

Leyden, J.J., McGinley, K.J., Holzle, E., Labows, J.N., and Kligman, A.M. 1981. The microbiology of the human axilla and its relationship to axillary odour. *Journal of Investigative Dermatology*, 77: 413–416.

Li, W., and Liu, M. 2002. Method and composition for preventing sweat-related odor. US patent 6 426 061 B1. Li W. and Liu M., New York.

Mahajan, A., and Pai, N. 2010. Simultaneous isolation and identification ofphytoconstituents from Terminalia chebula by preparative chromatography. *J. Chem. Pharm. Res.*, 2(5): 97–103.

Manosroi, A., Jantrawut, P., Akihisa, T., Manosroi, W., and Manosroi, J. 2010. In vitro anti-aging activities of *Terminalia Chebula* gall extract. *Pharmaceutical Biology*, 48: 469- 481.

Mao, J.W., and Murphy, L. 2001. Durable freshness for textiles. *AATCC Review*, 1: 28–31.

Maruthappan, V., and Sakthi Shree, K. 2010. Hypolipidemic activity of haritaki (*Terminalia Chebula*) in atherogenic diet induced hyperlipidemic rats. *Journal of Advanced Pharmaceutical Technology and Research,* 1(2): 229–235.

McQueen, M.K., Xu, Y., and Mah, T. 2013. In vivo assessment of odour retention in an antibacterial silver chloride-treated polyester textile. *The Journal of The Textile Institute*, 104(1): 108–117. doi: 10.1080/00405000.2012.697623

McQueen, R.H., Laing, R.M., Niven, B.E., Delahunty, C.M., and Wilson, C.A. 2007. Odour retention on apparel fabrics: Development of test methods for sensory detection. *Textile Research Journal*, 77(9): 645–652.

Middleton, E.J., and Kandaswami, C. 1994. The impact of plant favonoids on mammalian biology: Implications for immunity, in ammation and cancer, In J.B. Harborne (ed.), *The Flavonoids: Advances in Research Science* (pp. 619–652) 1986. Chapman & Hall.

Morais, D.S., Guedes, R.M., and Lopes, M.A.. 2016. Antibacterial approaches for textiles: From research to market. *Material*, 9: 498. doi: 10.3390/ma9060498

Morais, D.S., Guedes, R.M., and Lopes, M.A. 2016. Antibacterial approaches for textiles: From research to market. *Materials (Basel, Switzerland)*, 9(6): 498. doi: 10.3390/ma906049-8

Moulari, B., Pellequer, Y., and Lboutounne, H. 2006. Isolation and in vitro antibacterial activity of astilbin, the bioactive flavanone from the leaves of Harunganamadagascariensis Lam. Ex Poir. (Hypericaceae). *Journal of Ethnopharmacology*, 106: 272–278.

Murali, Y.K., Chandra R., and Murthy, P.S. 2004. Antihyperglycemic effect of water extract of dry fruits of *Terminalia Chebula*In experimental diabetes mellitus. *Indian Journal of Clinical Biochemistry*, 19(2): 202–204.

Nahar, K. 2020. A review on anti odor finishing of textiles, its application method and commercialization. *ACTA TECHNICA CORVINIENSIS – Bulletin of Engineering*, XIII (1): 79–84.

Naik, G.H., Priyadharsini, K.I., Naik, D.B., Ganghabhagirathi, R., and Mohan, H. 2004. Studies on the aqueous extract of *Terminalia Chebula* as a potent antioxidant and a probable radioprotector. *Phytomedicine*, 11: 530–538.

Naqvi, H.R., Asif, M.Y., Rehman, A.B., and Ahmad, M. 2010. Evaluation of antibacterial properties of *Terminalia Chebula* retz. *Pakistan Journal of Pharmacology*, 27(1): 29–35.

Okoli, A.S., Okeke, M.I., Iroegbu, C.U., and Ebo, P.U. 2002. Antibacterial activity of Harungana madagascariensis leaf extracts. *Phytotherapy Research*, 16: 174–179.

Palza, H. 2015. Antibacterial polymers with metal nanoparticles. *International Journal of Molecular Sciences*, 16: 2099–2116. doi: 10.3390/ijms16012099.

Parthiban, M., Srikrishnan, M.R., and Viju, S.. 2011. Studies on odor retention and antibacterial efficacy of non woven fabrics for technical textile applications. *International Journal of Pharmacy and Life Sciences*, 2(10): 1137–1141.

Quatrale, R.P., Coble, D.W., Stoner, K.L., and Felger, C.B. 1981. The mechanism of antiperspirant action by aluminum salts. II. Histological observations of human ecrine sweat glands inhibited by aluminumchlorohydrate. *Journal of the Society of Cosmetic Chemists*, 32: 107–136.

Rahman, M.A., Ahsan, T., and Islam, S. 2010. Antibacterial and antifungal properties of the methanol extract from the stem of argyreiaargentea. *Bangladesh Journal of Pharmacology*, 5: 41–44.

Raja, A.S.M., Thilagavathi, G., and Kannaian, T. 2010. Synthesis of spray dried polyvinyl pyrrolidone coated silver nanopowder and its application on wool and cotton for microbial resistance. *Indian Journal of Fiber and Textile Research*, 35: 59–64.

Raja W., Pandey S., and Agrawal, R.C. 2011. Studies on the anticlastogenic effect of *Terminalia Chebula* extract on cyclophosphamide-induced micronucleus formation and chromosome aberrations in Swiss albino mice. *International Journal of Genetics*, 1(2): 13–17.

Rathinamoorthy, R. 2016. Investigations on odour retention properties of textile, Doctoral thesis, Anna University, India.

Rathinamoorthy, R., and Thilagavathi, G. 2013. Optimisation of process conditions of cotton fabric treatment with *Terminalia Chebula* extract for antibacterial application. *Indian Journal of Fiber and Textile Research*, 38: 293–303.

Rathinamoorthy, R., and Thilagavathi, G. 2014. Characterisation and *in-vitro* evaluation of *Terminalia Chebula* extractfor Antibacterial potential. *International Journal of Pharmacy and Pharmaceutical Sciences*, 6(2): 927–933.

Rathinamoorthy, R., and Thilagavathi, G. 2014. Effect of antibacterial finish on odour control properties of apparel fabric. *Journal of Textile, Apparel Technology and Management (JTATM)*, 9(1): 1–15.

Rathinamoorthy, R., and Thilagavathi, G. 2016. GC-MS analysis of worn textile for odour formation. *Fibers and Polymers*, 17(6): 917–924. doi: 10.1007/s12221-016-5891-3.

Rathinamoorthy, R., and Thilagavathi, G. 2017. Axillary odour studies on alkali treated knitted polyester fabric. *International Journal of Clothing Science and Technology*, 29(2): 251–261. doi: 10.1108/IJCST-05-2016-0057.

Rathinamoorthy, R., and Thilagavathi, G. 2020. Statistical optimization of *Terminalia Chebula* fruit extraction for improved antibacterial activity against odour- causing bacteria. *Proceedings of the National Academy of Sciences*, 90: 365–374.

Rathinamoorthy, R., Thilagavathi, G., Brindha, S., Gayathri, P., Poornakala, N.S., and Pradeep, B. 2014. Odour control studies on apparel fabrics finished with Methanol extract of *Terminalia Chebula*. *Fibers and Polymers*, 15(8): 1669–1676.

Rathinamoorthy, R., Udayakumar, S., and Thilagavathi, G. 2011. Antibacterial efficacy analysis of PunicaGranatum L. Leaf, Rind And *Terminalia Chebula*Fruit extract treated cotton fabric against five most common human pathogenic bacteria. *International Journal of Pharmacy and Pharmaceutical Sciences*, 2: 1147–1153.

Saleem, A., Husheem, M., Harkonen, P., and Pihlaja, K. 2002. Inhibition of cancer cell growth by crude extract and the phenolics of *Terminalia Chebula* Retz. Fruit. *Journal of Ethnopharmacology*, 81: 327–336.

Shrama, A., Patel, V.K., and Ramteke, P. 2008. Shigellocidal activity of some medicinal plants used in folklore remedies by tribals of Mahakoshal region of central India. *Natural Product Radiance*, 7(5): 426–436.

Sharma, P., Prakash, T., Kotresha, D., Ansari, M.A., Sahrm, U.R., Kumar B., Debnath, J., and Goli, D. 2011. Antiulcerogenic activity of *Terminalia Chebula*fruit in experimentally induced ulcer in rats. *Pharmaceutical Biology*, 49(3): 262–268.

Shastri, J.P., Rupani, M.G., and Jain, R.L. 2012. Antibacterial activity of nanosilver-coated socks fabrics against foot pathogens. *The Journal of The Textile Institute*, 103(11): 1234–1243. doi: 10.1080/00405000.2012.675680

Shegadeh, N., and Kligman, A.M. 1963. The bacteria responsible for axillary odor. II. *Journal of Investigative Dermatology*, 4(1): 3.

Shelley, W.B., Hurley, H.J. Jr., and Nichols, A.C. 1953. Axillary odor: Experimental study of the role of bacteria, apocrine sweat and deodorants. *Review of Dermatology and Syphilis*, 68: 430–446.

Strauss, J.S., and Kligman, A.M. 1956. The bacteria responsible for apocrine odor of investigative dermatology. *Journal of Investigative Dermatology*, 27(2): 67–71.

Suchalatha, S., and Shyamala Devi, C.S.. 2004. Protective effect of *Terminalia Chebula*against experimental myocardial injury induced by isoproterenol. *Journal of Experimental Biology*, 42(2): 174–178.

Sun, H.X., Xie, Y., and Ye, Y.P. 2009. Advances in saponin-based adjuvants. *Vaccine*, 27(12): 1787–1796.

Tamhane, M., Thorat, S., Rege, N., and Dahanukar, S. 1997. Effect of oral administration of *Terminalia Chebula* on gastric emptying: An experimental study. *Journal of Postgraduate Medicine*, 43: 12–13.

Tayal, S., Duggal, S., Bandyopadhyay, P., Aggarwal, A., Tandon, S., and Tandon, C. 2012. Cytoprotective role of the aqueous extract of *Terminalia Chebula* on renal epithelial cells. *International Brazilian Journal of Urology*, 38(2): 204–214.

Ueda, M., Tokimitsu, I., and Masatoshi, A. 1990. Deodorant for axillary odor. EU patent 0 433911 A1. Kao Corp, Tokyo.

Walia, H., Kumar, S., and Arora, S. 2011. Comparative antioxidant analysis of hexane extracts of *Terminalia Chebula*Retz. Prepared by maceration and sequential extraction method. *Journal of Medicinal Plants Research*, 5(13): 2608–2616.

Walter, N., McQueen, R.H., and Keelan, M. 2014. In vivo assessment of antibacterial-treated textiles on skin microflora. *International Journal of Clothing Science and Technology*, 26(4): 330–342. doi: 10.1108/ijcst-12-2012-0078

Wasif, A.I., and Laga, S.K. 2009. Use of nano silver as an antibacterial agent for cotton. *AUTEX Research Journal*, 9(1): 5–13.

Xu, Y. 2012. Odour evaluation on antibacterial treated fabrics: An assessment of test methods, Masters' thesis, University of Alberta, Edmonton, Alberta.

8 Fragrance Finishing on Textile Substrate for Odour Control

Hemamalini Thillaipandian and
Giri Dev Venkateshwarapuram Rengaswami
Department of Textile Technology, Anna University, Chennai, India

CONTENTS

8.1 INTRODUCTION

Textiles are considered to be the second skin of humans, and they have the ability to absorb or adsorb unpleasant odours from the environment or the wearer's body odour that is produced through the physicochemical process of the human body. Unpleasant odour creates social embarrassment and reduces the self-confidence of the wearer. Odourants are defined as a substance that offers a sense of smell. The odour of the substance is generally sensed through the olfactory organs and gives a signal to the brain. Aromachology is the science that studies the relationship between fragrance and human psychology. Each body has its odour footprint representing its own physiological condition, such as stress, infection, age, and physical and hygiene practices.

DOI: 10.1201/9781003141426-8

The major sources of production of odorous volatile compounds in the human body include the armpit, sternum, anogenital area, axilla, and feet region due to the presence of bacterial growth, which breaks long-chain fatty acids into volatile organic chemicals, such as aldehyde, acid, amine, alcohol, hydrocarbon, ketones, sterols, sulfur compounds, and terpenoids, thereby generating human body odour. The bacteria not only create a bad odour but also cause hygiene problems to the wearer and deteriorate the textile structure because it is composed of a polysaccharide backbone (McQueen 2011). Hence, the textile fabric should not only offer protection from the external environmental factors but also result in reduction of the growth of microorganisms and pathogens causing bad odour. The ideal odour-resistant fabric should be able to release fragrance, thereby reducing bad odour and inhibiting the growth of volatile odour causing bacteria without causing damage to the fabric. Hence, odour resistant fabric can be designed as fabric intended to release fragrance and also to inhibit the growth of microorganisms (Buschmann, Dehabadi, and Wiegand 2015). Antibacterial properties can be cultivated in the fabric using various chemical finishing agents, such as triclosan, quaternary ammonium compounds, metal, and metal salts, which are considered to be poisonous to the environment. Bad odour from the fabric is initially controlled by usage of herbs, namely turmeric, which deteriorates the aesthetic performance of the fabric and paved the way for the development of topical odourants, such as antiperspirant and deodorant, to mask malodour (Callewaert, Lambert, and Van de Wiele 2017). The latter offered a temporary solution upon usage of a high level of fragrance. Thus, essential oil is the natural aromatic material that meets the requirement for an ideal odourant-resistant fabric as it is composed of flavours that release fragrance, and it acts as a nonpoisonous antibacterial agent, which makes the oil suitable for application in biomedical textiles (Pattnaik et al. 1997). This chapter summarizes the methods to apply essential oil to have sustained and controlled release of fragrance from fabric to reduce malodour.

8.2 MALODOUR CAUSING SOURCES

Body odour communicates nonverbal signals as the bacteria present in the body utilize sweat to produce malodour. The origin for the odour from the human body includes the breath, sweat, urine and faecal matter, blood, and underlying diseases. The quality of the breath depends on the exogenous and endogenous condition as the exogenous condition includes the smell inhaled from the environment, such as cigarette smoke, whereas the endogenous system includes the smell produced by humans' own system, such as gas released from the lungs to the environment. The other factors, such as urine, faecal matter, and blood, release the malodour due to complex volatile compounds released through the metabolic pathway, which can be used for the diagnosis of the diseases (Shirasu and Touhara 2011). Sweat is the major source of odour production in the human body, caused due to a thermophysiological condition. Eccrine and apocrine are the sweat glands present throughout the human body, especially with a high density of glands in the palms, soles, and armpits. Eccrine glands are responsible for the production of a watery substance to maintain the thermal comfort of humans, whereas the apocrine glands are found near the armpit and axillary region and produce viscous sweat, which produces malodour. The composition of sweat in the eccrine gland includes 99% water and other substances such as amino acids, ions, lactic acid, glycerol, urea, peptides and proteins. The amino acid secreted through the sweat of a healthy human leads to odour formation as the bacteria breaks the protein structure into volatile compounds (Baker 2019). *Brevibacterium linens* and *Staphylococcus epidermidis* are the bacteria responsible for the production of body odour as it breaks methionine into methanethiol and leucine into isovaleric acid at the feet. Socks and shoes therefore have bad odour. The bacteria, namely *Propionibacterium acnes*, *Pityrosporum folliculitis,* and *Acne vulgaris,* are more prone to cause inflammation, resulting in skin infection (Ara et al. 2006). Micrococcaceae family, aerobic diphtheroid (mainly *Corynebacterium*), anaerobic diphtheroid *(Propionibacterium*), and yeast (*Pityrosporum*) are the bacteria responsible for inflammation on the skin caused in the region of the apocrine glands, whereas *Corynebacteria, Stapphylococci,* and *Propionibacteria* are responsible for the production of odour from the eccrine glands. The bacteria found on the skin are responsible for the breakdown of metabolites, thereby causing malodour (Jha 2017; Kanlayavattanakul and Lourith 2011).

8.3 MECHANISM OF ODOUR CONTROL FROM THE FABRIC

The release of odour from the fabric can be inhibited by the application of finishes. These finishes act on the odour causing substance by reduction of bacterial growth on the fabric. The major requirement of the finishing agents used for odour control is long lasting, nontoxic, and user friendly. The foul smell from the fabric can be controlled using either passive or active techniques. In the passive technique of odour control, gas and moisture from the wearer and the fabric are absorbed using odourants, such as zeolite and activated carbon, whereas in an active method of reduction, the growth of bacteria/microbes on the fabric is inhibited, thereby reducing the odour release from the fabric (Nahar 2020).

8.4 FRAGRANCE OIL FOR FINISHING ON THE FABRIC

The odour caused due to environmental and physicochemical mechanisms of Homo sapiens should be controlled or reduced to increase their healthy social status. The odour can be temporarily controlled by employing perfumes or reagents added during the washing process, but the latter ways provide temporary odour control; hence, sustained and controlled release of fragrance from the human second skin (fabric) is a necessity. Essential oil is a strong volatile complex mixture available in abundance, with various sources of bioactive ingredients. These oils are produced by the aromatic plants as their metabolites; thus, the composition of the ingredients may vary based on the source. The major drawback of essential oil is poor stability during processing and storage, causing degradation and loss of fragrance. The essential oil can be classified based on the source of extraction, such as leaves, seeds, fruits, flowers, and roots, as shown in Figure 8.1. Each type of fragrance oil possesses its own medical and fragrance property, making it suitable for various technical applications (Bakry et al. 2016). Aromatherapy is the term used in the process of extracting the essential oil from the plants for health and well-being. The essential oils are extracted from the

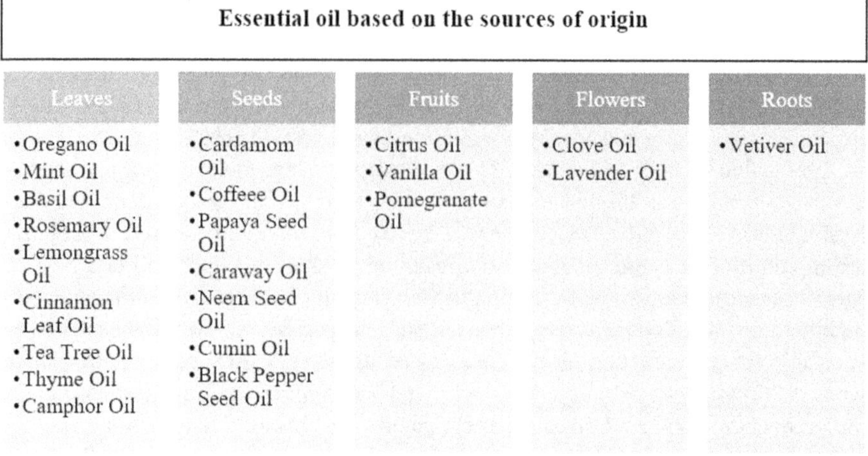

FIGURE 8.1 Essential oil extraction based on the sources.

plants using distillation, steaming, and solvent extraction techniques. The essential oil is composed of less than 5% vegetal dry matter and complex volatile compounds, which can be identified using gas chromatography and GC-mass spectrometry (GC-MS). The major chemical composition of essential oils includes terpene hydrocarbons and oxygenated compounds. Terpene hydrocarbons include monoterpene hydrocarbons and sesquiterpenes, which can be defined based on the number of isoprene units. In monoterpene hydrocarbons, two isoprene structures are present, whereas for sesquiterpenes, a single structure of isoprene is present, which is used for anticarcinogenic and anti-inflammatory applications (Mejri 2010). The essential oil is composed of several oxygenated compounds, as shown in Figure 8.2. It can be finished on the textile substrate based on the demand of release for anti-odour and antibacterial applications.

8.5 METHODS OF MANUFACTURING MALODOUR CONTROLLING TEXTILE SUBSTRATE

The odour control on the fabric is essential in the area of technical textiles, namely medical, home, and personal care textiles. The odourant finishes can be applied to the fabric using physical bonding, chemical bonding, and encapsulation technique (Perinelli et al. 2020). This chapter summarizes the method of application of odourants to fabric, such as physical coating and encapsulating the essential oil using a host-guest mechanism and within the polymer matrix using various microencapsulation techniques, as well as the material used for encapsulating the essential oil on the fabric to increase the durability and serviceability of the fabric.

8.5.1 Surface Treatment of Odourant on the Fabric

The odorous material can be coated on the fabric for the release of fragrance from the substrate to ensure the well-being of the wearer. The material can be coated on

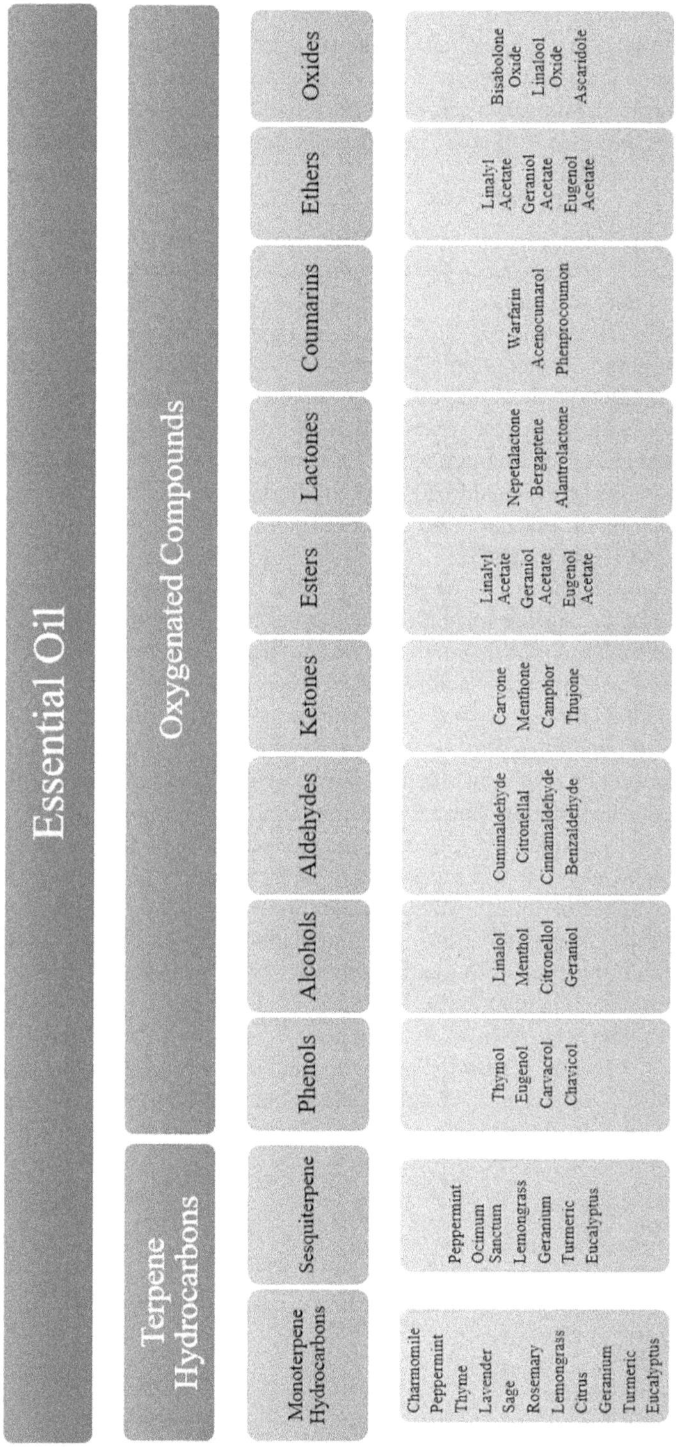

FIGURE 8.2 Essential oil based on major chemical composition.

the fabric either using the padding or exhaustion process by treating the fabric with fragrance materials. In the padding technique, a pair of positively loaded rollers apply pressure on the fabric, which decides the amount of final add-on value to the substrate. Several factors affect the efficiency of the padding mangle, such as the number of nips and dips, the concentration of liquor, and the pressure acting between the nip of padding rollers (Stewart 1967). The number of rollers based on a dip and nip requirement can be increased sequentially either horizontally or vertically based on the end application of the textile substrate. The padded fabric can be subjected to further operations, as shown in Figure 8.3. In the case of the exhaustion technique, the fabric was subjected to liquid loaded onto the machine, which offers higher affinity to the chemical groups present in the substrate, but the limitation of the process is energy consumption. The liquid consists of chemicals and auxiliaries that have substantivity to the fabric and can be applied using the exhaustion process, whereas in the padding process, chemicals with low or poor substantivity can be applied with the aid of an external agent. The temperature of the bath, material-to-liquid ratio, duration in the liquid bath, and condition of the substrate are the major influencing factors in finishing the fabric using the exhaustion process (Cay, Tarakçioğlu, and Hepbasli 2009).

Due to the volatile nature of fragrance material, the release of odour from the fabric can be lost during the manufacturing, storage, or washing stages. An ideal fragrance finished fabric will have a slow and controlled release of odourant from the fabric to increase the serviceability of the end product. Alcohol dehydrogenases is an enzymatic oxidation agent found in bacteria and yeast during the fermentation process. The function of this enzyme is to convert the alcohol group in the cellulosic structure of cotton fabric into an aldehyde group, which offers a higher affinity for polyelectrolyte nanoparticles. The latter is prepared using cationic polymer (chitosan) loaded with jasmine oil by electrostatic interaction with an anionic catalyst, such as tripolyphosphate. It was reported that chitosan nanoparticles were found to interact with oxidized cellulose, which was confirmed through FTIR spectra by the formation of imide groups. It was also reported that the fragrance lasted in the fabric even after 30 washing cycles, so the finished fabric has potential for application in the area of mosquito repellant fabric (Ammulu et al. 2019).

Fragrance coating was given to the cotton and aloe vera knitted fabric using lavender oil as an odourant. The oil, along with the binder, was applied to the fabric through the exhaustion process and was dried to study the laundering durability and sensorial intensity of the treated fabric. It was reported that aloe vera fabric exhibited

FIGURE 8.3 Process of padding the fabric with chemicals.

higher laundering durability for 5 washes and fragrance intensity for 25 days, which was confirmed by human experts through scratching the fabric. In contrast, the treated cotton fabric possessed laundering durability for 4 washes, and fragrance finishing lasted in the fabric for 20 days. It was concluded that aloe vera treated with lavender oil offered odour release properties compared to the cotton fabric and had the potential for wall hanging applications (Vasanth Kumar et al. 2012).

The cotton fabric was treated with ingredients such as citric acid, chitosan, disodium hydrogen phosphate, and silicone softener using padding and the exhaustion process to study the efficiency of antimicrobial and fragrance release from the two processed fabrics. The material-to-liquid ratio for padding mangle was set to 1:5, whereas in the exhaustion process, it was 1:30. The lavender oil, which was insoluble in water, was homogenized using propyl alcohol and sprayed on the fabric using the spray gun. The antimicrobial and fragrance release were found to be higher for padded fabric compared to exhaustion finished fabric. The antimicrobial activity of the treated fabric was due to the presence of positively charged chitosan on the cotton fabric, which then reacts with negatively charged bacteria, causing cell lysis of the bacterial cell wall (Li and Zhuang 2020). Citric acid crosslinks with the chitosan by the formation of chitosan citrate with the carboxylic acid residue, which in turn reacts with cellulose as fixation sites on treatment at the higher temperature. It was reported that the higher efficiency of padded fabric was due to host molecules within the structure compared to the exhaustion process. The antimicrobial property of the treated fabric offered inhibition to the growth of bacteria of 79% even after 3 wash cycles (Karolia and Mendapara 2007).

The structure and mass of the fabric decided the odour formation as air and water permeability were limited in the dense fabric, causing bacterial contamination. Cotton knit (rib and interlock) and the woven fabric were coated with fruit extract Terminalia chebula to reduce odour formation by controlling bacterial growth on the fabric. The pad-dry-cure method was adopted for the application of fruit extract, and the formation of the odour was tested by subjective assessment using SNV 19651. It was reported that the odour intensity decreased in the order of rib, interlock, followed by plain woven fabric, as the lighter structure resulted in better evaporation of moisture, thereby reducing the odour. The fabric was tested against the bacteria, which is more to multiply in the axilla region of the human body. The bacteria taken for the study were *Staphylococcus aureus*, *Corynebacterium* Sp, *Bacillus licheniformis*, *Pseudomonas* Sp, *Corynebacterium acnes,* and *Escherichia coli* using the disk diffusion method. The developed fabrics offered maximum zone of inhibition to the growth of bacteria, thereby reducing odour formation in the fabric (Rathinamoorthy and Thilagavathi 2014). Eucalyptus Cinerea is a medicinal plant that offers deodorant property when it is encapsulated on the fabric. Microcapsules of the plant extract were prepared by blending the extract with sodium alginate and spraying the solution onto the calcium chloride solution. The solution reacts with calcium ions, which result in the hardening of microcapsules. The latter was coated on the bamboo fabric using exhaust techniques with citric acid as a binder. The prepared fabric was found to have anti-odourant property and also offers protection against the infection causing bacteria (Ramya and Maheshwari 2019).

Activated carbon is an amorphous material rich in carbon content with other inorganic residues, such as nitrogen, oxygen, and sulphur, which acts as an odourant through an adsorption mechanism when treated on the fabrics. The charcoal can be used as an odourant material to absorb the odour releasing from wound, colour, and formaldehyde removal in wastewater treatment plants due to its adsorption characteristics (Mohan, and Bhaarathidhurai 2020). The activated carbon was coated and pigment printed on the cotton and polyester fabric to study the efficiency of the processing technique. The treated fabric was tested for odour control in an onion fragrance environment using the human panel as per ISO standard and portable electronic nose equipment. It was reported that the coating fabric was found to be superior when compared to pigment printed fabric due to uniform distribution and presence of activated carbon on the entire surface of the coated fabric, whereas in pigment printed fabric, the activated carbon was present on the localized region of the fabric. It was also reported that increasing the concentration of activated carbon reduced the intensity of onion fragrance due to its adsorption properties on both cotton and polyester fabrics (Eza et al. 2012).

Skin infection due to bacterial growth causes bad odour, which can be reduced by the coating of materials, which hinders the growth of such microorganisms. Silver nanoparticles and gold nanoparticles were synthesized using pine gum solution using silver nitrate and chloroauric acid. Pine gum is the complex polysaccharide exudated from the bark of the pine tree, which acts as a reducing agent as it is composed of various resin acids, namely abietic acid. Cotton and leather fabric were coated with prepared silver and gold nanoparticles to reduce the growth of odour causing bacteria on the substrate. A zone of inhibition study was used to analyze the efficiency of the coated substrate as the silver nanoparticle incorporated on cotton and tanned leather substrate offered higher inhibition against the odour causing bacteria with the zone of 13 mm (Velmurugan et al. 2014). Silver nanoparticles were reported as a strong antibacterial agent as it infiltrates into the cell wall of odour causing bacteria, thereby promoting cell lysis and controlling the release of bad odour. Silver nanoparticles were coated on the sock fabrics, namely cotton, nylon, and cotton/nylon blend. The fabric was tested against various gram-negative and gram-positive bacteria derived from the odour and foot ulcer causing bacteria, namely *Sarcina lutea, Klebsiella pneumoniae, Pseudomonas aeruginosa,* and *Staphylococcus epidermidis*. It was reported that silver-incorporated nylon socks offered strong antibacterial activity against the tested bacteria, except *Klebsiella pneumoniae,* compared to cotton fabric. It was also found that increasing the concentration of silver nanoparticles resulted in 80% inhibition for the growth of *Sarcina lutea* bacteria (Mashat and Awad 2014).

Zeolite is an aluminosilicate mineral found in abundance and used for various technical applications, such as medicine, construction, agriculture, environmental protection, gas, and petrochemical treatment. The polymer offers a microporous structure with the crystalline component of clinoptilolite, which forms a tetrahedron structure with other impurities, montmorillonite and mordenite, which acts as an effective adsorbent for absorbing volatile chemical compounds from the environment. The polymer was treated with surface mercerized cotton and alkali-treated polyester fabric in the form of nanoparticles to study the efficiency of coated fabric for UV protection after treating with fluorescence whitening agents, and antimicrobial activity after treating with an antibiotic, namely azithromycin. It was reported that cotton fabric offered higher UV resistance compared to polyester fabric and also promoted wound healing activity by inhibiting the growth of microorganisms in the in-vitro condition (Grancarić, Tarbuk, and Kovaček 2009).

Superhydrophobic cotton fabrics were prepared with the fragrance of lemon oil with the idea of slowing down the release of fragrance by creating a low surface energy film layer. The production of hydrophobic cotton fabric involved the preparation of fragrance nanoparticles followed by coating with polyurethane polymer as a film layer. The fragrance nanoparticles were prepared by dissolving polystyrene in hexadecyl trimethyl ammonium bromide, which was homogenized in an aqueous ethanol solution. The prepared solution was treated with tetraethoxysilane in order to yield hollow mesoporous silica nanoparticles, to which lemon oil was added and autoclaved. The fragrance nanoparticles were encapsulated on the cotton fabric with an aid of waterborne polyurethane and polyoxyethylene ether using the pad-dry-cure technique. The surface-treated fabric was found to be superhydrophobic with the contact angle of 166° with a self-cleaning effect as it did not uptake the dye from the water bath compared to pristine fabric. It was also reported that the release of fragrance was slow on treating the fabric with hydrophobic film but resulted in an increase in surface roughness of the fabric (Xue et al. 2016).

Mosquito repellant finishing was attempted on the fabric since the garment acts as the second skin to humans. The finishing agent applied for the prevention of mosquito-borne diseases includes permethrin and cypermethrin, which was reported as it caused a toxic reaction against the skin and also caused harm to aquatic life. Cymbopogon flexuosus (CF) oil was applied to nylon gauze fabric as it acts as a replacement for synthetic and chemical finishes. Microcapsules were applied to the gauze fabric, utilizing the layer-by-layer technique as it avoids the treatment at higher temperature, as in the pad-dry-cure technique. The microcapsule was prepared with the ratio of surfactant to oil of 1:1.5, and polyelectrolyte multilayer was prepared by treating the gauze fabric in cationic polyelectrolyte followed by the solution containing the anionic polyelectrolyte and microcapsules. The treated fabric was tested against the bacteria, namely *Staphylococcus aureus* and *Klebsiella pneumonia,* before and after the washing process. It was reported that the bacterial reduction was around 90% before washing, and the efficiency turned half after 15 washing cycles. It was also reported that the number of polyelectrolyte layers and concentration of microcapsules has no effect on the release of oil from the fabric (Bhatt and Kale 2019).

8.5.2 Adsorption of the Odourants on the Fabric

Essential oil can be held on the surface of the fabric by means of host-guest mechanism, such as cyclodextrin and dendritic polymers, which offers functional groups that bind the microcapsules and the fabric functional groups, which can slow down the release of fragrance from the fabric in a controlled manner rather than burst release. Cyclodextrin is a cyclic oligosaccharide that forms an inclusion complex with the hydrophobic substance as the guest molecules that fit into the host cavity without the formation of a covalent bond. It is made of a rigid conical structure with the hydrophilic group facing outwards and hydrophobic groups inwards. The structure of cyclodextrin can accommodate a wide range of polar solvents, such as alcohols, acids, amines, and non-polar compounds, namely aliphatic and aromatic hydrocarbon within the internal structure, whereas the outer hydrophilic groups react with aqueous solvents (López, de Vries, and Marrink 2011). Thus, cyclodextrin is widely used for various technical applications, such as drug delivery applications, extraction of cholesterol, food, and pharmaceutical applications. The different fragrance oils, such as lavender, eucalyptus, peppermint, jasmine, clove, and cedarwood oils, were applied on the cotton fabric with and without the presence of cyclodextrin to study the release of fragrance. It was reported that fragrance release was found to decrease in the order of peppermint, clove, cedarwood, lavender, eucalyptus, and jasmine when applied on the cotton fabric without the application of the crosslinking agent. It was found that the concentration of cyclodextrin was fixed to 9–11% as there were no cavities available for inclusion bonding. Without the crosslinker, fragrance release from the fabric was high; on coating with cyclodextrin, the release was slow and controlled due to the formation of inclusion complex with the fabric and hydrophobic oils (Khanna, Sharma, and Chakraborty 2015).

Essential oils have a volatile nature, and the fragrance is released in a burst mechanism. The release of fragrance from fabric has been designed based on body stimuli responses with the aid of hydrogel containing poly (N-methylene bisacrylamide), chitosan, and cyclodextrin. Essential aromatic oils were reported to be unstable at a particular pH; they may undergo an oxidation reaction or be lost due to their volatile nature during processing. To overcome the limitation of the nature of essential oils, the oils were entrapped in the hydrogel structure with the aid of a crosslinking agent, especially cyclodextrin, followed by padding on the cotton fabric. A crosslinking agent was utilized to bind the essential oil onto the hydrophilic structure of hydrogel and to control the release of the antimicrobial agent. It was reported that hydrogel particles of 350 nm were deposited on the surface of the fabric, and increasing the temperature released the essential oil from the hydrogel matrix. The developed substrate was tested against gram-positive and gram-negative bacteria, which offered a zone of inhibition of 1.5 mm, which finds a potential application in healthcare hygiene textiles (Štular et al. 2019). The limitation of cyclodextrin fragrance finished fabric is the complete release of odour when the substrate comes in contact with the aqueous medium, whereas with coacervation and the spray drying technique, the release of fragrance cannot be controlled to increase the durability of the fabric.

Dendrimers are macromolecules that contain a symmetric branching unit around the central molecule or polymeric core. They are also called cascade molecules that possess architectural motif. The structure of the dendrimers is composed of the central core, which is the combination of atoms or a single atom, and dendrons, the branched structure of the atoms around the core. Dendrimers are widely used for various biomedical applications, such as drug, anticancer, and gene delivery (Abbasi et al. 2014). The utilization of the dendrimers for technical applications is due to their properties, such as monodispersity, spherical three-dimensional morphology, highly branched structures offering a large number of functional end groups, and room between each branch within the structure. Thyme oil was encapsulated within the structure of polypropylene imine dendrimer, which was applied onto the structure of cotton fabric with the help of melamine resin as a binder using the pad-dry-cure technique. To study the efficiency of the dendrimer, the thyme oil was treated on the cotton fabric with melamine resin. The fabric was studied for antibacterial activity using the colony counting method, and anti-odour characteristic was tested using an electronic nose. The developed fabric with dendrimer surpassed the growth of bacteria, such as *Escherichia coli* and *Staphylococcus aureus,* to 2% and 3.4%, respectively. It was also reported that the fragrance of the essential oil was lost in the fabric without dendrimers after washing, whereas the dendrimer-treated fabrics offered a slow release of fragrance, which was confirmed using an electronic nose, and they also possessed antibacterial activity, even after 30 days of exposure to the free air (Fard, Shakoorjavan, and Akbari 2018).

8.5.3 Encapsulation of Odourant on the Fabric

The controlled release of fragrance or flavor from the textile is a demand from the user's perspective. Fragrance oil can be caged onto the fabric structure using a crosslinking agent, such as cyclodextrin; due to the high volatile nature of the oils, the odour can be dissipated over a period. Another way of applying the oil is spraying or airing of perfumes, which gives temporary treatments to remove the malodour. The oil can be adsorbed onto the surface of the fiber in the form of metal oxide films, but the limitation of the method is the deterioration of softness and handle of the fabric. Microencapsulation is the technique that allows the controlled release of chemicals by encapsulating the fragrance material within the polymer matrix. The structure of the microcapsule is made of two components: core and sheath material. The core material is an active ingredient that is volatile, whereas the sheath material is used to cover or protect the core material. The technique is widely used for a broad range of applications, such as textile, pharmaceutical, biotechnology, medical, agriculture, and waste management areas (Jyothi et al. 2010). Thus, the fragrance agent can be applied to the fabric by employing spray drying, coacervation, emulsification, in-situ polymerization, interfacial polymerization, ionotropic gelation technique, and microcapsules/nanocapsules to slow down the release of odour from the fabric to increase the serviceability of the end product.

8.5.3.1 Spray Drying of Odourant on the Fabric

Spray drying is a technique adopted widely for encapsulating fragrance particles on the surface of the fabric by means of supplying the essential oil along with hot air. The emulsion is atomized at a higher temperature, which leads to evaporation of water and the formation of core oil substance at a relatively a fast rate. The technique is readily available, low cost, and easy to reproduce, but it is not suitable for essential oil as it suffers a limitation due to the volatile nature of aroma oils and the possibility of undergoing oxidation when the oils are treated at higher temperatures and exposed to light, resulting in destruction of the product (Silva et al. 2014). Spray coating was adopted for fixing the essential oils on the oleophilic cavities of β-Cyclodextrin. The sol-gel was prepared using two precursors, 3-glycidyloxypropyltrimethodoxysilane and tetraethoxy orthosilicate, to which cyclodextrin was added and applied on the cotton and cotton/polyester woven fabric using a padding technique. Upon drying the treated fabrics, the essential oils – eucalyptus, lemon, and lavender oil – were sprayed as the forced spraying technique enhanced the penetration of the oils into the cavity of the crosslinking agent, thus producing inclusion complex on the fabric. It was reported that the fabric offered fragrance property retained on the fabric, even after 6 washes, due to the application of cyclodextrin and sol-gel formation on the fabric. It was also found that the strength of the fabric increased due to the application of the final add-on to the fabric, which increased the structural integrity of the fabric compared to untreated fabric. Polyester/cotton fabric offered the higher fragrance property compared to cotton fabric due to higher filling of cyclodextrin in the pores of sol-gel, thereby increasing the durability of fragrance finishing (Ghosh and Chipot 2015).

8.5.3.2 Coacervation

Coacervation is a phase separation process to separate the colloidal-rich phase from the liquid medium. There are two types of coacervation: simple and complex. In the simple coacervation technique, the single polymer is deployed to form coacervates by either dehydration or addition of nonsolvent to the polymer to precipitate it from the medium. In the case of the complex coacervation technique, the polyelectrolyte nature of polymers is deployed, resulting in the formation of ionic interaction between them to form coacervate (Timilsena et al. 2019). The complex coacervation technique involves five stages – polymer dissolution, emulsion preparation, coacervate formation, hardening of coacervate, and drying – as shown in Figure 8.4.

Lemon essential oil was encapsulated in gelatin and yeast dead cells and was coated on the cotton fabric using the pad-cure technique. The coacervate was prepared using oil/water emulsion in which gelatin and lemon essential oil were homogenized, followed by the addition of Arabic gum, and the slurry was maintained at a pH of 8–9 by the addition of sodium hydroxide. The coacervate was hardened using glutaraldehyde.

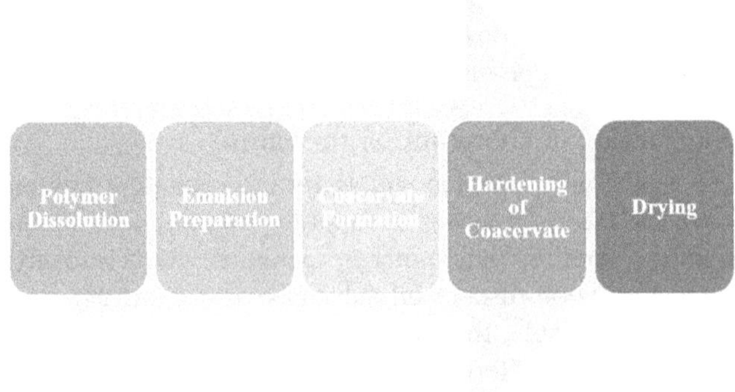

FIGURE 8.4 Process flow for the preparation of coacervate.

In the case of microcapsule based on yeast, lemon oil was added to inactivated yeast and centrifuged. The resultant was washed and applied on the fabric with and without the auxiliary agent. It was reported that the concentration of glutaraldehyde influences the fabric properties, as an increase in the concentration of glutaraldehyde increases the stiffness of the fabric, resulting in the aggregation of capsules, but on the other hand, increased washing fastness. The gelatin Arabic gum-based capsules offered increased fragrance durability compared to yeast cells encapsulated with lemon essential oil. The electronic nose was used to evaluate the oil content in the fabrics before and after laundering. The change in radar signal was noticed after the second wash cycle, indicating the loss of fragrance from the fabric, which was found to be undetectable by the human judges (Specos et al. 2010).

Citronella oil microcapsules were prepared using a complex coacervation technique using two different polymers: gelatin and gum arabic. Three emulsions were prepared: one by dissolving gelatin in water, the second by citronella oil in sodium lauryl sulfate, and the third by gum Arabic in water. These solutions were blended to form uniform microcapsules. The latter was applied on the wool fabric using butanetetracarboxylic acid (BTCA), which acts as a crosslinking agent as it contains zero traces of formaldehyde and sodium hypophosphite as a catalyst by means of the pad-dry-cure technique. The crosslinking agent increased the interaction between the fabric and microcapsules and also lead to an esterification reaction between the wool fabric and microcapsules. The slow release of citronella from the wool fabric showed the Fickian mechanism, implying that the oil content has some interaction with the polymeric chain of wool fabric, causing the slow release of fragrance from the structure (Bezerra et al. 2019).

Lime essential oil (*Citrus aurantifolia*) is used widely for therapeutic applications due to the presence of terpenes such as α-thujene, α-pinene, camphene, sabinene, β-pinene, myrcene, α-terpinene, p-cymene, linalool, and d-limonene, which help in the healing of wounds by acting as an antibacterial agent. Lime oil microcapsules were prepared using a complex coacervate technique with two polymers such as gum arabic and chitosan. The prepared microcapsules were mixed with succinic acid and applied to the cotton fabric by means of immersion, followed by air drying and ironing. The microencapsulated fabric was tested against bacteria, such as *Escherichia coli, Bacillus cereus, Salmonella typhimurium,* and *Staphylococcus aureus,* using the disk diffusion method. The microcapsules were found to be irregular in size of 15–300 μm, with an essential oil loading capacity of 2940 μL/g and loading efficiency of 82%. Cellulose esters were formed as the result of esterification with succinic acid, which aids in the attachment of microcapsules. The developed smart fabric offered a clear zone of inhibition against the tested bacteria, and the efficiency of antibacterial activity increases upon crushing the microcapsules by mechanical forces. It was reported that antibacterial activity retained on the microcapsule-treated cotton fabric, even after mild washing procedure at 500 rpm for 30 minutes (Wijesirigunawardana et al. 2018). Gelatin and sodium alginate were used to prepare the complex coacervate to encapsulate the lemon oil. The microcapsules were padded onto the surface of cotton fabric with the solution containing citric acid as the crosslinking agent, sodium dihydrogen phosphate dihydrate as the catalyst, followed by curing at high temperature. The substrate was tested against the two types of bacteria that cause infection on the skin: gram-positive bacteria (*Staphylococcus aureus, Staphylococcus epidermidis*) and gram-negative bacteria (*Escherichia coli and Klebsiella pneumoniae*). The microcapsules were found to be the size of 1.5 μm with an encapsulation efficiency of 84%. It was reported that the ionic interaction was found between the carboxylate group of sodium alginate and the amino group of gelatin, thereby indicating functional interaction between the polymers. The presence of carboxylate moieties in citric acid structure resulted in esterification reaction with a hydroxyl group in cellulose structure, thereby increasing the immobilization of essential oil. The substrate possessed antibacterial activity of 7 mm, which has potential application for antibacterial textiles (Julaeha et al. 2021).

Thyme oil (*Thymus vulgaris*) is widely used for treating several diseases due to its antiseptic, antimicrobial, antifungal, antioxidative, and antiviral properties. Microcapsules containing thyme oil were prepared using an emulsion containing gelatin/water and thyme oil to which gum arabic was added, and the resulting capsules were dried. The prepared coacervate was padded on the polyester nonwoven fabric, followed by air drying, and tested for antibacterial activity, *Escherichia coli, Staphylococcus aureus,* and *Candida albicans*. It was reported that the microcapsules were formed of size 25 micron, with the essential oil-loading capacity of 74%, and offered antibacterial activity suitable for the preparation of bandages and surgical gowns (Karagonlu et al. 2018).

8.5.3.3 In-situ Polymerization

The technique was adopted for the formation of wall material with the help of a solution containing the monomers and oligomers in a continuous phase at a single reactor without the reactive agent in the core material. The method utilizes suspension, emulsion, and dispersion polymerization techniques to produce a sheath material in order to protect the inner core essential oil (Nguon et al. 2018). Emulsification is one of the microencapsulation techniques that involves dispersing two immiscible liquids in the presence of an emulsifier/surfactant. The technique can be classified into two: oil-in-water and water-in-oil. The emulsifier plays a major role in the preparation of emulsion, with long-term durability as it interacts with the interfacial properties of two immiscible liquids. The emulsifier was composed of hydrophilic and lipophilic groups, in which lipophilic groups undergo hydrophobic interaction with the liquid, whereas the hydrophilic group reacts with the water-absorbing group of the other liquid, resulting in the formation of emulsion (Yamashita, Miyahara, and Sakamoto 2017). The odour control on the fabric was obtained by application of neem oil by means of emulsification technique. The produced fabric controls the growth of bacteria, such as *E. Coli* and *S. Aureus,* thereby controlling the odour. Azadirachtin is an active agent in the neem oil, which offers repellency against insects. Nanoemulsions were prepared by sonicating neem oil with surfactant, which was applied on the fabric in the presence of binder and resin. It was reported that the antimicrobial activity of the finished fabric lasted to 71% for *S. Aureus* and 65% for *E. Coli* after 15 washes (Sayed 2017).

Dimethylol dihydroxy ethylene urea (DMDHEU) is a cost-effective binding agent used for finishing cellulosic fabrics for various technical applications, namely crease and wrinkle resistant fabrics, but the limitation of the agent is the release of formaldehyde byproduct, which was reported as carcinogenic to humans (Schramm and Amann 2019). Jute batching oil, a petroleum-based lubricant added to the jute fibers during manufacturing into yarn, creates a kerosemic smell, which limits the fabric for home textile applications. Lavender oil offers a sedative fragrance that can be microencapsulated on the jute fabric for bed gowns, sheets, and curtain applications. Lavender oil is encapsulated in melamine formaldehyde pre-condensate with the addition of polyvinyl alcohol polymer, which acts as a stablishing agent to prevent agglomeration of micro globules. The microcapsules were attached to jute/cotton blend fabric by means of the pad-dry-cure method with three different types of binders: DMDHEU, polyurethane, and polyacrylate emulsion. The fabric was pretreated with sodium chlorite to possess the affinity for binder and microcapsules to encapsulate. The fragrance-incorporated fabric was evaluated for aroma by subjective terms using human judges as per the standard ISO 6564-1985. The wash durability of the finished fabric was analyzed using ASTM D 2960-05. It was reported that the polyacrylate binders offered better performance by retaining a large number of aroma microcapsules compared to DMDHEU and polyurethane. The fabric offered fragrance property for 10 washing cycle with the gradual decrease in aroma intensity (Biswas et al. 2015).

The fabric that was treated with microcapsules using padding mangle was generally cured at higher temperatures to enhance fixing of binders onto the fabric. Due to the volatile nature of fragrance oil, the treatment at higher temperature results in the loss of smell from the fabric. Thus, the curing method and curing temperature have an impact on the application of fragrance oil to the fabric. Lemongrass oil was microencapsulated on the cotton fabric using three binders: DMDHEU, polyurethane, and polyacrylate emulsion. The selection of DMDHEU as a binder was used due to excellent wrinkle recovery, polyurethane for its elastic and handle property, and acrylic binders for improved wash fastness. The padded fabrics were cured at different curing methods: conventional hot air stenter, infrared lamp, and microwave oven. It was reported that the fabric cured in the microwave oven resulted in rupture of the capsule wall, which in turn evaporated the aroma from the fabric compared to conventional stenter and infra-red lamp. The fabric thus produced possessed excellent repellency against insects and also offered multifunctional property, which was produced through a single finishing technique (Li et al. 2008).

Microencapsulation technique comprises of sheath material for encapsulating the core essential oil for the release of fragrance from the substrate to avoid volatilization and deterioration of the essential oil. Melamine formaldehyde resin was widely used for encapsulating chemicals that are volatile in nature due to its resistance to heat and water, nontoxicity, outdoor weatherability, and unlimited colorability, making it suitable for application in a dry and wet environment (Fei et al. 2015). The aromatic oils were volatile in nature; hence, the aroma can be protected using a stabilizing agent through the microencapsulation technique. The mint aroma was encapsulated through melamine formalin material in the form of microcapsules. These capsules were applied on the surface of the bleached cotton fabric with an aid of acrylic resin using the impregnation and exhaustion technique. The coated fabrics were tested for laundering and rubbing durability to study the efficiency of microcapsule adhesion on the surface of the fabric. It was reported that microcapsule adhered on the impregnated fabric was higher compared to the exhaustion process, which can be due to the application of mechanical force. The microcapsules were physically bound onto the surface of the fabric, indicating that the capsules have no affinity towards the fabric. The odour retained on the fabric after 10 washing cycles, which was tested as per the standard ISO 105 C01. The wet rubbing fastness was found to damage the microcapsules on the surface of the fabric, but the capsules within the structure remained in the protected position, thereby increasing the durability of the fabric (Monllor, Bonet, and Cases 2007).

To increase the washing durability of the fabric, microcapsules can be encapsulated on the structure of the textile substrate in two ways. The first way includes the usage of a polymeric binder, which forms the film, whereas the second way uses the crosslinking agent, which binds the microcapsules and functional group on the fabric structure. Melamine formaldehyde resin was used to encapsulate the essential oil due to resistance to acids, alkali, and abrasion, and the resin can be thermoset with good mechanical properties (He et al. 2019). Sage (*Salvia ocinalis*) and rose (*Rosa damascena*) oils were encapsulated into the melamine formaldehyde resin and coated on the structure of cotton and cotton/polyester woven fabric for the application of cosmetic textiles. The fabric was surface modified by means of bleaching and cationising to enhance the interaction between the polymer chain and microcapsules. The essential oil-encapsulated microcapsules were applied on the surface-modified fabric with an aid of a binder using the pad-dry-cure method, and the developed fabrics were tested for biocompatibility using human dermal fibroblast cells. It was reported that the developed samples offered permeability to air and water vapour, thereby making the substate suitable for home clothing application. The essential oil fragrance was found to remain on the fabric even after 5 washing cycles and provided biocompatibility suitable for the cosmetotextile industry (Stan et al. 2019).

Durable fragrance finishing of fabric was carried out by attaching the fragrance chemicals on the surface of the fabric, which has the advantage of controlled release of fragrance from the fabric. Fragrances applied after treatment have the limitation of poor wash fastness, whereas those applied during the fibre/filament spinning process depend on the type and volume of fragrant material. Polyester fabric was coated with lavender oil-loaded polyvinyl acetate nanoparticle by means of the pad-dry-cure technique. The preparation of nanoparticles involved the dissolution of polyvinyl acetate in ethyl acetate containing lavender oil, and polyvinyl alcohol was dissolved in water. The solutions were blended to form an emulsion from which polyvinyl acetate nanoparticle with lavender oil in the core was precipitated. To compare the release of fragrance, the fabric was coated with lavender oil through melamine formaldehyde condensate and microencapsulated. The controlled odour release was studied for polyvinyl acetate and melamine formaldehyde nanoparticle-loaded polyester fabric for 16 days. It was reported that polyvinyl acetate nanoparticles have controlled release compared to melamine formaldehyde as the latter was damaged due to an external frictional condition (Sohn et al. 2007).

8.5.3.4 Interfacial Polymerization

Interface free radical polymerization technique is used to produce a polymer using two reactive monomers as the polymerization takes place between the oil and water interface. The reaction takes place between the monomers and initiators, which exist in different phases, as the initiators trigger the polymerization reaction between the monomers to form polymers with higher encapsulation efficiency (Giro-Paloma et al. 2016). The polymerization technique was adopted for a single monomer, butanediol dimethacrylate, in the nitrogen atmosphere to produce poly (butanediol dimethacrylate). The initiator used for the polymerization of monomer was ammonium pyrosulfite. The fragrance oil (dementholished peppermint) was encapsulated in the core of the polymer shell (poly (butanediol dimethacrylate)) which was applied on the cotton fabric with an aid of penetrant and polyurethane binder with a pickup percentage of 90%. It was reported that the fragrance content on the fabric decreased with washing cycles. The curing temperature and time have the influence on the laundering durability, as the fabric with the microcapsules treated at 160°C for 2 minutes offered longer durability by retaining the nanoparticles by creating higher adhesive stress between the microcapsule and fabric (Zhao et al. 2016).

Neroline was encapsulated in the polyurethane microcapsules by means of the interfacial polymerization reaction. The microcapsule preparation involves the dissolution of hexane diisocyanate and neroline in toluene, and cyclodextrin and dibutyltindilaurate in polysorbate/ water solution. The two different solutions were mixed at a temperature of 80°C, followed by a centrifugation process in order to remove the microcapsules of the size 29 μm. The latter was applied on the cotton fabric with the help of a polyurethane crosslinking agent and dodecylamino diglycidyl ether of isosorbide, which acts as a cationic catalyst through impregnation, followed by air drying technique. The catalyst promoted the adhesion of the microcapsules on the surface of the fabric, thereby preventing the agglomeration of the capsules on the substate. It was reported that the formation of small particles on the surface of fabric offered higher washing resistance, thereby increasing the potential of the substrate for functional textiles (Abdelkader et al. 2019).

8.5.3.5 Ionotropic Gelation Technique

Ionotropic gelation technique is used to encapsulate the active ingredients within the polyelectrolyte polymeric matrix. The sheath material is formed with two different types of polymers with counter ionic nature, which interact with each other, thereby encapsulating the core material (Patil, Chavanke, and Wagh 2012). Cationic natural polysaccharide found in crustacean shells was used in the polymer matrix for encapsulating the rose fragrance due to its nontoxicity and biodegradability. The chitosan polymer was composed of glucosamine and acetyl glucosamine units linked together by a glycosidic bond, which is mucoadhesive in nature. The usage of an adhesive agent was eliminated by the formation of an ionic gelation reaction between cationic chitosan and anionic tripolyphosphate. The cotton fabric was finished using rose fragrance and chitosan-rose fragrance finishing on the substrate. The deposition of chitosan treated with rose fragrance nanoparticle was higher in the fabric, which can be due to the formation of hydrogen bonding between the hydroxyl group of chitosan and cotton fabric. It was reported that washing the fabric increased the fragrance release, with loss of the main component of rose oil, which was confirmed through GC-MS. Microencapsulation of rose fragrance oil within the chitosan capsule delayed the release of fragrance compared to fabric finished with rose fragrance (Hu et al. 2011). The readily available microcapsules of strawberry odour were coated on the polyester fabric of knitted and woven form using the pad-dry-cure method with an aid of chitosan as film former. Chitosan increased the durability of the fabric by controlling the release of fragrance. The woven fabric offered a slow release of odour compared to knitted structure due to its bulkiness and interlacement of threads (Islam, Troynikov, and Padhye 2012).

Ionic gelation technique was adopted to produce an activated charcoal-based microsphere for antibacterial, antifungal, and antiodour application. The efficiency of the charcoal-coated and charcoal microcapsules on the surface of cotton and bamboo nonwoven fabric was studied. The charcoal was dispersed in tween 20 and sodium alginate solution, to which calcium chloride was sprayed using sprayer, which resulted in the formation of microcapsules. The finishing of the fabric was carried out by spraying the prepared microcapsules, which were dispersed in an emulsion binder using a manual sprayer with a material-to-liquid ratio of 1:20. The solution was uniformly coated and cured at vaporizing temperature. It was reported that the fabric offered a zone of inhibition of 21 mm and 26 mm against *Escherichia coli* and *Staphylococcus aureus,* and 52 mm against the fungal agent *Aspergillus Niger.* Organoleptic Evaluation method was adopted for anti-odour measurement, using human judges, and shoe insole was used as test material. It was reported that the charcoal microcapsule-incorporated cotton and bamboo nonwoven fabric can be utilized for hygiene care applications (Pragadheeswari and Sangeetha 2017).

8.5.3.6 Microcapsules/Nanocapsules

The limitation of the microencapsulation technique is the size of the nanoparticles, which have the tendency to fall off from the fabric, thereby causing difficulty in providing the sustainable fragrance release with the fabric. The rose fragrance was homogenized in deionized water by the addition of a wetting agent. Butyl cyanoacrylate was added to the solution, and the polymerization was carried out at a pH of 7 with the addition of sodium hydroxide. Emulsion containing rose fragrance nanoparticle was coated on the cotton fabric using the impregnation technique. It was found that the rose fragrance particle adhered to the fabric was the size of 51.4 nm. The crystallinity of the oil-incorporated fabric was found to be lower compared to the pristine cotton fabric, which indicates the presence of fragrance oil on the fabric as the oil possess no crystal structure. The release of fragrance from the nanoparticle-incorporated fabric was found to be sustainable for 50 washing cycles compared to the coating of rose oil on the fabric, which was confirmed through testing using an electronic nose (Hu et al. 2011).

The fragrance encapsulated in the polymer matrix ranging from 10 nm to 1000 nm offers a higher advantage compared to microcapsules (3–800 μm) due to its higher surface area, which offers stronger adhesive force on the fabric. Smaller size of the capsules ensures uniform coating of products, thereby requiring a large amount of physical force in order to damage the covering, followed by the release of fragrance. Nanocapsules offer a slow release of fragrance compared to microcapsules, thereby increasing the durability of the textiles (Ahangaran, Navarchian, and Picchioni 2019). Zein, a natural hydrophobic polymer obtained from corn starch, was used for encapsulating the fragrance oil, such as pink fruity fragrance and white floral fragrance. The nanocapsules were then coated on cotton, polyester, and cotton/polyester blend (20/80) fabrics, along with fabric softener, by immersing at various concentrations of fragrance capsules and then dried using natural drying at room temperature. The pink fruity fragrance was composed of phenyl ethyl alcohol, terpineol, and geraniol as a major component of the fragrance agent, whereas white floral fragrance was composed of phenyl ethyl alcohol, benzoyl acetate, and citronellol as the main component. Zein polymer was dissolved in aqueous ethanol containing surfactant, to which fragrance oil was added, and the nanoparticles were dried and applied on the fabric to study the sustained release of fragrance from the textiles. It was reported that cotton fabric offered long-lasting fragrance release for 30 days compared to blend and polyester fabric. It was also concluded that the size of the pink fruity and white floral capsules was 219 nm and 301 nm, and the thermal stability of the fragrance increased upon encapsulation technique (Pithanthanakul et al. 2021).

Nanocapsules were prepared with the core and sheath structure for encapsulating fragrance oil for increasing the serviceability of the fabric. The sheath of the nanocapsule was prepared using methyl methacrylate-styrene copolymer, which is nontoxic, is chemically inert, and offers increased strength. Cologne essential oil was made into an emulsion by mixing with the polymer and emulsifying agent and then padded onto the fabric. It was reported that the nanoparticle-coated cotton fabric retained fragrance even after 16 washing cycles, implying that better adhesion of nanocapsules resulted in delayed release of fragrance (Liu et al. 2015). Lavender oil was coated on the three different fabrics – cotton, cotton/polyester, and polyester elastane fabric – in the form of nanocapsules by means of crosslinking with and without copolymer ethyl methacrylate-co-methacrylic acid. Nanocapsules comprising lavender oil were coated on the fabric using padding mangle, and it was found that the capsule adhered to the fabric was in the particle size of 15 nm and 40 nm. It was reported that copolymer chemically interacts with the fabrics to increase the durability of the product, as the hydroxyl group of the cotton fabric would form a bond with the carbonyl and hydroxyl group of a copolymer. In the case of polyester, the carbonyl group reacts with the hydroxyl group of copolymers, whereas in elastane fabric, a hydrogen bond would form between the hydroxyl and amino groups of elastane structure with the carbonyl group of the copolymer. It was concluded that cotton fabric treated with crosslinked lavender oil-coated nanocapsule offered higher durability and sustained release of fragrance compared to blended fabrics (Saade et al. 2021).

8.6 CONCLUSION

The release of fragrance from a textile substrate in a sustained fashion is desired to increase the serviceability of the product in the market. The addition of antimicrobial agents can prevent the growth of microorganisms that release malodour from the textile substrate. The method of manufacturing aromatherapeutic textiles includes physical coating, chemically crosslinking with aroma agents, or encapsulating the fragrance material to control the release of chemicals from the substrate. Essential oils are widely used for fragrance finishing of textiles as they offer dual properties: antibacterial activity and releasing fragrance to control malodour. The challenge in the entire process is adopting essential oil onto the fabric due to its volatile nature. Moreover, the characterization of fragrance-enriched substrates is still a greater challenge. At present, subjective evaluation using human trials and an electronic nose are used for characterizing the developed textile substrate for the diffusion of odour from the fabric, and a laundering test is carried out for the durability of the fragrance within the structure. The development of fabrics with durable fragrance will add another dimension to home and medical textiles.

REFERENCES

Abbasi, E., Aval, S.F., Akbarzadeh, A., Milani, M., Nasrabadi, H.T., Joo, S.W., Hanifehpour, Y., Nejati-Koshki, K., and Roghiyeh, P.-A. 2014. Dendrimers: Synthesis, applications, and properties. *Nanoscale Research Letters*, 9(1): 1–10. doi: 10.1186/1556-276X-9-247.

Abdelkader, M.B., Azizi, N., Baffoun, A., Chevalier, Y., and Majdoub, M. 2019. Fragrant microcapsules based on β-Cyclodextrin for cosmetotextile application. *Journal of Renewable Materials*, 7(12): 1347–1362. doi: 10.32604/jrm.2019.07926.

Ahangaran, F., Navarchian, A.H., and Picchioni, F. 2019. Material encapsulation in poly (Methyl methacrylate) shell: A review. *Journal of Applied Polymer Science*, 136(41): 1–21. doi: 10.1002/app.48039.

Ammulu, M.A., Tammina, K., Bonigala, B., Vemuri, P.K., Podha, S., and Ronda, S.R.. 2019. Fabricating multifunctional nanoparticles bonded to enzymatically oxidized fabrics for their various applications. *Indian Journal of Fibre and Textile Research*, 44(4): 381–388.

Ara, K., Hama, M., Akiba, S., Koike, K., Okisaka, K., Hagura, T., Kamiya, T., and Tomita F. 2006. Foot odor due to microbial metabolism and Its control. *Canadian Journal of Microbiology*, 52(4): 357–364. doi: 10.1139/W05-130.

Baker, L.B. 2019. Physiology of sweat gland function: The roles of sweating and sweat composition in human health. *Temperature*, 6(3):211–259. doi: 10.1080/23328940.2019.1632145.

Bakry, A.M., Abbas, S., Ali, B., Majeed, H., Abouelwafa, M.Y., Mousa, A., and Liang, L. 2016. Microencapsulation of oils: A comprehensive review of benefits, techniques, and applications. *Comprehensive Reviews in Food Science and Food Safety*, 15(1):143–182. doi: 10.1111/1541-4337.12179.

Bezerra, F.M., Lis, M., Carmona, Ó.G., Carmona, C.G., Moisés, M.P., Zanin, G.M., and Moraes, F.F. 2019. Assessment of the delivery of citronella oil from microcapsules supported on wool fabrics. *Powder Technology*, 343: 775–782. doi: 10.1016/j.powtec.2018.11.001.

Bhatt, L., and Kale, R.D. 2019. Lemongrass (Cymbopogon Flexuosus Steud.) wats treated textile: A control measure against vector-borne diseases. *Heliyon*, 5(12): e02842. doi: 10.1016/j.heliyon.2019.e02842.

Biswas, D., Chakrabarti, S.K., Saha, S.G., and Chatterjee, S. 2015. Durable fragrance finishing on jute blended home-textiles by microencapsulated aroma oil. *Fibers and Polymers*, 16(9): 1882–1889. doi: 10.1007/s12221-015-4829-5.

Buschmann, H.J., Dehabadi, V.A., and Wiegand, C. 2015. *Medical, Cosmetic and Odour Resistant Finishes for Textiles: Functional Finishes for Textiles: Improving Comfort, Performance and Protection.* Woodhead Publishing Limited. doi: 10.1533/9780857098450.1.303.

Callewaert, C., Lambert, J., and Van de Wiele, T. 2017. Towards a bacterial treatment for armpit malodour. *Experimental Dermatology*, 26(5): 388–391. doi: 10.1111/exd.13259.

Cay, A., Tarakçioğlu, I., and Hepbasli, A. 2009. Assessment of finishing processes by exhaustion principle for textile fabrics: An exergetic approach. *Applied Thermal Engineering*, 29(11–12): 2554–2561. doi: 10.1016/j.applthermaleng.2008.12.032.

Eza, T.S.M., Wan Ahmad, W.Y., Omar, K., and Noor Ahmad, M. 2012. The activated carbon anti-odour treated fabrics by coating and pigment printing. *ISBEIA 2012 - IEEE Symposium on Business, Engineering and Industrial Applications*, 12: 690–695. doi: 10.1109/ISBEIA.2012.6422978.

Fard, P.A., Shakoorjavan, S., and Akbari, S. 2018. The relationship between odour intensity and antibacterial durability of encapsulated thyme essential oil by PPI dendrimer on cotton fabrics. *Journal of the Textile Institute*, 109(6): 832–841. doi: 10.1080/00405000.2017.1376820.

Fei, X., Zhao, H., Zhang, B., Cao, L., Yu, M., Zhou, J., and Yu, L. 2015. Microencapsulation mechanism and size control of fragrance microcapsules with melamine resin shell. *Colloids and Surfaces A: Physicochemical and Engineering Aspects*, 469: 300–306. doi: 10.1016/j.colsurfa.2015.01.033.

Ghosh, S., and Chipot, N. 2015. Embedding aromatherapy essential oils into textile fabric using β-Cyclodextrin inclusion compound. *Indian Journal of Fibre and Textile Research*, 40(2): 140–143.

Giro-Paloma, J., Martínez, M., Cabeza, L.F., and Inés Fernández, A. 2016. Types, methods, techniques, and applications for microencapsulated phase change materials (MPCM): A review. *Renewable and Sustainable Energy Reviews*, 53: 1059–1075. doi: 10.1016/j.rser.2015.09.040.

Grancarić, A.M., Tarbuk, A., and Kovaček, I. 2009. Nanoparticles of activated natural zeolite on textiles for protection and therapy. *Chemical Industry and Chemical Engineering Quarterly*, 15(4): 203–210. doi: 10.2298/CICEQ0904203G.

He, Y., Yao, S., Hao, J., Hong, W., Zhu, L., Tian, S., Sun, Y., and Lin, J. 2019. Synthesis of melamine-formaldehyde microcapsules containing oil-based fragrances via intermediate polyacrylate bridging layers. *Chinese Journal of Chemical Engineering*, 27(10): 2574–2580. doi: 10.1016/j.cjche.2018.10.023.

Hu, J., Xiao, Z.B., Zhou, R.J., Ma, S.S., Li, Z., and Wang, M.X. 2011. Comparison of compounded fragrance and chitosan nanoparticles loaded with fragrance applied in cotton fabrics. *Textile Research Journal*, 81(19): 2056–2064. doi: 10.1177/0040517511416274.

Hu, J., Xiao, Z., Zhou, R., Ma, S., Wang, M., and Li, Z. 2011. Properties of aroma sustained-release cotton fabric with rose fragrance nanocapsule. *Chinese Journal of Chemical Engineering*, 19(3): 523–528. doi: 10.1016/S1004-9541(11)60016-5.

Islam, S., Troynikov, O., and Padhye, R. 2012. New automotive fabrics with anti-odour and antimicrobial properties. *Sustainable Automotive Technologies*, 2012: 81–89. doi: 10.1007/978-3-642-24145-1_12.

Jha, S.K. 2017. Characterization of human body odor and identification of aldehydes using chemical sensor. *Reviews in Analytical Chemistry*, 36(2): 1–16. doi: 10.1515/revac-2016-0028.

Julaeha, E., Puspita, S., Eddy, D.R., Wahyudi, T., Nurzaman, M., Nugraha, J., Herlina, T., and Anshori, J.I. 2021. Microencapsulation of lime (Citrus Aurantifolia) oil for antibacterial finishing of cotton fabric. *RSC Advances*, 11(3): 1743–1749. doi: 10.1039/d0ra09314a.

Jyothi, N.V.N., Muthu Prasanna, P., Sakarkar, S.N., Surya Prabha, K., Seetha Ramaiah, P., and Srawan, G.Y. 2010. Microencapsulation techniques, factors influencing encapsulation efficiency. *Journal of Microencapsulation*, 27(3): 187–197. doi: 10.3109/02652040903131301.

Kanlayavattanakul, M., and Lourith, N. 2011. Body malodours and their topical treatment agents. *International Journal of Cosmetic Science*, 33(4): 298–311. doi: 10.1111/j.1468-2494.2011.00649.x.

Karagonlu, S., Başal, G., Ozyıldız, F., and Uzel, A. 2018. Preparation of thyme oil loaded microcapsules for textile applications. *International Journal of New Technology and Research*, 4(3): 263122.

Karolia, A., and Mendapara S. 2007. Imparting antimicrobial and fragrance finish on cotton using chitosan with silicon softener. *Indian Journal of Fibre and Textile Research*, 32(1): 99–104.

Khanna, S., Sharma, S., and Chakraborty, J.N. 2015. Performance assessment of fragrance finished cotton with cyclodextrin assisted anchoring hosts. *Fashion and Textiles*, 2(1): 1–17. doi: 10.1186/s40691-015-0042-9.

Li, J., and Zhuang, S. 2020. Antibacterial activity of chitosan and its derivatives and their interaction mechanism with bacteria: Current state and perspectives. *European Polymer Journal*, 138: 109984. doi: 10.1016/j.eurpolymj.2020.109984.

Li, S., Lewis, J., Stewart, N., Lei, Q., and Boyter, H. 2008. Effect of finishing methods on washing durability of microencapsulated aroma finishing. *Journal of the Textile Institute*, 99(2): 177–183. doi: 10.1080/00405000701489701.

Liu, C., Liang, B., Shi, G., Li, Z., Zheng, X., Huang, Y., and Lin, L. 2015. Preparation and characteristics of nanocapsules containing essential oil for textile application. *Flavour and Fragrance Journal*, 30(4): 295–301. doi: 10.1002/ffj.3245.

López, C.A., de Vries, A.H., and Marrink, S.J. 2011. Molecular mechanism of cyclodextrin mediated cholesterol extraction. *PLoS Computational Biology*, 7(3): 1–11. doi:10.1371/journal.pcbi.1002020.

Mashat, B., and Awad, A. 2014. The impact of nanosilver particles on foot - Odor producing microbes. *Current World Environment*, 9(3): 615–622. doi: 10.12944/cwe.9.3.09.

McQueen, R.H. 2011. *Odour Control of Medical Textiles: Handbook of Medical Textiles*. Woodhead Publishing Limited. doi: 10.1533/9780857093691.3.387.

Mejri, M. 2010. Aromatherapeutic textiles. *Acta Horticulturae*, 853: 439–446. doi: 10.17660/actahortic.2010.853.54.

Mohan, A.V., and Bhaarathidhurai, D. 2020. Activated carbon: A textile material. *International Journal of Innovative Science and Research Technology*, 5(September): 829–832.

Monllor, P., Bonet, M.A., and Cases, F. 2007. Characterization of the behaviour of flavour microcapsules in cotton fabrics. *European Polymer Journal*, 43(6): 2481–2490. doi: 10.1016/j.eurpolymj.2007.04.004.

Nahar, K. 2020. A review on anti odor finishing of textiles, its application method and commercialization. *Acta Technica Corviniensis-Bulletin of Engineering*, ISSN: 2067–3809 :79–84.

Nguon, O., Lagugné-Labarthet, F., Brandys, F.A., Li, J., and Gillies, E.R. 2018. Microencapsulation by in situ polymerization of amino resins. *Polymer Reviews*, 58(2): 326–375. doi: 10.1080/15583724.2017.1364765.

Patil, P., Chavanke D., and Wagh, M. 2012. A Review on Ionotropic gelation method: Novel approach for controlled gastroretentive gelispheres. *International Journal of Pharmacy and Pharmaceutical Sciences*, 4(SUPPL. 4): 27–32.

Pattnaik, S., Subramanyam, V.R., Bapaji, M., and Kole, C.R. 1997. Antibacterial and antifungal activity of aromatic constituents of essential oils. *Microbios*, 89(358): 39–46.

Perinelli, D.R., Palmieri, G.F., Cespi, M., and Bonacucina, G. 2020. Encapsulation of flavours and fragrances into polymeric capsules and cyclodextrins inclusion complexes: An ipdate. *Molecules (Basel, Switzerland)*, 25(24): 1–33. doi: 10.3390/molecules25245878.

Pithanthanakul, U., Vatanyoopaisarn, S., Thumthanaruk, B., Puttanlek, C., Uttapap, D., Kietthanakorn, B., and Rungsardthong, V. 2021. Encapsulation of fragrances in xein nanoparticles and use as fabric softener for textile application. *Flavour and Fragrance Journal*, 36(3): 365–373. doi:10.1002/ffj.3648.

Pragadheeswari, R., and Sangeetha, K. 2017. Functional properties of activated carbon treated textile material. *International Journal for Scientific Research & Development*, 5(04): 1766–1768.

Ramya, K., and Maheshwari, V. 2019. Effect of fragrance finish of the 100% bamboo knitted fabrics finished with herbal extracts.

Rathinamoorthy, R., and Thilagavathi, G. 2014. Effect of antimicrobial finish on odor control properties of apparel fabric. *Journal of Textile and Apparel, Technology and Management*, 9(1): 1–15.

Saade, H., Díaz De León-Gómez, R., Treviño, M.E., Rodríguez-Fernández, O.S., Soriano-Corral, F., Borjas, J.J., Castellanos, F., Enríquez-Medrano, F.J., Valdez, J.A., and López, R.G. 2021. Performance of cotton, cotton-polyester, and polyester-elastane fabrics impregnated with ultrafine polymeric nanoparticles loaded with lavender oil. *Journal of Nanomaterials*, 6665327: 1–10. doi: 10.1155/2021/6665327.

Sayed, U. 2017. Application of essential oils for finishing of textile substrates. *Journal of Textile Engineering & Fashion Technology*, 1(2): 42–47. doi: 10.15406/jteft.2017.01.00009.

Schramm, C., and Amann, A. 2019. Of cellulosic material modified by a hydrolyzed di-carboxylic acid based alkoxysilane / Melamine finishing system. *Cellulose*, 26(7): 4641–4654. doi: 10.1007/s10570-019-02391-3.

Shirasu, M., and Touhara, K. 2011. The scent of disease: Volatile organic compounds of the human body related to disease and disorder. *Journal of Biochemistry*, 150(3): 257–266. doi: 10.1093/jb/mvr090.

Silva, P.T., Fries, L.L.M., de Menezes, C.R., Holkem, A.T., Schwan, C.L., Wigmann, É.F., de Oliveira Bastos, J., and de Bona da Silva, C. 2014. Microencapsulation: Concepts, mechanisms, methods and some applications in food technology. *Ciência Rural*, 44(7): 1304–1311. doi: 10.1590/0103-8478cr20130971.

Sohn, S.O., Lee, S.M., Kim, Y.M., Yeum, J.H., Choi, J.H., and Ghim, H.D. 2007. Aroma finishing of PET fabrics with PVAc nanoparticles containing lavender oil. *Fibers and Polymers*, 8(2): 163–167. doi: 10.1007/BF02875786.

Specos, M.M., Escobar, G., Marino, P., Puggia, C., Victoria Defain Tesoriero, M., and Hermida, L. 2010. Aroma finishing of cotton fabrics by means of microencapsulation techniques. *Journal of Industrial Textiles*, 40(1): 13–32. doi: 10.1177/1528083709350184.

Stan, M.S., Chirila, L., Popescu, A., Radulescu, D.M., Radulescu, D.E., and Dinischiotu, A. 2019. Essential oil microcapsules immobilized on textiles and certain induced effects. *Materials*, 12(12): 1–15. doi: 10.3390/ma12122029.

Stewart, N.D. 1967. Padding, drying, steaming and baking stages in the application.

Štular, D., Šobak, M., Mihelˇciˇc, M., Šest, E., Iliˊc, I.G., Jerman, I., Simonˇciˇc, B., and Tomšiˇc, B. 2019. Proactive release of antimicrobial essential oil from a 'Smart' cotton fabric. *Coatings*, 9(242): 1–18. doi: 10.3390/coatings9040242.

Timilsena, Y.P., Akanbi, T.O., Khalid, N., Adhikari, B., and Barrow, C.J. 2019. Complex coacervation: Principles, mechanisms and applications in microencapsulation. *International Journal of Biological Macromolecules*, 121: 1276–1286. doi: 10.1016/j.ijbiomac.2018.10.144.

Vasanth Kumar, D., Boopathi N., Karthick N., and Ramesh P. 2012. Aesthetic finishes for home textile materials. *International Journal of Textile Science*, 2012(3): 5–9. doi: 10.5923/j.textile.20120103.01.

Velmurugan, P., Lee, S.M., Cho, M., Park, J.H., Seo, S.K., Myung, H., Bang, K.S., and Oh, B.T. 2014. Antibacterial activity of silver nanoparticle-coated fabric and leather against odor and skin infection causing bacteria. *Applied Microbiology and Biotechnology*, 98(19): 8179–8189. doi: 10.1007/s00253-014-5945-7.

Wijesirigunawardana, P.B., Gayani, B., and Perera, K. 2018. Development of a cotton smart textile with medicinal properties using lime oil microcapsules. *Acta Chimica Slovenica*, 65(1): 150–159. doi: 10.17344/acsi.2017.3727.

Xue, C.H., Deng, L.Y., Jia, S.T., and Wei, P.B. 2016. Fabrication of superhydrophobic aromatic cotton fabrics. *RSC Advances*, 6(109): 107364–107369. doi: 10.1039/c6ra23558a.

Yamashita, Y., Miyahara, R., and Sakamoto, K. 2017. *Emulsion and Emulsification Technology: Cosmetic Science and Technology: Theoretical Principles and Applications*. Elsevier Inc. doi: 10.1016/B978-0-12-802005-0.00028-8.

Zhao, D., Jiao, X., Zhang, M., Ye, K., Shi, X., Lu, X., Qiu, G., and Shea, K.J. 2016. Preparation of high encapsulation efficiency fragrance microcapsules and their application in textiles. *RSC Advances*, 6(84): 80924–80933. doi: 10.1039/c6ra16030a.

9 Textile and Other Odours: A Focus on Third-hand Smoke and Laundry Odour

R. Rathinamoorthy
Department of Fashion Technology, PSG College of Technology, Coimbatore, India

G. Thilagavathi
Department of Textile Technology, PSG College of Technology, Coimbatore, India

CONTENTS

9.1 INTRODUCTION

The odour adsorption property of textiles is highly related to the individual's cleanliness and hygiene. The apparel fabrics mainly receive odour from the body and secondly from the environment. As the other chapters of this book discuss body odour and textile interaction, this chapter focuses on other environmental odours and

DOI: 10.1201/9781003141426-9

textiles. Airborne contaminants usually carry semi-volatile organic compounds from the environment and deposit them on the fabric surface via a different mechanism that leads to odour formation. The transport of airborne chemicals to skin through textile/fabric mediation is of great concern in recent time (Chen-Chou Wu et al. 2019). Textile fibers are very prone to catching odour components from the environment due to their chemical structure (active groups) and physical structure (compactness and other physical properties). The odour sorption characteristics of textile fabrics are highly correlated to the fabric and to the air distribution coefficient (Cao et al. 2017). The type of reactive sites and types of functional groups on the surface of the fiber decide the mechanism of interaction. In the case of structure, it aids a physical interaction with the textiles through compactness, surface roughness, and other parameters (Obendorf et al. 2006). The most commonly used fibrous materials, like polyester, cotton, and wool, were proven for their odour retention characteristics by relating their physical and chemical properties (McQueen et al. 2008).

During daily activities, textile materials are exposed to various odourous materials, like food-related and laundry-related odours. Adhesion of these odours creates unpleasant smells to the wearer and also creates health hazards in some cases. Hence, this chapter details the various other possible odours textile material will interact with. The first part of this chapter details the smoke odour and textile interaction. It details the importance of second-hand and third-hand smoke and its impact on textiles and wearers' health. The second part of this chapter deals with the most common laundry odour, and the formation of laundry odour and the role of washing machines and drying are detailed. The role of textile material in this context is also detailed. The last part of this chapter depicts other potential odours, like cooked food, fish, vegetables, and pet odours in the household environment.

9.2 TOBACCO/CIGARETTE SMOKE AND TEXTILES

Out of several potential environmental odours, tobacco smoke is of high interest as it creates serious health issues for nonsmokers. Tobacco smoke contains several carcinogenic chemicals that lead to cancer (U.S. EPA 1992). Cigarette smoke/odour harms the human lungs as it contains more than 4000 chemicals, out of which around 69 are carcinogenic and at least 250 chemical substances develop potential health hazards (Brunnemann and Hoffmann 1991). Environmental tobacco smoke (ETS) generally indicates the tobacco content in the indoor air after the end of smoking, without any active smokers. It is known as second-hand smoking (SHS). The smoke or components of the tobacco further cling to the surfaces of furniture, walls, home textiles, and apparel and pollute the air for months, which is known as third-hand smoke (THS) (Chien, Chang, and Liu 2011: Nilsen and Nilsen 1997). THS further can react with the dust or oxidants in the atmosphere upon ageing and develop a secondary pollutant, which can also create serious health hazards (Matt et al. 2008). One of the most important pollutants that results from THS and atmospheric nitrous acid is tobacco-specific nitrosamines (TSNA), which is very well known for its carcinogenic nature (Matt et al. 2011). Further, hydrogen cyanide, butane, toluene, formaldehyde, and even radioactive polonium-210 and polycyclic aromatic hydrocarbons (PAHs) were found and reported in THS (Talbot and Palmer

TABLE 9.1
Comparison between second-hand odour and third-hand odour. (From Chi-Yung Cheng et al. (2016). Reprinted with permission)

	Second-hand smoke (SHS)	Third-hand smoke (THS)
Source	Composed of the mainstream smoke exhaled by active smokers and sidestream smoke expelled from the lit tobacco product	A residual contamination from tobacco smoke that remains after the cigarette is extinguished; usually clings to hair, skin, clothing, and furniture surfaces
Major compounds	Benzene, toluene, styrene, acetone, xylene, naphthalene, formaldehyde, polonium-210, phenol, PAHs, nicotine, 3-ethenylpyridine, carbon monoxide, and N-nitrosamine	Nicotine, 3-ethenylpyridine, phenol, cresols, formaldehyde, tobacco-specific nitrosamines, naphthalene, xylene, styrene, benzene, and toluene (Basically, the compounds of THS are very similar to SHS but at lower concentration)
Related disease	Lung cancer, oral cancer, asthma, chronic obstructive pulmonary disease (COPD), and coronary heart disease	Allergic symptoms, asthma, and altering brain and lung development in children

2013). Table 9.1 compares the various issues developed by SHS and THS (Chi-Yung Cheng et al. 2016).

Organic substances with different natures, such as very volatile (VVOCs), semi-volatile (SVOCs), and volatile organic compounds (VOCs), are the major components of ETS (Daisey 1999). Further, it also contains particulate matter (PM) and associated organics, as well as gas-phase inorganic chemicals. The intensity of these chemical compounds as a THS depends mainly on the sorption and emission characteristics of the indoor or clothing material. Out of different materials, household fabrics and clothing materials play a vital role in the re-emission of major ETS components, such as nicotine, naphthalene, 3-ethenylpyridine (3-EP), and phenol (Elkilani et al. 2003). A study analysed the ETS uptake by clothing material using a simple method by exposing the textiles to cigarette smoke in a closed container. Fabrics such as wool, linen, cotton, silk, rayon, acetate, and polyester were used in this study, and the weight addition immediately before and after ETS exposure was analysed. The results reported the smoke sorption in terms of weight addition; a higher mass change was noted in natural fibers like linen and wool, followed by cotton and silk. Out of all the natural fibers, silk was found to have lesser weight gain. In the case of man-made textiles, rayon and acetate fibers absorbed more ETS, with a higher weight gain, whereas polyester was noted as the least absorbing material. In this study, there was a different amount of weight gain noted with the wool fiber itself due to its structural difference. Higher sorption of ETS was noted with wool gabardine structure, followed by suiting material and crepe. Similarly, the researcher noted a difference between the wool gabardine and polyester gabardine structures. Hence, he reported that the chemical and physical structures of fabric are the main reason for the different sorption abilities of the fabric. Furthermore, he compared the moisture regain of the fibers and related it with ETS sorption. Fibers with higher moisture regain percentage (13.6–16%, wool) gained more weight than cotton (8.5%). Similarly, in man-made fiber also, rayon with 11.6–16% moisture regain absorbed more ETS than the polyester (0.4%) (Noble 2000).

The cigarette smoke absorption capacity of the cotton, linen, silk, acetate, and polyester fabric was evaluated by Ueta et al. (2010). The fabrics were exposed to cigarette smoke, and the smoke from the fabric was extracted using a needle extraction device and analysed through gas chromatography-mass spectrometry (GCMS). The results reported higher absorption of ammonia and pyrrole in cotton and linen fabric, followed by acetate, silk, and polyester. In the case of benzene and toluene, higher absorption was noted with acetate, followed by linen, cotton, silk, and polyester. Based on the absorption behaviour, the study pointed out the role of fiber chemical and physical structure differences. Further, Ueta et al. (2010) evaluated the VOC release behaviour of the textiles for 10 minutes and noted VVOCs released faster than VOCs. The emission of VOCs from fabric reduced with time. Piade, D'andres, and Sanders (1999) evaluated the nicotine sorption behaviour of different materials like glass, cotton, and nylon. They reported a higher absorption of cotton fabric over the other two sink materials they selected. It is noted that up to 1 mg of nicotine can be absorbed and reemitted from cotton material over a few hours.

In a similar study, a higher release of polar and volatile compounds (furfural, phenol, and benzonitrile) was noted from cotton and linen fabric than the polyester and acetate fabric. In the meantime, both synthetic and natural textiles released the same amount of nonpolar and nonvolatile compounds. The fibers from plant origin (cotton, linen) only released low molecular weight substances, whereas the animal-origin natural fibers like wool and silk released higher molecular-weight VOCs (Chien, Chang, and Liu 2011). When compared to the other fibers, wool fiber exhibited higher sorption and less emission compared to other structures. This is mainly attributed to the higher chemical sorption and physical entrapment of odourous substances in wool structure than other fiber (Cieslak 2006). Though both wool and cotton absorbed a similar amount of tobacco smoke, the reemission greatly differs between them. Chien, Chang, and Liu (2011) denied the weight-based method of evaluation proposed by Noble (2000), though it had a higher correlation with moisture regain. Concerning the cotton fabric, the fabric structure had a significant difference in odour emission. The knitted fabric released a higher amount of VOC than the cotton-woven fabric. The research did not identify any relationship between the fabric weight density and other parameters with odour emission. When the effect of fabric colour was analysed, there was some difference noted, but it is not significant. However, a cotton white fabric reemitted a significantly higher amount of gas than the black fabric. In the case of linen, a reverse relationship was noted concerning colour of the fabric. The exposure time also gave a significant impact on the odour emission. The fabrics exposed for 12 minutes had a higher amount of gas release than the fabric exposed for 8 min, except for formaldehyde.

Bahl et al. (2014) exposed 100% cotton terry cloth and 100% polyester fleece fabric for 114 and 257 hours over 12 and 10 months, respectively, to analyse the ageing effect. The exposed fabrics were extracted using aqueous and organic solvents and analysed for their VOC substances. The results showed that the THS chemicals remained on the selected fabric even after 1.5 years of last exposure. The aqueous extract of cotton fabric showed the presence of nicotine and its derivatives similar to the composition of saliva. The THS content in the polyester fabric significantly differs from cotton fabric. The VOCs were found both in the gas and solid phase at indoor temperature. The lower nicotine sorption of polyester than cotton terry cloth is mainly due to the surface chemistry of the fiber. As the free hydroxyl groups are higher in cotton fabric, it forms hydrogen bonds with the nicotine and its derivative (Senthilkumar et al. 2012). The oleophilic nature of the polyester repels the polar components more than the cotton (Bendak and El-Marsafi 1991). Bahl et al. (2014) demonstrated that textile materials are a good reservoir of THS chemicals. Wearing a fabric of 500 g that is exposed to the amount of exposure used by Bahl et al. (2014) will expose the wearer to 7,894 mg of nicotine/day and 32.7 mg of TSNAs/day. As per their study, mouthing a 5g of cotton textile with used exposure for 1 hour will be 529 mg of nicotine/day and 2.2 mg of TSNAs/day for a 12 kg toddler. The interaction mechanism of terry cloth and nicotine is illustrated in Figure 9.1. These exposure levels are similar to an active smoker's consumption. Based on the potentially harmful effect of THS exposures on textiles, the study recommended the immediate removal of contamination from textiles.

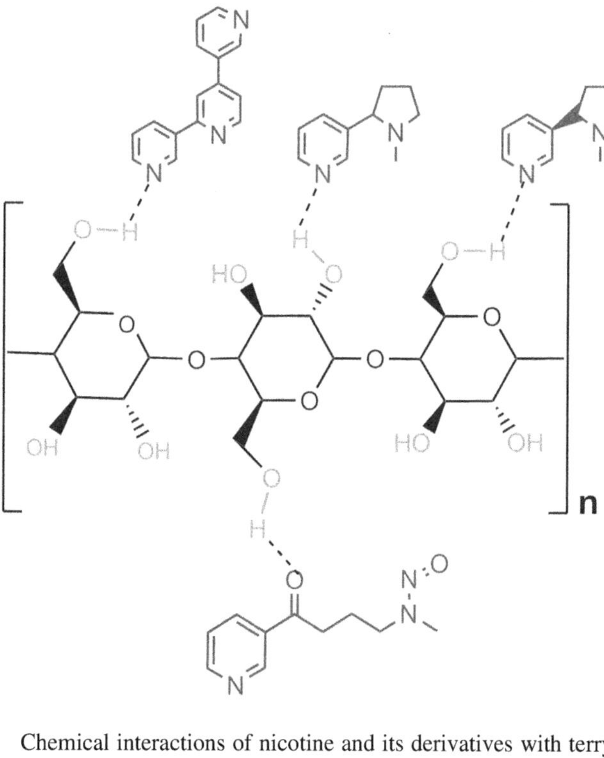

FIGURE 9.1 Chemical interactions of nicotine and its derivatives with terry cloth through hydrogen bonds. Terry cloth absorbs nicotine and related chemicals as these are polar and can form hydrogen bonds with the free hydroxyl groups in terry cloth. (From Bahl et al. (2014). Reprinted under Creative Commons license.)

To reduce the cigarette smoke odour (THS) sorption, retention, and release from the textile, the application of monochlorotriazinyl cyclodextrin (MCT-β-CD) as a finish was evaluated by Setthayanond et al. (2017). They measured the odour intensity using an E-Nose, and the results showed a significant reduction after MCT-β-CD treatment. Upon three different treatment concentrations, namely from 1%, 3%, and 5% of MCT-β-CD weight per volume of liquid used, all the concentrations reduced the THS release behaviour. Though an increase in MCT-β-CD concentration reduced the THS release percentage, there is not much difference noted between 3% and 5%. However, a THS release percentage reduced up to 65.5% for 3% MCT-β-CD treatment. While comparing the dyed and undyed fabric, dyed fabric showed a reduced amount of THS release than the undyed sample up to 15%. The THS odour emission results from different dyed and MCT-β-CD treated samples are provided in Table 9.2.

The fundamental reason for the reduction of THS release is the entrapment of odour molecules in the MCT-β-CD structure. MCT-β-CD engages the hydroxyl reactive sites in the cellulose chain of the cotton structure. The reaction permanently bonds the finish to cotton fabric through a covalent bonding as MCT acts as a reactive anchor. The complete fixing of MCT-β-CD helps the complexing reaction of CD with the THS from the sources, like nitrosamine and benzene, on the surface of the cellulose. This complex formation is expected to happen by noncovalent physical interactions, such as van der Waals forces, H-bonds, and hydrophobic attractions. Due to the hydrophobic nature of CD, it provides a suitable environment for the host material complexing. The positively charged b-cyclodextrin cavity forms a hydrogen bond with the nitrosamine molecules, by replacing oxygen in the -N-N=O, through a HO- bond with the rim of b-cyclodextrin and complexes the nitrosamine within the b-cyclodextrin. Similarly, the alkyl group was trapped into the wider rim of the b-cyclodextrin cavity. As the volatile components are not only trapped into the cavity of the b-cyclodextrin but also interact through hydrogen bond (Wang et al. 2004), the complexing phenomenon is noted as a strong option for the reduction of THS release from the textile.

TABLE 9.2

Odour emittance reduction performances (%) of MCT-β-CD treated cotton fabrics dyed with reactive dyes at 0.5% (on the weight of fabric) dye concentration (From Setthayanond et al. (2017). Reprinted with permission)

MCT-β-CD treated cotton fabric type	MCT-β-CD treated cotton fabric type		
	MCT-β-cyclodextrin		
	1%w/v	3%w/v	5%w/v
Undyed	26.52	51.41	53.61
Dyed with red X-6BN	41.94	58.76	65.04
Dyed with navy X-GN	33.05	59.17	62.15
Dyed with yellow X-4RN	35.23	65.11	65.51

9.3 THS AND FABRIC INTERACTION MECHANISM

The life of SHS constituents in the air is very short, whereas the life of THS components in the indoor environment is very long, from days to months. Hence, compared to SHS, THS can get ample time to interact with air pollutants and surface chemicals of different indoor surfaces. Indoor free radicals like nitrous acid (HONO) and hydrogen peroxide (H_2O_2), can develop reactive species indoors. The free oxygen and nitrogen-containing radicals, oxidants, and nitrosating species, along with long residence of THS, potentially form various hazardous by-products (Gómez et al. 2013). Nicotine is the main component of THS. While in contact with ozone, it oxidizes and creates by-products like carbonyls, amides, N-oxides, and carboxylic acids, which are more hazardous than nicotine itself (Sleiman, Destaillats, and Gundel 2013). Other than the oxidation reaction, the nitrosation of nicotine by HONO produces tobacco-specific nitrosamines indoors (TSNAs) like N-nitrosonornicotine (NNN), 4-(methylnitrosamino)-1-(3-pyridyl)-1-butanone (NNK), and 4-(methylnitrosamino)-4-(3-pyridyl)butanal (NNA), as represented in Figure 9.2 (Jacob et al. 2017). It is also to be noted that these tobacco-specific TSNAs were not identified in the fresh tobacco smokes but in the THS (Hecht et al. 1978).

Hecht et al. (1978) reported the formation of nitrosamines like N'-nitrosonornicotine, 4-(N-methyl-N-nitrosamino)-l-(3-pyridyl)-l-butanone, and 4-(N-methyl-N-nitrosamino)-4-(3- pyridy1)butanal were by the intermediacy of cyclin iminium salts. The initial addition of NO^+ to the amine forms three iminium species. These intermediate species can be hydrolysed to amines and correspondingly nitrosated, or these intermediate species can directly react with NO_2 and form the above-mentioned nitrosamines. This is one the most common mechanisms that happens in neutral pH. Sleiman et al. (2010) detailed a similar reaction mechanism between the cotton textile and nitrosation of nicotine. The researchers proposed three

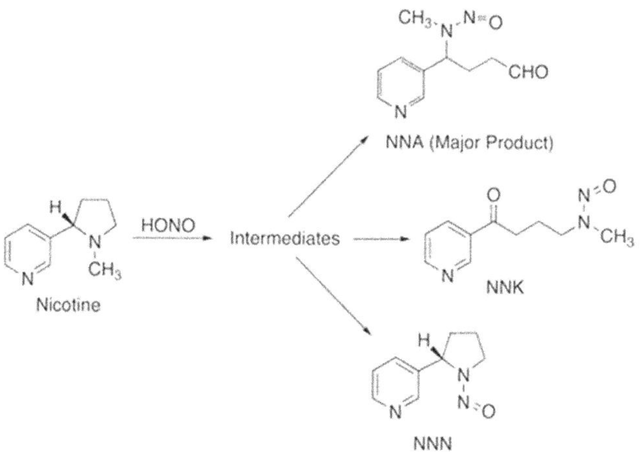

FIGURE 9.2 Formation of TSNAs from the reaction of nicotine and nitrous acid. (From Jacob et al. (2017). Reprinted under Creative Commons license.)

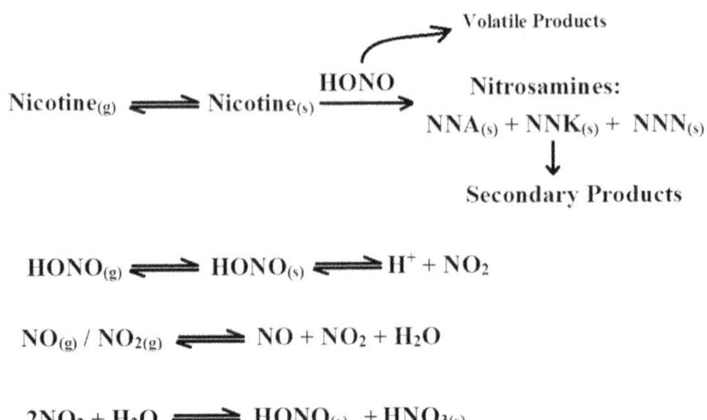

FIGURE 9.3 Physical-chemical processes involved in the formation of TSNAs on the surface of the cotton textiles. (See Sleiman et al. 2010.)

different pathways to convert nicotine into a different hazardous chemical like TSNA. NO^+ is identified as the main reactive species that removes one electron from nicotine and forms an unstable intermediate. The second NO^+ reacts with carbon atoms and forms iminium ions, and it forms TSNA with water and HONO. The researchers also noted some secondary products as a result of the above reaction in the gas phase (1-methyl-5-pyridin-3-yl-pyrrolidin-2-one (Cotinine), 4-(N-methyl-N-nitrosamino)-2-oxi-mino-1-(3- pyridyl)-1-butanone, methyl 3-pyridinecarboxylate (methyl nicotinate), N-methylnicotinamide and 1-methyl-5-(3 pyridinyl) pyrazole). The textile and nicotine interaction was reported in three different aspects, namely by direct adsorption of HONO, by heterogeneous disproportionation of NO_2, and also by surface-catalyzed reaction between NO and NO_2, as mentioned in Figure 9.3 (Sleiman et al. 2010).

9.4 POLYCYCLIC AROMATIC HYDROCARBONS (PAHS) FROM CIGARETTE SMOKE

Another important component of tobacco smoke is polycyclic aromatic hydrocarbons (PAHs). PAHs are responsible for creating carcinogenic condensates from cigarette smoke and are also known to be partially responsible for the tumorigenicity of the smoke (Wynder and Hoffmann 1967). Out of several PAHs present in cigarette smoke, benzo(a)pyrene was identified as a major component. A normal active smoker may consume approximately 1–5 microgram of benzo(a)pyrene from one pack of cigarettes, whereas a room highly polluted with cigarette smoke may release an amount of 22 ng per cubic meter of space (WHO, 1987; Menzie, Potocki, and Santodonato 1992). Tobacco smoke-exposed textiles like cotton, acetate, polyester, and silk were evaluated for their PAHs content, and their adsorption behaviours were analysed by Pahila et al. (2014). The results of the study showed that a higher amount of total naphthalene and alkyl naphthalenes was adsorbed on the polyester, and the lowest adsorption was noted with silk. When the benzo(a)pyrene concentration is considered, the acetate fiber showed higher adsorption than other selected textiles.

The research also reported a larger amount of lower molecular weight PAHs than the higher molecular weight PAHs in cotton, silk, acetate, and polyester. It was reported that the smoke odour was adsorbed the lowest in the silk fabric. The PAH adsorption on the silk was nearly half of the amount adsorbed in the polyester fabric. This is mainly due to polar amino acid chains in the chemical structure of the silk fiber, which has low affection towards the nonpolar PAHs. In the case of cotton fabric, the presence of more hydroxyl reactive sites on its surface forms a hydrogen bond with odourant molecules (Needles 1986). However, the case of synthetic textiles like acetate and polyester, due to their higher affinity towards nonpolar compoubnds, showed a higher amount of PAHs adsorption (Pahila et al. 2014).

9.5 HUMAN EXPOSURE OF THS FROM TEXTILES

As the effect of THS becomes a potential threat to nonsmokers, women, infants, and kids, the release behaviour of THS from textiles was evaluated by Hammer et al. (2011). In their study, cotton textile was analysed for its THS releasing behaviour as it is one of the most common textiles worldwide. The cotton fabric was exposed to cigarette smoke and incubated on a human adult's full-thickness skin with phosphate-buffered saline (PBS) and artificial sweat. The permeation study from the textile revealed that after 24 h incubation, all the samples showed a permeation of nicotine through the skin, but the diffusion percentage largely depended upon the type of solvent used. In the case of PBS, the nicotine permeation through the textile was noted to be very high (50%), more than the artificial sweat-incubated cotton fabric. In the artificial sweat-treated textile, around 54% of nicotine remained on the textile, and only 5.7% nicotine permeated through the skin model. The researchers also analysed the cytotoxicity of smoke-exposed textiles' sweat extracts on skin fibroblast. Based on the amount of exposure, all the smoke-exposed textiles showed cytotoxicity against the fibroblasts checked. Further, the results also reported a higher nicotine release from sweat extract-influenced zebrafish development. These findings showed a higher negative effect of THS residue from the textile samples on skin fibroblasts (Hammer et al. 2011).

Matt and co-workers analysed nicotine sorption of various indoor surfaces. Specifically, their analysis of infants' exposure to nicotine was significant. The concentration of continine, a bioindication for nicotine exposure, in urine samples of infants with smoking parents were noted to be higher than the nonsmoking parents. Infants whose parents smoked outside the home also showed an impact on this study. The urine samples indicated a lower concentration of continine, but still, the percentage was many times higher than the nonsmoking parents (Matt et al. 2004). The nicotine level in the urine sample was highly correlated to the nicotine level in the living room, bedroom, and dust on the surfaces. THS extracts on *in-vitro* studies showed a concentration-oriented response, based on the exposure time. The exposure did not kill the cells; however, it lead to damaged DNA, alteration of cytoskeleton, and inhibition of motility. This may not have an immediate effect on health but it will affect the health in due time. As the THS exposure affects the DNA, specific attention is needed or else it may lead to cancer, like cigarette smoke (Jacob et al. 2017). In a human exposure study conducted by Matt et al. (2011), it was noted that the house previously occupied by a smoker can expose a higher THS to the new nonsmoker occupant. The nicotine levels in the fingertip and urine samples were higher than the occupants who moved into a house that was previously occupied by a nonsmoker. The results were evident even after the houses were cleaned and kept idle for 2 months after the smoker moved on. The results were also true in the case of the hotel environment. The results of a nonsmoker staying in a hotel with no smoking ban showed a higher level of nicotine content in the fingertips and urine than the occupant who stayed in a hotel room with a complete smoke ban. A significant level of nicotine increase was noted with the participants who stayed in the most polluted hotel rooms than the smoking-banned hotels (Matt et al. 2014). The other long-term human exposure study results showed that even after the smoker quit smoking, the nonsmoking cohabitant showed a significantly higher level of cotinine and NNAL for up to 6 months. At this time, no additional tobacco was used in the habitant's home, and no active smoking person cohabitated. Though the results showed a significant reduction in cotinine and NNAL content in the first week after quitting smoking (compared to the time when an active smoker habituated), a higher cotinine and NNAL level persisted for 6 months (Matt et al. 2016). These findings showed strong evidence for the long life of THS over SHS and its importance to human health.

The *in-vivo* animal study was performed with mice that were exposed to THS textile materials like curtains and floormats. The exposed mice were analysed *in-vivo* after an exposure duration of 6 months. The results showed that the THS-exposed mice showed alcoholic fatty liver disease, a condition where prolonged exposure causes fibrosis, cirrhosis, and cancer (Martins-Green et al. 2014). THS exposure to the lungs leads to cellular infiltration due to the pro-inflammatory cytokines, and the walls of alveoli are disrupted significantly more often than control samples. THS-exposed mice showed hyperactive behaviour and impaired healing. The changes in the metabolism of the liver were one of the main data points that has a direct implication on humans. The researcher compared the results of the animal study with humans and reported that humans, specifically children in the smoking or SHS or THS environment, will have a significant health risk in terms of both long and short period (Martins-Green et al. 2014). Figure 9.4 represents the potential health impact of THS on the mouse model.

Among the human population, infants and children are most vulnerable to getting exposed and affected by THS (Matt et al. 2011; Matt et al. 2004; Winickoff et al. 2009). Winickoff et al. (2009) mentioned that around 88 million nonsmokers in the United States, above 3 years old, who live with a smoker in the home, have a higher amount of nicotine metabolite and TSNAs in their urine. Infants and young children are most affected by THS. Children and infants crawl and do frequent hand-to-mouth activities. This is identified as one of the major reasons for their higher exposure to THS. They touch, crawl, and move all over the house, more than an adult, which ultimately exposes them to furniture, home textiles, and other surfaces, including dust. This kind of exposure results in higher oral and dermal inhalation of THS (WHO Children's Health and the Environment 2008; US-EPA 2008). In addition to that, the following factors also increase the risk of THS for children and infants,

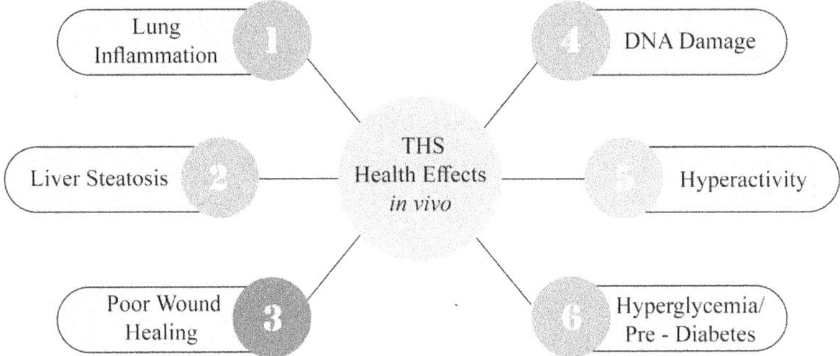

FIGURE 9.4 Potential health effects of THS exposure using a mouse model of exposure that mimics exposure of humans in their homes. (From Jacob et al. (2017). Reprinted under Creative Commons license.)

- Their skin is thinner than adult skin; hence, a smaller exposure leads to higher inhalation.
- Children mostly play on the floor, which leads to exposure to surfaces and dust. Further, their short stature exposes them to more dust than adults.
- Infants and children breathe more air than adults; this makes them inhale more THS pollutants.
- The larger surface-area-to-body-weight ratio was also reported as one of the reasons for higher THS exposure for children than adults.
- The internal body parts and organs are in a premature state and developing rapidly. Hence, the detoxifying effects of pollutants are different than for adults. These reasons are strongly reported in the literature for the higher vulnerability of children to THS (WHO Summary of Principles for Evaluating Health Risks in Children Associated with Exposure to Chemicals 2011; Matt et al. 2004, 2011; Jacob et al. 2017).

9.6 CONTROL OF THS EXPOSURE

As THS smoke exposure occurs as a result of reemitted smokes from all the absorbents ranging from furniture, floors, along with home textile materials, no preventive measures were reported from the textile side. As reported in the previous section, limited studies reported the possibility of controlling the emission of THS from textiles (Setthayanond et al. 2017). Frequent laundering of home textiles and clothing, mopping of hard surfaces, and frequently scrubbing walls and ceilings will reduce the deposits of THS in an indoor environment (Kristeen Cherney 2018). Quitting smoking in an indoor environment was often referred to as the only potential solution to completely protect humans from exposure to THS (American Academy of Pediatrics 2017; Drehmer et al. 2017). Studies frequently reported that along with a smoking ban, proper knowledge spreading among the public and patients will help to control THS in public (Drehmer et al. 2017). Northrup et al. (2016) reported the urgency of policy updates and educating the public regarding the THS issue due to its relatively longer persistence and ineffective means of complete cleaning. A set of documents and awareness pamphlets were developed and made publically available through a program called Clinical Effort Against Secondhand Smoke Exposure (CEASE). They suggest that these can be made available in child care centers, in schools, by landlords, and by management of different companies to educate the labourers/public (Drehmer et al. 2017). Several policies implemented by several states in the United States to control THS and SHS were consolidated by Lisa et al. (2016). The necessity of implementing health care policies and strategies to avoid ETS should be developed and communicated with targeted groups who are exposed to THS frequently (Kuo and Rees 2019). Though several points were discussed, it seems there is no standard method or technology to reduce or control THS in public places and indoor environments. Often, frequent laundering was reported as an effective method in reducing THS, but no studies reported on the complete removal of THS from clothing. Several other medical researchers repeatedly insisted that complete elimination of THS can be attained only by avoiding or quitting smoking at locations.

9.7 LAUNDRY ODOUR AND TEXTILES

Laundering is performed on textiles to remove soiling from the textiles, but at the same time, the laundering is expected to remove unwanted odours generated in the textile materials. The use of bleach and higher temperature plays an important role in odour removal from textiles. Recent knowledge on energy consumption practices reduced the average washing temperature. Hence, the laundry odour problem also is rising among users. Munk et al. (2001) mentioned that the formation of odour in the laundered textile is mainly associated with the sweat residue in the textile and its associated bacterial metabolism. In comparison with the fresh sample and soiled samples, it is reported that the human secretion on soiled fabrics showed a higher odour after laundering. The laundry process was ineffective to completely remove the odour from cotton textiles. Results of the study measured ketones, specificcally aldehydes like octanal (E)-2-octenal, methional, (Z)-2-nonenal, (E,Z)-2,6-nonadienal, and (E,E)-2,4-decadienal as much contributed odour after laundering than the low molecular organics acids. Though organic acids from sebum and skin secretion were noted as important for odour formation after laundering, their roles were noted as less important with dilution factors (Munk et al. 2000). The wet and dust-like odour in the laundered cloth was mainly related to the formation of 4-methyl-3-hexenoic acid (4M3H), a fatty acid associated with the human sweat odour (Takeuchi et al. 2012). Though sweat residue present in laundered clothes is mentioned as one of the reasons for odour formation, the studies also reported other possible ways of contamination. The first is washing-related contamination through textile-to-textile interaction or textile-to-washing-machine interaction, which will cause a significant microbial load to the textile and create odour on the textiles (Dirk et al. 2019). Further, it is also a known fact that odour removal or generation in the textile is highly fiber type-dependent (McQueen et al. 2008, 2007; Rathinamoorthy et al. 2014). For instance, researchers proved a higher or effective removal of odour molecules from the cotton textile compared to the synthetic textiles (Abdul-Bari et al. 2018; McQueen et al. 2014). In contrast to the natural fibers, synthetic textiles showed higher odour retention abilities due to their hydrophobic structure and oleophilic or soil absorption nature. This is also reported as one of the major reasons for higher odour buildup and the development of malodour after laundering (Klepp et al. 2016). Washing machines are a potential source of contamination as they accumulated different pathogens from different wash loads via biofilm inside the drum, which can detach during the next wash and re-contaminate the fresh textile (Lucassen et al. 2014). The second is related to environmental exposure of textiles after the laundering process (Nagoh et al. 2005).

9.8 WASHING MACHINES ON LAUNDRY MALODOUR

Washing machines are identified as one of the major sources of odour formation in textiles after laundering. Poor washing process, lower temperature, and types of soiling in the textiles lead to the formation of biofilms on the surface of the washing machine. This biofilm acts as a reservoir of potential pathogens and re-contaminates the textiles in subsequent washes (Gattlen et al. 2010). The machine washing process has a significant effect on odour removal from textiles. However, the effectiveness of such action is mainly based on the laundry additives and process conditions. For example, the detergent type has a significant impact; a bleach-based detergent showed an effective odour removal property compared to nonbleach detergents. A similar effect was also noted with a liquid detergent compared to powder-based detergents (Morris et al. 1984). The water temperature, wash load quantity, and frequency of washing also showed a significant impact on the laundry malodour development. Researchers also reported the trend of using nonbleach detergents and lower or cold temperature washes as the main reason for the malodour formation (Herreweghen et al. 2020). When additives like fabric softeners are evaluated, the role in malodour development is very significant in the case of synthetic textiles and their blends, whereas in the case of cotton or natural fibers, they did not show any malodour after a softener used during washing (Laitala, Boks, and Klepp 2011).

A detailed study on the impact of washing machine laundry malodour formation was performed by Stapleton et al. (2013). Research works reported that the wash drum, rubber seal, product drawer, coin trap, and sump of the machine were noted as potential sources of biofilm formation. Brand-new machines and also used machines were found to contain a certain group of odour-forming bacteria. Sterilized cotton fabrics (terry towel and fleece) were washed using a cold cycle without detergent with local water. The fabrics were removed from the machine at different time intervals, and sensorial odour analysis and microbial analysis were performed. Table 9.3 represents the olfactory gradings of different samples recovered from different washing machines. The grading was done on a 10-point scale, where 0 represents *no odour* and 10 represents *extreme malodour.*

The odour analysis by subjects showed that the odour intensity of the fabric increased with the length of storage. In the case of different washing machines, a higher odour intensity was always noted with the used washing machine compared to the brand new machines (control). They noted a higher odour intensity in the fabrics

TABLE 9.3
Olfactory grading of fabrics washed in the different washing machines and at different times (Olfactory grading based on the average of three different panelists) (From Stapleton et al. (2013). Reprinted with Permission)

Hours	Fabric type and machine	Washing Machine 1	Washing Machine 2	Washing Machine 3	Washing Machine 4
0	Terry-towel	0.7	0.0	3.3	2.3
	Fleece	0.7	0.7	3.7	2.0
	Machine	0.0	1.0	6.7	5.0
4	Terry-towel	0.0	1.0	5.2	3.7
	Fleece	0.7	0.7	4.3	3.0
	Machine	1.3	1.7	7.0	5.5
24	Terry-towel	0.3	1.7	6.0	5.0
	Fleece	0.7	2.0	4.5	3.8
	Machine	2.0	2.7	8.2	6.0
28	Terry-towel	1.3	2.0	6.0	4.3
	Fleece	0.7	2.3	5.2	4.3
	Machine	2.7	2.5	8.3	6.7
48	Terry-towel	1.7	2.0	6.7	5.0
	Fleece	0.7	2.3	6.0	4.0
	Machine	3.0	2.8	8.7	7.0
52	Terry-towel	1.3	2.0	7.0	5.7
	Fleece	1.3	2.3	6.0	4.0
	Machine	3.0	2.8	9.0	7.5

laundered at a used machine at "0" storage time, meaning immediately after the washing. However, this was not the case in the control machines. When the bacterial isolation studies were performed, the results showed a wide variety of 62 different bacterial strains. Out of those, *Pseudomonas putida, Pseudomonas fluorescens, Pseudomonas aeruginosa, Microbacteriumoxydans, Brevundimonasdiminuta, Brevundimonasvesicularis, Ochrobactrumanthropii, Sphingobacteriumspiritvorum, Sphingomonaspaucimobilis, Stenotrophomonas maltophilia, Micrococcus luteus,* and *Paenibacillus sp.* were reported in the biofilm formed in the washing machine (Gattlen et al. 2010). When the washing machine types considered only *Pseudomonas oleovorans, S. maltophilia, S. spiritvorum, Rhodococcuserythropolis, Sphingomonas sp.*, and *Shewanellaputrefaciens* were found in control machines, and in the used machine, almost all bacterial strains were noted. On the isolation of fabric, the researchers also found a similar amount and type of bacterial from the washing machine drum and rubber seal. This confirms the transfer of bacteria from machine to cloth. On the headspace GC-MS analysis of fabric, irrespective of the machines used, all the samples showed some common peaks for dimethyl disulphide, 3-methyl- 1-butanol, and isobutyric acid, respectively. Other than this, the research also showed various odour-forming VOCs from the fabric extracts through chromatography analysis. The results confirmed that the environmental odour in the washing machine has a significant impact on laundry malodour. Further, a higher correlation between the bacterial count and malodour formation was obtained. This confirms the interrelationship between the bacterial growth in the machine and fabric with fabric malodour after laundering (Stapleton et al. 2013).

Callewaert et al. (2015) evaluated the transmission of microorganisms from soiled cloth to new materials. The research evaluated the influent, effluent water samples along with the textile samples. The result showed that the washed cotton samples still contained all the types of bacterial strains as the unwashed ones but with lower abundance. The effluent sample consisted of a mixture of microorganisms related to the skin secretions and also from the influent water. Concerning the unused cotton fabric, a higher abundance of all skin-related species like *Enhydrobacter, Acinetobacter, Corynebacterium, and Staphylococcus* sp., along with the *Pseudomonas sp.,* which is from the biofilm formed in the washing machine, was noted. These results confirm the effective transfer of microbial load from one fabric to a new fabric through the laundry process. Out of all the strains noted, a higher abundance of the *Moraxellaceae* family was noted in the cotton sample, which is dominated by *Enhydrobacter sp.* and *Acinetobacter sp.* Further, a higher amount of *Corynebacterium, Staphylococcus, Propionibacterium, and Micrococcus sp.* was also reported as a representative of skin-born microflora. The study reported that the washing machines mix skin bacteria and bacterial strains from textiles and washing machine contamination and spread it to different unsoiled clothing. The use of lower temperature, nonbleach detergent, and poor maintenance of the machine was also reported as the main reason. The use of more plastic components in the washing machines also aids the biofilm formation and subsequent contamination of washing machines.

This was further confirmed by other researchers, who identified *Klebsiella oxytoca* with newborn babies in a particular hospital. However, no such isolate was reported in the country earlier. Hence, by a complete surface analysis inside the hospital, they found that the *Klebsiella oxytoca* originated from the sinks, washing machine drum, and rubber seal. They reported that the washing machines could be a reservoir of potential bacterial species. After the study, the sources/washing machines were removed from the hospital, and professional laundry services were used for the laundry purpose. From that, the transmission of *Klebsiella oxytoca* was restricted. Though several studies reported the ineffectiveness of the laundry process in removing bacterial strains from contaminated textiles (Heudorf et al. 2017; Sasahara et al. 2011; Yoh et al. 2010), this is the first study that reported the transmission of bacterial strains from the washing machine to humans (Schmithausen et al. 2018). Next to the machine components, the stagnant water in the outlet tubing of the washing machine was also noted and reported as the main source of microorganism growth and subsequent odour formation. The study reported a range of VOCs in the water streams from the washing machine (Hammond 2013). Table 9.4 represents the various odourous VOCs (mainly from the human body) identified in textiles after the washing process and before washing, as consolidated and reported by Herreweghen et al. (2020).

TABLE 9.4

VOCs detected in textiles before and after washing after a wear trial (From Herreweghen et al. (2020). Reprinted under Creative Commons attribute Licence)

Volatile Group	Odorous Volatiles	Involved Microorganisms	Analysis before and/or after washing, FD-value
Fatty acids	Ethanoic acid (acetic acid)	Propionibacterium sp., Staphylococcus sp.	After wearing
	Propanoic acid	Propionibacterium sp., Staphylococcus sp.	After wearing
	2-methylpropanoic acid (isobutyric acid) Butanoic acid	Bacillus subtilis	After wearing
	2-methylbutanoic acid	Bacillus subtilis	After wearing
	3-methylbutanoic acid	Bacillus subtilis	After wearing and after washing
	(isovaleric acid)	Staphylococcus epidermidis	After wearing and after washing
	3-methyl-2-hexenoic acid	Corynebacterium sp., Micrococcus sp.	After wearing
	4-methyl-3-hexenoic acid (4M3H)	Moraxella osloensis	After wearing
	5-methyl-4-hexenoic acid		After wearing
	3-methyl-3-hydroxy-hexanoic acid	Corynebacterium bovis, Corynebacterium jeijeikum, Corynebacterium striatum	After wearing
	6-heptenoic acid4-methyloctanoic acid		After washing
	4-ethyloctanoic acid		After washing; FD > 256
Steroid compounds	5-_-androstenol		After wearing
	5-_-androstenone		After wearing
	5-_-androst-2-en-17-one	Staphylococcus sp., Corynebacterium sp.	After wearing and after washing
	5-_-androst-16-ene-3-one		After wearing
Sulfur compounds	3-methyl-3-sulfanyl-hexan-1-ol(3M3SH)	Staphylococcushaemolyticus, Staphylococcus hominis	After wearing
	Dimethyl disulphides		After washing
	Dimethyl trisulphides		After washing
	Benzyl mercaptan	Corynebacterium sp.	After wearing
Ketones	1-hexen-3-one		After washing
	2-heptanone2-octanone		After wearing
	1-octen-3-one		After wearing and after washing
	2-nonanoneMedium-chain ketones (undetermined)		After wearing

TABLE 9.4 (Continued)
VOCs detected in textiles before and after washing after a wear trial (From Herreweghen et al. (2020). Reprinted under Creative Commons attribute Licence)

Volatile Group	Odorous Volatiles	Involved Microorganisms	Analysis before and/or after washing, FD-value
Esters	Ethyl-2-methylpropanoate		After washing; FD > 256
	Ethyl butanoate		After wearing and after washing; FD > 256
	Methyl-3-methyl-hexanoa		After wearing
	Methyl laurate		After wearing
	Methyl myristate		After wearing
	2-Aminoacetophenone		After wearing; FD > 256
	Diethyl phthalate		After wearing
Aldehydes	Methional		After washing
	Hexanal		After washing
	(Z)-4-heptenalOctanal		After washing; FD > 256
	(E)-2-octenal		After washing
	Cis/trans-2-nonenal(E,Z)-2,4-nonadienal](E,Z)-2,6-nonadienal		After wearing and after washing; FD > 256
	Decanal, (E,E)-2,4-decadienal		After washing
	(E)-4,5-epoxy-E-2-decenal4-methoxybenzaldehyde		After wearing; FD > 256
	Medium-chain aldehydes (not determined)		After washing; FD > 256
Alcohols	Oct-1-en-3-ol		After washing
	2-Nonanol		After wearing
	1-Decanol		After washing
	1-Dodecanol		After washing; FD > 256
	2-Phenylethanol		After wearing and after washing; FD > 256
	2-Methoxyphenol (guaiacol)		After wearing and after washing
Others	Naphthalene		After wearing

9.9 LAUNDRY ODOUR FROM DRYING/ENVIRONMENT

The laundry process removes microorganisms incompletely based on the type of laundry cycle (time, temperature, and speed), type of detergent, and other laundry aids used in the process. A study that analysed bed linens and hospital linen reported that the complete removal of microorganisms in industrial laundry can be achieved only at a longer duration (more than 13 min) and higher washing temperature (60°C) (Walter and Schillinger 1975). While evaluating the effectiveness of the US-based and Europe-based detergent, Munk et al. (2001) reported that detergent concentration, washing temperature, and composition of the detergent, such as bleach content, greatly influence bacterial survival on cloth after laundering. US-based detergent showed higher survival of gram-negative bacterial strains in the cloth, even after laundering. In a subsequent study, they evaluated the survival of the gram-negative strains on drying using tumble dry. The results showed that the drying process carried out at a normal time (15 min, 42°C) did not kill the gram-negative strains. Only extended drying temperature and time (60 min, 53°C) was able to kill the bacterial strain. This is one of the main reasons for after laundry malodour generation in the textile. Upon the analysis of cotton and polyester textiles, researchers extracted the odourous substance on the textile that is developed as a part of identified microorganisms. Hence, they concluded, when the odour builds upon every washing, finally the clothing started to generate malodour to the level of unpleasantness (Munk et al. 2001).

A similar study performed by Wiksell et al. (1973) analysed the survival of *Escherichia coli*, T3 bacteriophage, *Staphylococcus aureus,* and *Bacillus stearothermophilus* spores on polyester/cotton blended fabric after the washing and drying process. They used both lower and higher temperatures in both the washing and drying processes. The results of the study reported that wash temperature and type of cycle played an important role in microbial survival and odour formation. The study also proved that the washing process removed a large number of microorganisms (gram-positive) via washing machine water that cross contaminates the fresh clothing. The drying process significantly reduced the bacterial count on textiles; however, the study reported two possibilities for the reduction. Either the death of microorganisms or improved adhesion after drying might be the reason for the lower recovery of strains after drying. After drying, out of the selected species, only *Bacillus stearothermophilus* survived on the textile. In situations where the fabric was dried at a lower temperature or in an indoor situation, the cloth was still wet and developed a dirty, dust-like malodour. This odour is unpleasant, and several researchers reported such odours on the cloth after washing and drying (Hiromi Kubota et al. 2012; Nagoh et al. 2005). The development of 4-methyl-3-hexenoic (4M3H) acid in the laundry was reported as one of the major odour-causing components in the laundered cloth. Though the amount reported is small, it contributes to the overall malodour (Takeuchi et al. 2012). An analysis was performed by Kubota et al. (2012) using random samples (bath and hand towels, T-shirts) collected from Japan. The indoor dried samples were incubated and isolated for the existing bacterial load. Based on the 4M3H odour formation ability, only mild and heavy malodour-developing bacterial strains were taken forward for the analysis. Out of all the bacterial strains found, *Moraxella osloensis* was noted in large quantities (2×10^2 CFU/Cm^2) in the laundered fabric even after laundry and drying. This was the first time *Moraxella osloensis* was reported in clothing after washing and drying and as a potential species to produce malodour. On an *in-vitro* analysis, the odour formation characteristics of *Moraxella osloensis* were confirmed by the researchers. They reported that the presence of dirt, sebum, and sweat components are necessary for *Moraxella osloensis* to form malodour; however, no such mechanisms have been proposed so far. From the literature, it can be understood that 4M3H can be produced from linalool, which is present in laundered clothing and is produced from dirt in the clothing by other strains. Hence, the study mentioned that there could be an alternative path of odour formation, but *Moraxella osloensis* is one of the main reasons for malodour generation in clothing after washing and drying (Hiromi Kubota et al. 2012).

Nagoh et al. (2005) evaluated the odourous compounds in indoor dried textiles using solid phase microextraction and solvent extraction method. In this research, one set of samples was dried indoors with 90% RH, and another set was dried at 50% RH in the outdoor environment. The samples were analysed using GC-MS and noted that indoor (90% RH) dried samples were mostly contaminated with medium-chain aldehydes, medium-chain alcohols, and ketones. On the same sample, low levels of short-chain amines (represents the N content) and also some sulfide components were reported as odour causing elements. The presence of a broad peak in GC-MS analysis also confirmed the fatty acids as a source of malodour development in the indoor hung fabrics. The further fraction showed the existence of isovaleric acid, medium-chain fatty acids, octanoic acids, methyl-octanoic acid, and ethyl-octanic acid as a source of malodour in the indoor hung fabrics. In the case of the outdoor (50% RH) hung sample, very minimal peaks were noted in the GC-MS analysis compared to the indoor hung material. The research reported that the malodour generated in the textile is highly associated with the drying time and humidity of the environment. The second most important aspect of malodour generation in the fabric is residual soils in the cloths. When the malodour analysis was performed with used and brand-new unused cloths, even at higher relative humidity no malodour formation was noted with new material. This finding confirms the effect of soils in malodour generation after washing and drying. Considering this study, the main soils with the textile are from body secretions. Several other researchers also detailed the role of these components on odour formation of textiles (McQueen et al. 2008; Rathinamoorthy et al. 2014). These findings were in line with the results of McQueen et al. (2014), who anlaysed the effect of repeated laundering on odour build-up on knitted textiles. In their research, they used cotton and polyester T-shirts with 5% elastane content for the analysis. The worn samples after the wear trail experiment were analysed for their odour character after the laundering process. A higher odour intensity was noted with polyester fabric after washing, followed by cotton. The chemical component analysis reported more carboxylic acid in both fabrics after laundering; however, a higher amount of soiling was noted with polyester textiles. The research results confirmed that though the laundry process removed few soils, based on the types and laundry parameters, the presence of soils is possible, even after washing. They also reported that upon repeated use and washing the remaining soil slowly builds up on textiles and becomes the reason for laundry odour.

The fresh laundry odour in the laundered fabric is a combined action of clean laundry, the detergent type used, and the number of microorganisms in the fabric. The type of drying process also had great importance on the fabric smell and feel (Morris et al. 1984). To analyse the effect of outdoor drying on laundry malodour, researchers experimented with laboratory-washed fabric by drying it outdoors in sun, outdoors in shade, and indoors. The results of the study indicated that when the samples dried in sunlight, the formation of fresh smell in the laundered clothes was evident compared to the other indoor dried samples. The outdoor dried samples showed higher peak intensities of carboxylic acids, aldehydes, and ketones those components specifically have a pleasant smell. The room-dried sample and shade-dried one did not show any smell compared to the sun-dried one. The potential mechanism reported was the ozonolysis of alkenes present in the clothing. When the sunlight is exposed, the alkenes react with ozone and convert it into carbonyl and a Criegee intermediate. These intermediates react with water on the surface of the wet towel and degrade by producing hydroxy-alkyl hydroperoxides (Sheps et al. 2017). This reaction produces a most stable product on the wet clothing, which was further enhanced by the larger surface area of the towel. This Criegee intermediate then reacts with the cellulose network in the cotton clothing and forms ethers, which are stably bound on the surface of the clothing for a longer period. The results showed that the surface water in the wet cloth played the main role in the interaction of outdoor VOCs. Though the reaction mechanism between the cloth and VOCs was not clear whether they are of photochemistry or oxidisation by ozone, the exposure to the sun enhances the presence of aldehydes in the sunlight-exposed fabric, which produces a fresh washed smell on the clothes. The precursor for the oxidisation must be present in the fabric or they must be adhering to the clothing during the wet stage before the drying (Silvia Pugliese et al. 2020). The overall mechanism of odour formation after the laundry is provided in Figure 9.5.

In summary, it can be understood that the laundry odour is mainly associated with the soiling level in the cloth and also the laundry parameters. Mainly, the improper cleaning of body odour-related soils remains in cloth and is bio-transformed as odourous substances by microbes. The washing machine and drying method and temperature also play a vital role in laundry odour formation. Literature suggests a higher temperature washing with bleach-containing detergent for odour-free laundry. The washing machine was noted as one of the sources for microbial transfer from laundry to clothing. It performs inter-textile microbial transfer and also loads the textile with microbial load obtained from influent water and washing machine surfaces. Hence, an effective cleaning of the washing machine can reduce laundry odour. In the case of the drying method, machine dried or line dried, environmental humidity has a significant impact on odour formation. The sunlight-dried textiles produced a fresh laundry smell compared to shade- and indoor-dried textiles. Hence, it is necessary to dry in sunlight for improved laundry fragrance.

9.10 OTHER ODOURS IN TEXTILES

9.10.1 Food Odour

The most common contamination of clothing materials is food-related household odours. Mild and pleasant odours like coffee and fragrance from various fruits are mostly accepted by humans as they cause a relaxing and soothing feeling. In contrast, strong, intense vegetable odours (garlic, onion etc), meat odours (fish), and other food odours will adhere on the surface of the textile and create unpleasant odour to the

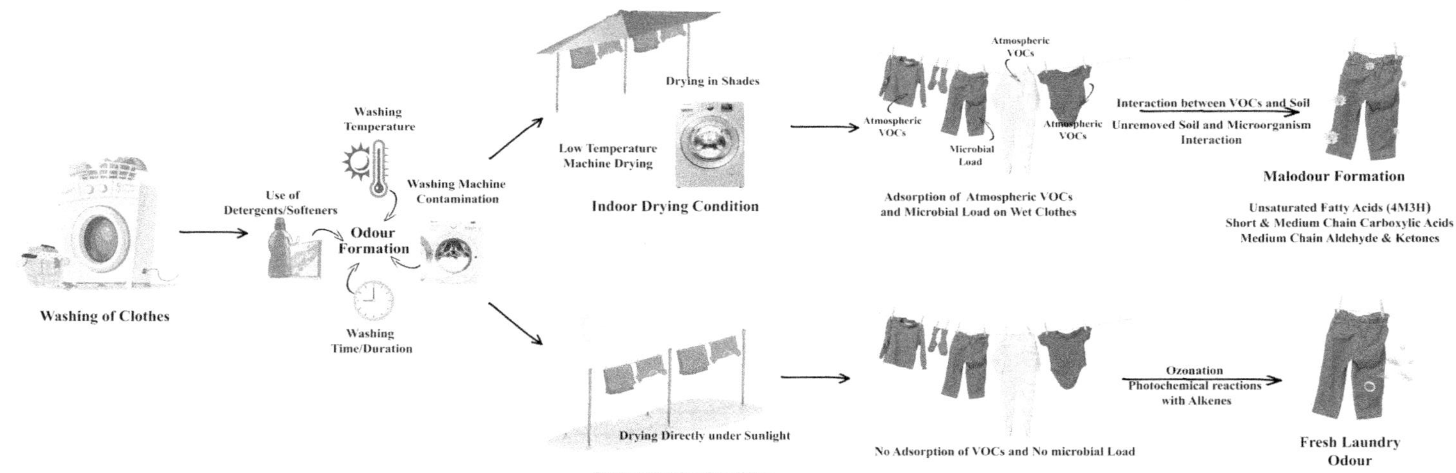

FIGURE 9.5 Mechanism of laundry odour formation.

wearer and environment, even days after the exposure. Odours developed from food stuff differ in their characteristics, which are widely based on the method of cooking, raw materials used for the food, and other add-ons. Hence, it is very difficult to narrow down the entire range of food smells. Complete detail of various foods (around 450 types) prepared in the world and their various odourous VOCs were reported by Boelens (2000). Various types of general (indoor and outdoor) odours that usually clings to home textiles and apparel are provided in Figure 9.6.

A study analysed the adsorbent characteristics of chafing dish soup, a type of hotpot in China, on textile fabrics like cotton, wool, and polyester. The soup is generally made of meat, vegetables, and spice like pepper, cinnamon, onions, ginger, garlic, and fennel. The higher volatile organic substance makes the odour of the dish easy to cling to the cloths. Due to poor adsorption in the preliminary study, cotton and polyester samples were excluded from the study, and woolen material was utilized. The results were evaluated after exposing the fabric using a box and a tent. The results showed a significant increment in the odour adsorption of fabric with an increment in the spice content. Further, the adsorption device (environment like closed box or tent) and size of the sample exposed shows a higher relation with odour intensity. The steam distillation and GC-MS analysis of exposed textile showed around 99 major volatile compounds, out of which the most abundant aroma components were 4-isopropyl-benzaldehyde, di-isobutyl phthalate, cedrol, lauryl alcohol, and diphenylamine. The analysis confirmed the higher adsorption characteristics of woollen textiles with cooking or food aromas (Mi Li et al. 2018).

9.10.2 Fish Odour

Aquatic life forms generally contain amino acids and amines, and they will degrade or break down into odourous trimethylamine. This is noted as the main reason for

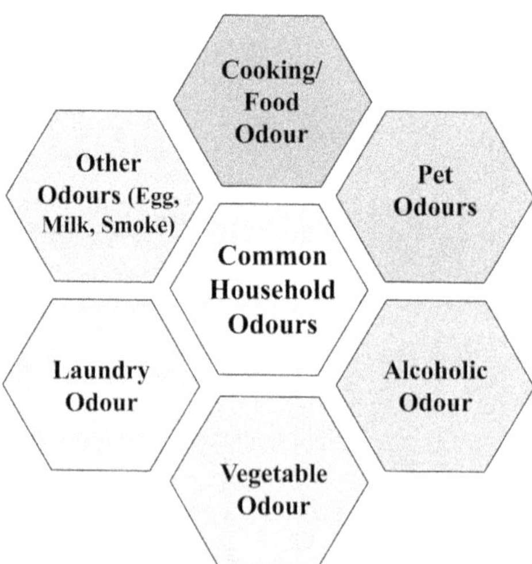

FIGURE 9.6 Common odour sources in indoor and outdoor environments.

the fishy smell in the kitchen or food (if improperly cleaned) sometimes (Anon 2020; PhattaraBoraphech and PaitipThiravetyan, 2015). Fish odour is mainly comprised of different VOCs, based on the species type. Aldehydes are the major type of VOC identified in salmon (Refsgaard et al. 1999), and in the case of whitefish, it is reported that the alcohols and aldehydes are major components. In the case of catfish, along with aldehydes, geosmin and 2-methyl isoborneol (1,6,7,7-tetra methyl bicycloheptan-6-ol) were reported. However, a higher concentration and abundance were noted with the aldehyde VOCs in all the fish types (Hammond 2013). These VOCs are very common in body odour and resin-finished textile (with lower abundance) and create a pungent odour in clothing (Michael 2007). Hence, exposure of cotton (natural) fibers to such an environment with a fishy smell will deposit a large number of VOCs on the surface of the textile and create a fishy odour from apparel. No direct studies reported on the control of fish odour (from kitchen or market or seafood) or interaction mechanism separately. However, a few studies used the fish as an odourant material and developed odour adsorbent material to control the release of the odour (Tülay Gülümser 2017). Other researchers used the fish odour in the cotton fabric as a strong odour and evaluated the effectiveness of the VOC extraction through the head phase of GC-MS methods (Xiaoning, Hao and Qing 2011). Based on analysis of food waste that contained fish, cabbage, rice, potatos, and apples, researchers reported more sulfur-containing components like hydrogen sulfide, and dimethyl disulfide. The presence of carboxylic acids like ethanoic acid, propanoic acid, butanoic acid, 2-methyl propanoic acid, pentanoic acid, 3-methyl butanoic acid, and hexanoic acid was also reported from the food waste. This fish sample alone contained amines as a source of odour. The findings revealed domestic food wastes mainly consist of carboxylic acids and sulfur as a source of the odour (Kamiya and Ose 1984).

9.10.3 Vegetable Odour

Vegetable odour is also one of the types of household odours. Out of several vegetables used in the kitchen, onion and garlic are the most specific items as far as odour is concerned. Commonly perceived, onion odour is mainly associated with bodily secretion and its bacterial biotransformation. The process develops sulfur-containing thioalcohols on the surface of sweaty clothing and damp towels and creates malodour. However, specific onion odour development depends on the skin microflora of the individual (Luis Villazon 2021). In general, onion contains hydrogen sulfide, n-Propyl mercaptan, methyl disulfide, Isopropyl alcohol, n-Propyl disulfide, methyl trisulfide, trisulfide acetaldehyde, acetone, and methyl ethyl ketone as its major components (Carson and Wong 1961). The interaction mechanism of cotton and onion-based aldehydes is detailed based on its polarity. The cotton fabric sorbs more aldehydes than synthetic textiles like polyester due to its higher polarity. Polyester attracts more aromatic odours and fatty acids (Prada, Curran, and Furton 2011). In a study conducted to analyse the odour masking effect of activated carbon from different sources, onion odour was used to represent body odour. Cotton and polyester fabrics were exposed to onion odour before and after activated carbon treatment (as print). The odour sorption on the fabric was measured using human olfactometry and also using

E-nose. The results showed higher sorption of odour in the cotton fabric than the polyester for the above-mentioned reasons (Eza et al. 2014).

Similar to onion, garlic is another food item that has a distinct smell and is most of the time noted as unpleasant on different occasions. Though consumption of garlic is effective in improving immunity, it also possesses some negative effects, like unpleasant breath and body odour after consumption (Amagase et al. 2001). To be specific, garlic consumption significantly affects the axilla odour of humans (Fialova, Roberts, and Havlicek, 2016). Garlic mainly produces allicin when it is crushed, and it is responsible for the typical odour of garlic. Shukla and Kalra (2007) reported that the main VOCs created by allicin were sulfur compounds. Due to its unstable nature, allicin transforms into sulfides (mono-, di– and tri-), ajoene, and vinyldithiines. Among the different VOCs, sulfide components induce a strong and onion-like powerful odour through surface deposits. Often these components (dimethyl disulfide, dimethyl trisulfide) are detected in the axilla-worn textile as a major source of body odour (Chamila, Ian, and John 2016). Similar to onion odour, garlic odour can also be removed from the textile with a suitable laundering process. However, no detailed studies exist in the textile discipline for directly representing the interaction mechanism of these components.

9.10.4 Pet Odour

Dog and cat odours are the general pet odours that can be found in household objects, surfaces, fabrics, and clothing. Like humans, animals also secrete bodily fluids like sweat, and it creates a distinct odour. Animals use these odours to identify other animals and humans, and also to scent mark their territory. Like humans, dogs also have sweat glands in places where they don't have hair. However, a major odour arises from the dog when it is wet. The higher intensity of the odour of wet hair is due to the evaporation of VOCs from dog skin along with the water molecules. Coats made of dog hair released 16 different odours at the wet stage; the odours are mostly strong sulfur (fecal) odours like dimethyl trisulfide and phenol, p-cresol, mushroom (1-octen-3-ol), fruity (2-nonanone), floral (â-damascenone), and earthy smelling (2,3-diethyl-5-methylpyrazine) odours and branched or complex aldehydes (Young et al. 2002). The addition of water content increased the dog hair odour multiple times than the dry situation, and this is generally referred to as a "wet dog" odour. Similarly, domestic cats also use their odour for scent marking and identification. The major components in the cat's urine are sulfur-containing amino acid and 2-amino-7-hydroxy-5,5-dimethyl-4-thiaheptanoic acid, which is known as felinine. These components slowly degrade into the volatile 3-mercapto-3-methylbutan-1-ol (MMB) (Miyazaki et al. 2006). Both cats and dogs generally had few VOCs like sulfur-based components and short-chain amino acids in common. The interaction of sulfur, aldehydes, carboxylic acids, and other aforementioned VOCs are already detailed in previous studies (McQueen et al. 2007; 2008; Munk et al. 2000; 2001; Callewaert et al. 2015; Rathinamoorthy and Thilagavathi 2016). Due to the simple and easy removal of these odours from (home) textiles, it is considered as less important than the body odour. Hence, though many studies reported on the fragrance and odour analysis of indoor or household environments, studies on textiles were found to be meagre.

9.11 SUMMARY

This chapter summarises the interaction mechanism of cigarette smoke as THS on textile material. The review shows that textile fabrics can hold the smoke and potential carcinogenic material for days and sometimes for months. Exposure to such hazardous chemicals causes serious health issues to infants and toddlers, even in the absence of active smokers. The findings of the research showed the importance of textiles and clothing in transferring the pollutants to humans. THS that directly comes in contact with skin can cause serious negative health issues for adults, also. In the case of laundry odour, low temperature and shorter cycle washing are the major causes for improper soil removal from textiles and cause laundry odour. Further, the washing machine could be an effective reservoir in microbial transmission from the previous wash or from contaminated cloth to other fresh fabrics. Similarly, the role of the drying method and environment (like environmental relative humidity and shade drying, machine drying, and sun-drying) was also noted as significant in odour formation. Drying the laundered clothes in the open sun causes ozonisation and photochemical reaction with cloth and creates a "fresh laundry" smell to cloth. Though other household odours can also cling to textiles, lower intensity of odour, their easy-to-remove nature, and lower health hazards make those odours less important than the previously discussed odours. However, future studies on the interaction between those household odours and textiles will help us develop new odour-resistant materials that do not cause health hazards to humans, even after exposure to environmental odours.

REFERENCES

Abdul-Bari, M.M., McQueen, R.H., Nguyen, H., Wismer, W.V., de la Mata, A.P., and Harynuk, J.J. 2018. Synthetic clothing and the problem with odor: Comparison of nylon and polyester fabrics. *Clothing and Textiles Research Journal*, 36: 251–266.

Amagase, H., Petesch, B.L., Matsuura, H., Kasuga, S., and Itakura, Y. 2001. Intake of garlic and its 460 bioactive components. *Journal of Nutrition*, 131(3): 955S–962S.

American Academy of Pediatrics. 2017. How parents can prevent exposure to thirdhand smoke. https://www.healthychildren.org/English/health-issues/conditions/tobacco/Pages/How-Parents-Can-Prevent-Exposure-Thirdhand-Smoke.aspx (Accessed on July, 2020).

Anon. 2020. How to eliminate fish smells. Accessed May 15, 2021. https://www.downthecove.com/food-drink/getting-rid-of-fish-smells/

Bahl, V., Jacob, P. III, Havel, C., Schick, S.F., and Talbot, P. 2014. Thirdhand cigarette smoke: Factors affecting exposure and remediation. *PLoS ONE*, 9(10): e108258. doi: 10.1371/journal.pone.0108258

Bendak, A., and El-Marsafi, S.M. 1991. Effects of chemical modifications on polyester fibers. *Journal of Islamic Academy of Sciences*, 4(4): 275–284.

Boelens, M. 2000. *BACIS VCF Database of Volatile Compounds in Food*. Leffingwell & Associates.

Brunnemann, K.D., and Hoffmann, D. 1991. Analytical studies on tobacco-specific N-nitrosamines in tobacco and tobacco smoke. *Critical Reviews in Toxicology*, 2(4): 235–240.

Callewaert, C., Van Nevel, S., Kerckhof, F.-M., Granitsiotis, M.S., and Boon, N. 2015. Bacterial exchange in household washing machines. *Frontiers in Microbiology*, 6: 1381. doi: 10.3389/fmicb.2015.01381

Cao, J., Liu, N., and Zhang, Y. 2017. SPME-based Ca-history method for measuring svoc diffusion coefficients in clothing material. *Environmental Science & Technology*, 51: 9137–9145.

Carson, J.F. and Wong, F.F. 1961. The volatile flavor components of onions. *Agricultural and Food Chemistry*, 6(2): 140–143.

Chamila J.D., Ian A.F., and John R.D. 2016. Source, impact and removal of malodour from soiled clothing. *Journal of Chromatography A*, 1438(2016): 216–225.

Cheng, C.-Y., Huang, S.-S., Yang, C.-M., Tang, K.-T., and Yao, D.-J. 2016. Detection of third-hand smoke on clothing fibers with a surface acoustic wave gas sensor. *Biomicrofluidics*, 10: 011907. doi: 10.1063/1.4939941

Chien, Y.C., Chang, C.P., and Liu, Z.Z. 2011. Volatile organics off-gassed among tobacco-exposed clothing fibers. *Journal of Hazardous Materials*, 193: 139–148.

Cieslak, M. 2006. New approach to environmental tobacco smoke exposure and its relation to reemission process. *International Journal of Occupational Medicine and Environmental Health*, 19: 1–7.

Daisey, J.M. (1999). Tracers for assessing exposure to environmental tobacco smoke: What are they tracing? *Environmental Health Perspectives*, 107: 319–327.

Dirk, P.B., Schages, J., and Rehberg, L. 2019. Laundry and textile hygiene in healthcare and beyond. *Microbial Cell*, 6(7): 299–306.

Drehmer, J.E., Walters, B.H., Nabi-Burza, E., and Winickoff, J.P. (2017). Guidance for the clinical management of thirdhand smoke exposure in the child health care setting. *Journal of Clinical Outcomes Management: JCOM*, 24(12): 551–559.

Elkilani, A.S., Baker, C.G.J., Al-Shammari, Q.H., and Bouhamra, W.S. 2003. Sorption of volatile organic compounds on typical carpet fibers. *Environment International*, 29: 575–585.

Eza, T.S.M., Wan Ahmad, W.Y., Ahmad, M.R., Omar K., and Ahmad, M.N. 2014. Effectiveness of activated carbon produced from coconut and oil palm shells as anti-odour on textile fabrics. *Indian Journal of Fibre & Textile Research*, 39: 190–195.

Fialova J., Roberts S.C., and Havlicek, J. 2016. Consumption of garlic affects hedonic perception of axillary body odour. *Appetite*, 97: 8–15.

Gattlen, J., Amberg, C., Zinn, M., and Mauclaire, L. 2010. Biofilms isolated from washing machines from three continents and their tolerance to a standard detergent. *Biofouling*, 26: 873–882.

Gómez A.E., Amedro D., Afif C., Gligorovski S., Schoemaecker C., Fittschen C., Doussin J.F., and Wortham H. 2013. Unexpectedly high indoor hydroxyl radical concentrations associated with nitrous acid. *Proceedings of the National Academy of Sciences of the United States*, 110: 13294–13299.

Hammer, T.R., Fischer, K., Mueller M., and Hoefer, D. 2011. Effects of cigarette smoke residues from textiles on fibroblasts, neurocytes and zebrafish embryos and nicotine permeation through human skin. *International Journal of Hygiene and Environmental Health*, 214: 384–391.

Hammond, C.J. 2013. Chemical composition of household malodours: An overview. *Flavour and Fragrance Journal*, 28: 251–261.

Hecht, S.S., Chen, C.H.B., Ornaf, R.M., Jacobs, E., Adams, J.D., and Hoffmann, D. 1978. Chemical studies on tobacco smoke. 52. Reaction of nicotine and sodium nitrite: Formation of nitrosamines and fragmentation of the pyrrolidine ring. *The Journal of Organic Chemistry*, 43: 72–76.

Hecht, S.S., Chen, C.-H.B., Ornaf, R.M., Jacobs, E., Adams, J.D., and Hoffmann, D. 1978. Chemical studies on tobacco smoke. 52. Reaction of nicotine and sodium nitrite: Formation of nitrosamines and fragmentation of the pyrrolidine ring. *The Journal of Organic Chemistry*, 43(1): 72–76. doi: 10.1021/jo00395a017

Herreweghen, F.V., Amberg, C., Marques, R., and Callewaert, C. 2020. Biological and chemical processes that lead to textile malodour development. *Microorganisms*, 8: 1709; doi: 10.3390/microorganisms8111709

Heudorf, U., Gasteyer, S., Müller, M., Serra, N., Westphal, T., Reinheimer, C., and Kempf, V. 2017. Handling of laundry in nursing homes in Frankfurt am Main, Germany, 2016 - laundry and professional clothing as potential pathways of bacterial transfer. *GMS Hyg Infect Control*, 12: Doc20.

Jacob, P. III, Benowitz, N.L., Destaillats, H., Gundel, L., Hang, B., Martins-Green, M., Matt, G.E., Quintana P.J.E., Samet, J.M., Schick, S.F., Talbot, P., Aquilina, N.J., Hovell, M.F., Mao, J.-H., and Whitehead, T.P. 2017. Thirdhand smoke: New evidence, challenges, and future directions. *Chemical Research in Toxicology* 30(1): 270–294.

Kubota, H., Mitani, A., Niwano, Y., Takeuchi, K., Tanaka, A., Yamaguchi, N., Kawamura, Y., and Hitomia, J. 2012. Moraxella species are primarily responsible for generating malodor in laundry. *Applied and Environmental Microbiology*, 78(9): 3317–3324.

Kamiya, A., and Ose, Y. 1984. Study of odorous compounds produced by putrefaction of foods: V. Fatty acids, sulphur compounds and amines. *Journal of Chromatography*, 292: 383.

Klepp, I.G., Buck, M., Laitala, K., and Kjeldsberg, M. 2016. What's the problem? Odor-control and the smell of sweat in sportswear. *Fashion Practices*, 8: 296–317.

Kristeen Cherney. 2018. Thirdhand smoke: What you should know. Accessed on July, 2020. https://www.healthline.com/health/thirdhand-smoke

Kuo, H.W., and Rees, V.W. 2019. Third-hand smoke (THS): What is it and what should we do about it?. *Journal of Formosan Medical association*, 118(11): 1478–1479.

Laitala, K., Boks, C., and Klepp, I.G. 2011. Potential for environmental improvements in laundering. *International Journal of Consumer Studies*, 35: 254–264.

Lisa, A., Kasey, F., Jacob, H., Michelle, S., and Jiangang, C. 2016. Third-hand smoke: Old smoke, new concerns. *Journal of Community Health*, 41(3): 680–687. doi: 10.1007/s10900-015-0114-1

Lucassen, R., Blümke, H., Born, L., Fritz, A., Geurtz, P., Hofmann, N., Hoffmann, L., Steiner, R., Merettig, N., and Bockmühl, D. 2014. The washing machine as a source of microbial contamination of domestic laundry - a case study. *Househ Pers Care Today*, 9(5): 54–57.

Martins-Green, M., Adhami, N., Frankos, M., Valdez, M., Goodwin, B., Lyubovitsky, J., Dhall, S., Garcia, M., Egiebor, I., Martinez, B., Green, H.W., Havel, C., Yu, L., Liles, S., Matt, G., Destaillats, H., Sleiman, M., Gundel, L.A., Benowitz, N., Jacob, P. 3rd, Hovell, M., Winickoff, J.P., and Curras-Collazo, M. 2014. Cigarette smoke toxins deposited on surfaces: Implications for human health. *PLoS One*, 9: e86391.

Matt, G.E., Quintana, P.J., Fortmann, A.L., Zakarian, J.M., Galaviz, V.E., Chatfield, D.A., Hoh, E., Hovell, M.F., and Winston, C. 2014. Thirdhand smoke and exposure in California hotels: Nonsmoking rooms fail to protect non-smoking hotel guests from tobacco smoke exposure. *Tobacco Control*, 23: 264–272.

Matt, G.E., Quintana, P.J.E., Hovell, M.F., Chatfield, D., Ma, D.S., Romero, R., and Uribe, A. 2008. Residual tobacco smoke pollution in used cars for sale: Air, dust, and surfaces. *Nicotine and Tobacco Research*, 10(9): 1467–1475.

Matt, G.E., Quintana, P., Zakarian, J., Hoh, E., Hovell, M.F., Mahabee-Gittens, M., and Chatfield, D. 2016. When smokers quit: Exposure to nicotine and carcinogen persists from thirdhand smoke population. *Tobacco Control*. In Press. doi: 10.1136/tobaccocontrol-2016-053119

Matt, G.E., Quintana, P.J., Zakarian, J.M., Fortmann, A.L., Chatfield, D.A., Hoh, E., Uribe, A.M., and Hovell, M.F. 2011. When smokers move out and non-smokers move in: Residential thirdhand smoke pollution and exposure. *Tobacco Control*, 20: e1.

Matt, G.E., Quintana, P.J.E., Hovell, M.F., Bernert, J.T., Song, S., Novianti, N., Juarez, T., Floro, J., Gehrman, C., Garcia, M., and Larson, S. 2004. Households contaminated by environmental tobacco smoke: Sources of infant exposures. *Tobacco Control*, 13: 29–37.

Matt, G.E., Quintana, P.J.E., Destaillats, H., Gundel, L.A., Sleiman, M., and Singer, B.C. 2011. Thirdhand tobacco smoke: Emerging evidence and arguments for a multidisciplinary research agenda. *Environmental Health Sciences*, 119: 1218–1226.

McQueen, R.H., Laing, R.M., Brooks, H.J. L., & Niven, B.E. 2007. Odor Intensity in apparel fabrics and the link with bacterial populations. Textile Research Journal, 77(7): 449–456.

McQueen, R.H., Laing, R.M., Delahunty, C.M., Brooks, H.J.L., and Niven, B.E. 2008. Retention of axillary odour on apparel fabrics. *Journal of Text I*, 99(6): 515–523.

McQueen, R.H., Harynuk, J.J., Wismer, W.V., and Keelan, M. 2014. Axillary odour build-up in knit fabrics following multiple use cycles. *International Journal of Clothing Science and Technology*, 26(4): 274–290. doi: 10.1108/IJCST-05-2013-0064

Menzie, C.A., Potocki, B.B., and Santodonato, J. 1992. Exposure to carcinogenic PAHs in the environment. *Environmental Science andTechnology*, 26(7): 1278–1284.

Mi Li, S.Y., Zhong, Q., Guo, Y., Cheng, Y., Yu, H., Qian, H., Yao, W., and Xie, Y. 2018. Evaluation of adsorption and desorption of chafing dish odor on woolen fabric, *IOP Conference Series: Materials Science and Engineering*, 392: 032005 doi: 10.1088/1757-899X/392/3/032005

Morris, M.A., Prato, H.H., White, N.L. 1984. Line-dried vs. machine-dried fabrics: Comparison of appearance, hand, and consumer acceptance. *Home Economics Research Journal*, 13: 27–35.

Miyazaki, M., Yamashita, T., Suzuki, Y., Saito, Y., Soeta, S., Taira, H., and Suzuki, A. 2006. A major urinary protein of the domestic cat regulates the production of felinine, a putative pheromone precursor. *Chemistry & Biology*, 13: 1071–1079. doi: 10.1016/j.chembiol.2006.08.013

Michael. 2007. Removing odors & smells from clothing. https://organicclothing.blogs.com/my_weblog/2007/02/removing_odors_.html

Munk, S., Johansen, C., Stahnke, L.H., and Adler-Nissen, J. 2001. Microbial survival and odor in laundry. *The Journal of Surfactants and Detergents*, 4(4): 385–394. doi: 10.1007/s11743-001-0192-2

Munk, S., Münch, P., Stahnke, L., Adler-Nissen, J., and Schieberle, P. 2000. Primary odorants of laundry soiled with sweat/sebum: Influence of lipase on the odor profile, *Journal of Surfactants and Detergents,* 3(4): 505–515. doi: 10.1007/s11743-000-0150-z

Nagoh, Y., Tobe, S., Watanabe, T., and Mukaiyama, T. 2005. Analysis of odorants produced from indoor drying laundries and effects of enzyme for preventing malodor generation. *Tenside Surfactants Detergents*, 42: 7–12.

Needles, H.L. 1986. *Textile Fibers, Dyes, Finishes, and Processes: A Concise Guide*. Noyes Publications Mill Road, 227.

Nilsen, T., and Nilsen, O.G. 1997. Accumulation of nicotine in human hair during long-term controlled exposure to a low concentration of nicotine vapour. *Pharmacology and Toxicology*, 81: 48–52.

Noble, R.E. 2000. Environmental tobacco smoke uptake by clothing fabrics. *Science of the Total Environment*, 262: 1–3.

Northrup, T.F., Jacob, P., 3rd, Benowitz, N.L., Hoh, E., Quintana, P.J., Hovell, M.F., Matt, G.E., and Stotts, A.L. 2016. Thirdhand smoke: State of the science and a call for policy expansion. *Public Health Reports (Washington, D.C.: 1974)*, 131(2): 233–238. doi: 10.1177/003335491613100206

Obendorf, S.K., Liu, H., Leonard, M.J., Young, T.J., and Incorvia, M.J. 2006. Effects of aroma chemical vapor pressure and fiber morphology on the retention of aroma chemicals on cotton and poly(ethylene terephthalate) fabrics. *Journal of Applied Polymer Science*, 99: 1720–1723.

Pahila, I.G., Servidad, L.C., Nillos, M.G.G., and Taberna, Jr. H.S. 2014. Sorption of polycyclic aromatic hydrocarbons from tobacco smoke on different fabric materials. *AES Bioflux*, 6(1): 17–25.

Phattara Boraphech and Paitip Thiravetyan. 2015. Trimethylamine (fishy odor) adsorption by biomaterials: Effect of fatty acids, alkanes, and aromatic compounds in waxes. *Journal of Hazardous Materials*, 2(284): 269–277. doi: 10.1016/j.jhazmat.2014.11.014.

Piade, J.J., D'andres, S., and Sanders, E.B. 1999. Sorption phenomena of nicotine and ethenylpyridinevapors on different material in a test chamber. *Environmental Science & Technology*, 33: 2046–2052.

Prada, P.A., Curran, A.M., and Furton, K.G. 2011. The evaluation of human hand odor volatiles on various textiles: A comparison between contact and noncontact sampling methods. *Journal of Forensic Sciences*, 56: 866–881.

Pugliese, S., Jespersen, M.F., Pernov, J.B., Shenolikar, J., Nygaard, J., Nielsen, C.O.J., and Johnson, M.S. 2020. Chemical analysis and origin of the smell of line-dried laundry. *Environmental Chemistry*, 17(5): 355–363. doi: 10.1071/EN19206.

Rathinamoorthy, R., and Thilagavathi, G. 2016. GC-MS analysis of worn textile for odour formation. *Fibers and polymers*, 17(6): 917–924. doi: 10.1007/s12221-016-5891-3.

Rathinamoorthy, R., Thilagavathi, G., Brindha, S., Gayathri, P., Poornakala, N.S., and Pradeep, B. 2014a. Odour control studies on apparel fabric finished with Methanol extract of *Terminalia chebula*. *Fibers and Polymers*, 15(8): 1669–1676.

Refsgaard, H.H.F., Haahr, A.M., and Jensen, B. 1999. Isolation and quantification of volatiles in fish by dynamic headspace sampling and mass spectrometry. *Journal of Agricultural and Food Chemistry*, 47: 1114.

Sasahara, T., Hayashi, S., Morisawa, Y., Sakihama, T., Yoshimura, A., and Hirai, Y. 2011. Bacillus cereus bacteremia outbreak due to contaminated hospital linens. *European Journal of Clinical Microbiology & Infectious Diseases*, 30:219–226.

Schmithausen, R., Exner, M., Rösing, C., Savin, M., Hack, S., Kaase, M., Gebel, J., Engelhart, S., and Exner, D. 2018. The washing machine as a reservoir for transmission of ESBL-producing Klebsiella oxytoca in newborns. unpublished. 1–28. doi: 10.1101/354613

Senthilkumar, L., Ghanty, T.K., Kolandaivel, P., and Ghosh, S.K. 2012. Hydrogenbonded complexes of nicotine with simple alcohols. *International Journal of Quantum Chemistry*, 112: 2787–2793. doi: 10.1002/qua.23304

Setthayanond, J., Sodsangchan, C., and Suwanruji, P. 2017. Influence of MCT-β-cyclodextrin treatment on strength, reactive dyeing and third-hand cigarette smoke odor release properties of cotton fabric. *Cellulose*, 24: 5233–5250.

Sheps, L., Rotavera, B., Eskola, A.J., Osborn, D.L., Taatjes, C.A., Au, K., Shallcross, D.E., Khan, M.A.H., and Percival, C.J. 2017. The reaction of Criegee intermediate CH2OOwith water dimer: Primary products and atmospheric impact. *Physical Chemistry Chemical Physics*, 19: 21970–21979. doi: 10.1039/C7CP03265J

Shukla, Y., and Kalra, N. 2007. Cancer chemoprevention with garlic and its constituents. *Cancer Letters*, 247(2): 167–181. doi: 10.1016/j.canlet.2006.05.009

Sleiman, M., Destaillats, H., and Gundel, L.A. 2013. Solid-phase supported profluorescent nitroxide probe for the determination of aerosol-borne reactive oxygen species. *Talanta*, 116: 1033–1039.

Sleiman, M., Gundel, L.A., Pankow, J.F., Jacob, 3rd, P., Singer, B.C., and Destaillats, H. 2010. Formation of carcinogens indoors by surface-mediated reactions of nicotine with nitrous acid, leading to potential thirdhand smoke hazards. *Proceedings of the National Academy of Sciences*, 107(15): 6576–6581.

Stapleton, K., Hill, K., Day, K., Perry, J.D., and Dean, J.R. 2013. The potential impact of washing machines on laundry malodour generation. *Letters in Applied Microbiology*, 56: 299–306.

Takeuchi, K., Hasegawa, Y., Ishida, H., and Kashiwagi, M. 2012. Identification of novel malodour compounds in laundry. *Flavour and Fragrance Journal*, 27(1): 89–94. doi: 10.1002/ffj.2088

Talbot, L., and Palmer, J. 2013. Effects of smoking on health and anaesthesia. *Anaesthesia and Intensive Care*, 14: 107–109.

Tülay Gülümser. 2017. The role of microcapsules in masking bad odors of cotton fabrics. *IndustriaTextila*, 68(1): 275. doi: 10.35530/IT.068.04.1289

U.S. EPA. 1992. United States Environmental Protection Agency, Office of Research and Development, Office of Health and Environmental Assessment, Respiratory health effects of passive smoking. Washington, DC, EPA/600/6-90/006F, 525 pp.

Ueta, Y., Saito, Y., Teraoka, K., Miura, T., and Jinno, K. 2010. Determination of volatile organic compounds for a systematic evaluation of third-hand smoking. *Analytical Sciences*, 26: 569–574.

US-EPA. 2008. *Child-Specific Exposure Factors Handbook (Final Report)*. EPA/600/R-06/096F, U.S. Environmental Protection Agency.

Villazon, L. 2021. Why do I smell of onions after I've been working out?. Science Focus Magazine. https://www.sciencefocus.com/the-human-body/why-do-i-smell-of-onions-after-ive-been-working-out/

Walter, W.G., and Schillinger J.E. 1975. Bacterial survival in laundered fabrics. *Applied Microbiology*, 29: 368–373.

Wang, Y., Zhou, S.L., Xia, J.R., Xue, J., Xu, J.H., and Zhu, J.H. 2004. Trapping and degradation of volatile nitrosamines on cyclodextrin and zeolites. *Microporous Mesoporous Mater*, 75(3): 247–254.

WHO Children's Health and the Environment. Available at http://www.who.int/ceh/capacity/Children_are_not_little_adults.pdf (July 2008).

WHO Summary of Principles for Evaluating Health Risks in Children Associated with Exposure to Chemicals. 2011. Available at http://www.who.int/ceh/publications/health_risks_exposure_chemicals/en/

WHO. 1987. Polynuclear aromatic hydrocarbons (PAH). In: *Air Quality Guidelines for Europe. Copenhagen* (pp. 105-117). World Health Organization Regional Office for Europe. https://apps.who.int/iris/handle/10665/107364.

Wiksell, J.C., Pickett, M.S., and Hartman, P.A. 1973. Survival of microorganisms in laundered polyester-cotton sheeting. *Applied Microbiology*, 25: 431–435.

Winickoff, J.P., Friebely, J., Tanski, S.E., Sherrod, C., Matt, G.E., Hovell, M.F., and McMillen, R.C. 2009. Beliefs about the health effects of "thirdhand" smoke and home smoking bans. *Pediatrics*, 123: e74–e79.

Wu, C.-C., Wang, W.-J., Bao, L.-J., Shi, L., Tao, S., and Zeng, E.Y. 2019. Impacts of texture properties and airborne particles on accumulation of tobacco-derived chemicals in fabrics. *Journal of Hazardous Materials*, 369, 108–115.

Wynder, E.L., and Hoffmann, D. 1967. *Tobacco and Tobacco Smoke: Studies in Experimental Carcinogenesis*. Academic Press, 730.

Xiaoning, W., Hao, W., and Qing, L. 2011. Determination of cotton fabric fish odor by HS-GC-MS. *Journal of Textile Research*, 32(2): 68–72.

Young, L, Pollien, P, Roberts, D., and Cline, J. 2002. Compounds responsible for the odor of dog hair coat, WSAVA 2002 Congress. https://www.vin.com/doc/?id=3846472

Yoh, M., Matsuyama, J., Shime, A., Okayama, K., Sakamoto, R., and Honda, T. 2010. Can industrial laundry remove Bacillus cereus from hospital linen?. *KansenshogakuZasshi*, 84: 583–587.

10 Textile Cleaning and Odour Removal

Kirsi Laitala, Ingun Grimstad Klepp, and Vilde Haugrønning
Consumption Research Norway (SIFO), Oslo Metropolitan University, Oslo, Norway

CONTENTS

10.1 INTRODUCTION

Textiles have a central part in all cultures and as a ubiquitous good, clothing is a basic need and a personal item visible in public arenas. We are all dressed in social situations and use clothing to express ourselves visually but also to protect our

DOI: 10.1201/9781003141426-10

bodies from the external environment and to facilitate activities such as sports (Kaiser 1997, Lamb and Kallal 1992). In addition to appearance, clothes hide our bodies and can conceal, absorb, and retain the odours it emanates. Entwistle (2000) emphasizes that clothes exist as the body's boundary with the world. Clothes are seen as protecting the surroundings from the body's impurities or, conversely, as protecting the body from one's surroundings (Broby-Johansen 1953). The function of clothes includes to ensure the body is accepted in the various social situations we may encounter. Absence of body odour is an important part of this border (Klepp 2005). Cleanliness is deeply rooted in culture, associated with religion and morality, health, hygiene, and well-being (Klepp 2003, 2005, 2007). Cleaning helps to remove unwanted microorganisms and thus prevent diseases from spreading (Bloomfield and Scott 2003, Bockmühl 2017, Larson and Duarte 2001).

Different bodily secretions, such as sweat, urine, dandruff, and other discharges, are often considered culturally problematic (Mauss 1979). When clothes become 'contaminated' by them, this contamination may be visible in the form of stains, it may be invisible, or it may be sensed through other means, such as smell or touch. Today, bodily malodours are especially considered socially stigmatising (Classen, Howes, and Synnott 1994, Klepp 2005). When the odour is considered to be a problem, it is often associated with filth. Unpleasant odour, especially body odour, is understood to be closely related to uncleanliness and poor hygiene.

Consumers' textile maintenance practices today are characterized by frequent laundering (Ashenburg 2007, Shove 2003). Sociologist Elizabeth Shove (2003) discusses the goals of laundering concerning the concepts of sensation, disinfection, and deodorization. Klepp (2003) has looked at shifts in the reasoning behind laundering in Norway and discussed these changes in relation to aesthetic, hygienic, practical, and ritual aspects.

Since the 1950s, the focus on dirt as something external has weakened compared to the focus on dirt that comes from within, especially by increased attention on sweat and odours. This change is closely related to the increased emphasis on odour as opposed to purity (Klepp 2005). The demand for laundry to be whiter than white is no longer as absolute as it once was (Klepp 2003), especially since a higher share of textiles are dyed (Grimsvang and Klonteig 1985). In addition, the textile industry has focused on developing stain-repelling finishes for easy-care fabrics (Namligoz et al. 2009). Sensitivity to body odour has changed a lot in Western countries over the last 100 year, and it is still changing. The population has become much more aware of body smell than before, and the experience of perfume and other odours is different both in time and in different regions (Klepp et al. 2016). The textile industry has responded to this by developing garments with various odour control treatments (Höfer 2006; Patra and Gouda 2013).

Body odour, such as sweat, is the most common odour to be removed from clothing through laundering, and it has been shown that unpleasant odour is the most common reason for washing sports clothing (Damm 2011), but it is also an important reason for washing other textiles (McQueen et al. 2019, Miracle et al. 2020). Fibre content, fabric structure, and garment fit contribute to how quickly clothes get dirty and how much they emit odour. Increased access to air reduces the odour. Since microorganisms depend on moisture, a thick cotton towel that requires a longer time to dry will start smelling sour faster than a thin one.

Previous research indicates that many consumers have encountered problems with odour removal in laundering. A study in Norway indicated that 65% had experienced that laundry odour had not disappeared in the wash, 48% had experienced that clothes had received an odour after laundering, and 34% of respondents had experienced the washing machine starting to smell unpleasant (Laitala, Kjeldsberg, and Klepp 2012). The experiences related to laundry receiving an odour could be related to cross-contamination from other textiles or the washing machine being affected by the growth of microbial colonization (Nix, Frontzek, and Bockmühl 2015, Stapleton et al. 2013). This chapter discusses textile cleaning and odour removal based on research on the area and gives guidance on suitable maintenance methods for different materials and how to remove specific types of soils.

This chapter first discusses the various soils and related odours, followed by laundering aspects that impact the odour/soil removal and retention, including water, washing temperature, mechanical agitation, time, and contents of the used laundry chemicals. This is followed by a description of various maintenance methods, including washing with a machine or by hand and cleaning without the use of water, such as dry-cleaning and airing. The following sections discuss the cleaning of different materials based on their fibre contents. The next section focuses on environmental impacts, and the final section concludes and summarises this chapter.

10.2 SOILS AND ODOURS

One 3 kg laundry load contains on average more than 30 g soil, varying from 15 g for a lightly soiled load to 100 g or more in extreme cases (Laurent et al. 2007). In general, main stain categories in clothing are divided to water-soluble (polar) contaminations such as salts, sugar, perspiration, and urea; hydrophobic solvent-soluble stains, such as fat, grease, wax, and oil, where removal is conducted by detergents or dry-cleaning solvents and can be aided by heat; coloured stains, such as coffee, fruit juices, and red wine, that must be bleached away; and proteins, such as eggs and blood, that should be dissolved in cold water first because heat sets them, similarly to some carbohydrates, such as starch. The alkali environment and enzymes in detergents aid this process. In addition, there are solid particles, such as dust, scales of skin, sand, humus, and soot, that can be partly removed by mechanical means, such as brushing, in addition to laundering (Hasenclever 2007a; Klepp and Tobiasson 2019; Laurent et al. 2007).

The different odours in the laundry are based on various sources and chemical compositions. Sweaty body odour has been reported to be the most common type of odour detected in clothing, with a total of 88% of respondents noticing it regularly, followed by food odour (i.e., cooking smells) (57%) (McQueen et al. 2019). The odour generation mechanisms correlate with microbial participation (Takeuchi, Yabuki, and Hasegawa 2013). Sebaceous lipids, such as triglycerides and wax esters, sweat, and dead skin cells from our bodies, function as microbial nourishment.

Numerous studies have focused on the chemical composition of sweat and odorous components of laundry, and have identified several hundred volatile compounds (VCs) (see, for example, Curran et al. 2005, Curran et al. 2007, Gallagher et al. 2008, Takeuchi et al. 2012, Takeuchi, Yabuki, and Hasegawa 2013). These compounds include organic fatty acids, ketones, aldehydes, esters, alcohols, hydrocarbons, and nitrogen-containing compounds, and are discussed in more detail in Chapter 3 of this book. In the next section, we will discuss the various factors that impact cleaning efficiency of soils and odours.

10.3 CLEANING FACTORS

The cleaning result is dependent on solvents (water and types and amounts of laundry chemicals), washing temperatures, length of washing cycles, and mechanical agitation applied. If one of these elements is reduced, another has to be increased to achieve a similar cleaning result. This is often referred to as Sinner's Circle (Sinner 1960), with temperature, chemistry, mechanics, and duration as parts of the circle. Since water is also important in water-based cleaning methods, Stamminger later added it as a fifth parameter as an inner circle that is the combining element for the other four factors (Stamminger 2010). The right mix of these factors will depend on both what the textiles are made of and thus can withstand, and what kind of dirt needs removal.

10.3.1 WATER

Water is by far our of the most important textile cleaning solvents used when laundry is washed by hand, in a washing machine, or by professional wet-cleaning methods. It dissolves a lot of soils as well as detergents. Water also has an important function to carry out the soil once it is loosened (Johansson and Somasundaran 2007).

Water works in combination with the other factors of Sinner's Circle. The mechanical processing moves the water through the textile to remove the stains and odours, while the dirty water is removed from the clothes through centrifugation or twisting. Water washes more efficiently when it is heated, and more efficiently if it gets sufficient time to function. There needs to be enough water available for it to move through the textiles during the mechanical agitation and to dissolve the soils. The amount of water needed in laundering has reduced due to more efficient washing machines where technological improvements have taken place, especially due to the need to meet the energy labelling requirements (Boyano, Espinosa, and Villanueva 2020, Miracle et al. 2020).

The quality of water also impacts the cleaning result. Impurities in unfiltered water or reuse of grey water can impact the cleaning results. Its degree of mineral deposits impacts the cleaning results (Brown et al. 1991, Gotoh et al. 2016). Hard water that is high in minerals, such as calcium and magnesium, can cause the soil to build up on the textiles and cause harder feeling, grey or yellow hue, or stains appearing on dyed textiles (Okada, Fujii, and Okuyama 1976, Okada, Fujii, and Minagawa 1984). The mineral deposits react with the soap and form insoluble compounds (lime soaps), which appear as a sticky paste or deposit. Therefore, the calcium and magnesium need to be deactivated for the detergent to be able to function as a cleansing agent (Brown et al. 1991). Copper content of water is shown to substantially increase the formation of some known malodour molecules, which can be a problem, especially in areas where water runs through copper piping (Miracle et al. 2020). Consequently, more detergent is needed to get a clean result when washing with hard water, and detergent formulations include ingredients that react with the minerals (see Section 10.3.4).

10.3.2 Temperature

Cleaning with warmer water is more efficient on most soils than using cold water. Use of cool water can make odour removal from fabrics more difficult because sebum residues harden (the viscosity increases) at lower temperatures, such as 20°C (Burton 1970; Motwani, Rhein, and Zatz 2001). However, many protein-based soils (such as egg and blood) dissolve better in cold water and solidify and adhere when washed in warm water. This can be seen, for example, in the experiment conducted by Kruschwitz et al. (2013), where they compared among other things laundry cycles without any detergent at 30°C and 60°C. The cleaning result was better on average at 60°C, but not on all soil types. An example of how Sinner's Circle functions can be seen in the study by Laitala and Jensen (2010). They compared cleaning performance of laundering at 30°C and 40°C with different detergents. Some laundry cycles at 30°C got a better result than those at 40°C if a more efficient detergent was used. The cleaning performance was tested by measuring the reflection values of pre-soiled textile swatches after wash, where the level of removal of soils can be evaluated as a degree of whiteness. The difference in reflection value was on average only 1.1% higher at 40°C than at 30°C, indicating a very small difference in cleaning effect between the two temperatures.

Recommended laundering temperature will vary by textile types and is usually labelled in garments. In many countries, this labelling is obligatory. Too high temperature may damage the clothing and is discussed in section 10.5.

Higher temperatures and boiling can be used to sterilize laundry. Thermal disinfection sufficient to kill almost all bacteria, viruses, and yeasts can be achieved with a temperature of 85°C for 6 minutes in an industrial laundering process (Terpstra 2001). The elimination rates at temperatures below 60°C are not sufficient if disinfection is needed, and in these cases, additional disinfection methods will be needed, such as chemical disinfection (Eberle et al. 2007, Smith et al. 1987). These measures can be important in cases of epidemics or for particularly vulnerable user groups.

Munk et al. (2001) tested the survival and distribution of microorganisms after laundering at 30°C or 40°C according to North American and European conditions and detergents and found that the four bacterial pathogens in the test survived and transferred between textiles when washed with colour detergents, but did not survive laundering with bleach-containing detergents. The reduction was greater at European laundering conditions at 40°C and longer washing cycles. Since odour in the laundry is often connected to microbial activities, an increase in temperature or chemical disinfection aids removal. After washing at low temperature without bleach, the most common odorants from sweat still left in laundry included ketones, esters, and aldehydes, while organic acids were more readily removed (Munk et al. 2000). In another experiment, Denawaka, Fowlis, and Dean (2016) found that low-temperature laundering at 20°C did remove a large share of malodorous VCs from used socks, but laundering at 50°C was more efficient. For t-shirts, overall VC concentration reductions of between 25% and 98% were recorded on samples from before and after laundering at 20°C.

Microbial survival in the washing machine can lead to bacterial colonisation (formation of biofilms), that in turn contribute to malodour associated with the machine itself and cross-contamination of clothing washed in it (Gattlen et al. 2010; Munk et al. 2001). Bacterial colonisations have been detected especially around the detergent drawer, sump, and rubber seal (Stapleton et al. 2013). To avoid the formation of biofilms, it is recommended to wash occasionally at hot temperature or use washing machine cleaners. In many cases, consumers' laundering practices are affected by the selection between a high level of hygiene and sustainability considerations, as many aspects that increase the bacterial removal require more energy or more efficient chemicals (Terpstra 2001).

There are large national differences in average laundry temperatures. The average European washing temperature was 42.4°C in 2020 (A.I.S.E. 2020), but the average for Scandinavia was higher (46.2°C) and Southern Europe was lower (39.7°C) (A.I.S.E. 2017). In the United States, the average temperature is close to 30°C, whereas in Japan, it is most common to use unheated water around 20°C (Golden et al. 2010; Laitala et al. 2020; Pakula and Stamminger 2010). In general, higher temperature laundering is more efficient in odour removal, especially when the odours are connected to microbial growth.

10.3.3 Mechanical Agitation

Washing machines aid the physical removal of soils through tossing and turning the laundry in water, which aids the general cleaning efficiency through mechanical agitation (Honisch et al. 2016; Laitala, Boks, and Klepp 2011). Mechanical agitation varies between different washing machine programs. Most machines have as a minimum, a program for regular/cotton laundry, and gentle/delicate cycle, combined with different water-level choices or temperature settings. Additionally, many modern machines have more specific programs for various types of laundry, such as wool or jeans (Laitala and Vereide 2010).

The physical removal of soil is connected to the hygienic effect of the washing process, as a large share of microorganisms is removed with the dirt they are connected to. If laundering results in poor soil removal, it is also likely to have a poor hygienic quality (Terpstra 2001) and therefore also reduced removal of odorous components.

10.3.4 Time

One of the commonly used alternatives to reduce the energy demand of a washing machine is to reduce the washing temperature and compensate for it by extending the duration of the washing cycle to achieve the same cleaning result. Therefore, today's most energy-efficient eco-programmes programmes have a very long duration (Alborzi, Schmitz, and Stamminger 2016) that can last up to 4 hours (Boyano, Espinosa, and Villanueva 2020; Laitala and Vereide 2010).

10.3.5 Chemistry – Detergents

Detergents are used in laundering to improve the cleaning result. Various types of detergents are developed for different types of washing methods, such as hand-washing detergents, wool detergents, detergents for white/light textiles, and specific types for washing machines with vertical and horizontal axis drums (high degree of tumbling in the front-loading h-axis machines require low-sudsing detergents). Detergents are made as granules and liquids in regular and compact formulations, and more recently tablets and single-dose liquid laundry products (liquitabs) (Johansson and Somasundaran 2007). Liquids are better suited for short programs, wool washing, and hand washing because they dissolve more quickly in water. For regular machine washing, powder/granules are more suitable.

Detergents are labelled with a recommended dosage that ensures the optimal cleaning result. It varies based on water hardness, the size of the washing machine, the amount of clothes, and how dirty the laundry is. In hard water, detergent builders react with the calcium, and more detergent is needed for achieving the same cleaning result as in soft water (Brown et al. 1991). Too little detergent is unable to loosen all the dirt, and there are too little means to keep it floating in the washing water. The result is that the dirt re-attaches to the clothes. However, overdosing detergent is not equal to increased cleanliness, as the extra amount can cause additional foaming that will reduce the mechanical agitation and thus reduce the cleaning result (Garrett 2013). The detergent is also more likely to be retained in the textiles.

The contents of detergents vary, but some of the general components and their functions are listed here:

Surfactants constitute about 15–40% in household laundry detergent formulations (Koohsaryan, Anbia, and Maghsoodlu 2020). They modify and reduce the tension between the surfaces, such as those between air, water, fabric, and soils, based on their amphiphilic nature with both hydrophilic and hydrophobic parts in their structures (Koohsaryan, Anbia, and Maghsoodlu 2020). They are used to reduce the surface tension of the water, thus improving its wetting ability, loosen and disperse the soil, keep the soil floating in the washing water, as well as provide or control the suds (Nayak and Ratnapandian 2018). Detergents may contain more than one type of surfactants with different ionic properties. Anionic surfactants are most widely used, and are effective at cleaning oily soils and at suspending clay soils, but they are easily deactivated by calcium and magnesium in hard water (Bajpai and Tyagi 2007). Non-ionic surfactants are especially good at solubilizing and emulsifying grease/fat-based soils, and they are not deactivated by hard water (Bajpai and Tyagi 2007).

Enzymes catalyse the breakdown of stains. Specific enzymes are developed for removing different types of stains or other unwanted contaminants or ingredients. Lipase dissolves fat and oil, amylase attacks starchy stains, while cellulase can be used to remove loose cellulosic fibres from the surface to avoid fuzzing and pilling. Protease dissolves protein-based soils, such as blood, eggs, milk, and skin secretions. Detergents that contain this type of enzyme should not be used for wool or silk. Morales-Garcia et al. (2020) have demonstrated that addition of nuclease enzymes to detergent can aid removal of accumulated soils that give a dingy appearance in clothing, characterized by dull colours, yellowing, and an unpleasant malodour. Nucleases target extracellular DNA, that is a key component in soil matrices that binds proteins and polysaccharides and is likely to be a part of the tenacious deposits that persevere through a laundry washing cycles. The nuclease was shown to be effective when washing at 30–40°C. However, some enzymes can also contribute negatively to odour formation. For example, Munk et al. (2000) found that the presence of lipases in detergents increased malodour from aldehydes (one of the primary odorants in the laundry).

Enzymes have been developed to function at low washing temperatures down to 20ºC, but most work best at a temperature around 30–40°C. Enzymes are sensitive to heat, and many start to denaturate and lose their efficiency at temperatures above 70°C (Daniel, Dines, and Petach 1996; Lund et al. 2012).

Builders are added to detergents in order to enhance the detergency efficiency through decreasing the water hardness. They bind cations to soften the water, provide good buffering properties to the alkaline region, aid the surfactants in the soil dispersion and anti-redeposition, and improve the wetting and emulsification characteristics of detergent (Bajpai and Tyagi 2007). Builders used to consist mainly of phosphates, but since they cause eutrophication and have been banned in many countries, they are now increasingly replaced with more sustainable alternatives, such as zeolites (Koohsaryan, Anbia, and Maghsoodlu 2020).

Bleaching agents can be added to granulate detergents in order to bleach away difficult stains, but at the same time, they also bleach the fabric and should not be used on coloured textiles. Bleaching agents, such as sodium percarbonate, release hydrogen peroxide upon exposure to water, that will oxidate and break stain molecules into tiny parts that can be carried away by surfactants at the washing stage (Koohsaryan, Anbia, and Maghsoodlu 2020). In most detergents, it has no significant effect until at 60°C, but with help of a bleach activator, such as tetra-acetylethylenediamine (TAED), it gives a bleaching effect already at 40°C. Bleaching agents contribute to chemical disinfection of the laundry (Terpstra 1998) and therefore reduce odour formation.

Some detergent powders also include bacteriostats, such as hexachlorophene, to prevent growth of bacteria and/or fungi, which is beneficial for avoiding odour formation in low-temperature laundering (Koohsaryan, Anbia, and Maghsoodlu 2020).

Perfumes are added to detergents to cover the smell of the detergent and to add a smell to the laundry. They are known to cause allergic reactions and consist of many different chemicals with a possible negative effect on health and the environment.

Detergents can include many other ingredients such as preservatives and optical brighteners (fluorescent whitening agents) that reduce the yellowness of the fabric.

There exists a variety of alternative cleaning products, such as soap nuts, soapwort, laundry/wash balls, washing pellets, and laundry magnets. Studies on alternatives to detergents indicate that, in general, they are not as effective as regular detergents

(Kruschwitz, Augsburg, and Stamminger 2013; Laitala and Kjeldsberg 2012). Most of them result in similar cleaning results as when washing with water only. However, many consumers use these products and are satisfied with the cleaning result, indicating that water alone is an efficient cleaning agent. It also indicates that items may be washed even though they are not that dirty. We have not found any studies on odour removal efficiency of these alternative detergents.

Laundry odour can also be impacted by the use of additional laundry products, such bleaching agents or fabric softeners. As discussed previously, bleach contributes to chemical disinfection. Laundry can also be disinfected by using a hypochlorite bleach in the last rinse of the washing process, as is often used in Mediterranean countries (Terpstra 2001).

Fabric softeners (also called fabric conditioners) are used to reduce the harsh feeling of laundered products, but also due to their ability to give a scent to laundry and reduce static electricity (Crutzen 2006). Fabric softeners consist primarily of cationic surfactants with long hydrophobic hydrocarbon chains that get attached to the anionic charged wet fibres and form a layer on the surface (Toedt, Koza, and Cleef-Toedt 2005). Cationic surfactants called esterquats have been used since the 1990s (Braun and Stamminger 2011).

Many consumers use softeners due to the scent they add to the laundry, but in fact, they may have a negative contribution to odour. A comparison of odour intensity of between laundering with or without softeners showed that laundry washed with perfume-free softeners was more odorous than one washed without it. This was especially noticeable on synthetic samples (Laitala, Kjeldsberg, and Klepp 2012). The same study showed that consumers who used laundry softeners more often had problems with laundry receiving an unpleasant odour after wash. These results indicate that it is especially important to avoid softeners on polyester sports clothing.

10.4 MAINTENANCE METHODS

Methods of cleaning clothes vary greatly around the world. Removing odour by washing textiles in the washing machine with detergents is by far the most common, but in some regions, such as China, washing by hand is still important (Kruschwitz et al. 2014; Laitala, Klepp, and Henry 2017b, Laitala, Klepp, and Henry 2017a). There are also means to clean textiles without or with less water, such as airing, steaming, stain removal, physical removal (brushing, vacuuming, or beating, shaking and hanging outdoors), or dry-cleaning.

The finishing work with the laundry, drying, smoothing, folding, and cleaning in place, has major consequences for the result, the amount of work, and the use of energy.

10.4.1 Washing with a Machine

The type of washing machines used vary greatly between different regions. Horizontal axis drum machines are common in Europe, and in these machines, only the bottom of the washtub is filled with water, and the machine has an internal water heating system. In vertical axis machines, usually the whole tub is filled with water, and if warm water is used, it is usually heated externally. Therefore, there are great differences in water and energy consumption between the different types of machines, in addition to variations in ways of using them. Vertical axis top-loading machines are used mainly in America, Australia, and Asia, although the share of horizontal axis machines is rising in these markets as well (Pakula and Stamminger 2009; Wang et al. 2014). In some areas such as India, semi-automatic top-loading machine types called "twin tubs" are common. The first tub is for wash, rinse, and soak, while the second is for spin-drying the laundry.

Traditional vertical top-loading washing machines have an agitator in the centre of the tub that rotates back and forth, whereas the newer machines use an impeller or pulsator with rotating tubs and wash plates that toss clothes around in addition to jets and streams of water to rinse the laundry. The machines with an agitator are less suitable for washing delicate clothes because the aggressive action of the agitator can be hard on textiles. This also increases the shedding of fibres during washing and, hence, the number of synthetic microfibres in the effluent (Hartline et al. 2016). In addition to these differences in mechanical cleaning action, the way water is heated also varies. The European drum machines have an internal heating unit, whereas in other regions, use of either ambient or externally heated water is common (Golden et al. 2010). Replacement of washing machines with high-efficiency versions in both vertical and horizontal drum formats has enabled reduction in energy and water consumption (Chen-Yu and Emmel 2018).

10.4.2 Washing by Hand

Hand washing is especially common in rural areas of developing countries (Euromonitor International 2011, Gordon et al. 2009), but it is also common in some areas in addition to the use of washing in the machine, such as China (DuPont 2013; Kao 2010a). The Chinese consumers who washed by hand reported that the main reason was superior cleaning performance (70%), followed by less water required (about 53%) and less damage to clothes (about 47%). About 20% also preferred hand wash due to lower electricity costs (Kao 2010b). Heavy/large loads were preferably washed in the machine. The average time used on hand washing was rather short: 12% used less than 15 minutes, 63% used 15–30 minutes, and 25% used 30–60 minutes (DuPont 2013). According to Kao (2010b), all of the respondents used scrubbing technique with both hands to wash the laundry, but in addition, every fifth used a washboard and about 12% brushing. Some consumers in Western countries also wash laundry by hand, but it is quite seldom and in general for some specific clothing items only (Aalto 2003; Gwozdz et al. 2013).

10.4.3 Dry-Cleaning

Dry-cleaning is a process of cleaning garments with the help of chemical solvents, predominantly volatile organic solvents with perchloroethylene (PERC, C_2Cl_4), which is the most common solvent used in recent years (Troynikov et al. 2016). The dry-cleaning process is typically constituted of several steps, including washing, extracting, drying, spotting, and finishing (pressing/ironing).

Dry-cleaning is used on materials that shrink, felt, crease, or otherwise get damaged when washed with water. Apolar solvents like hydrocarbons are not absorbed by natural textile fibres because of the high polarity of the fibres but are effective in the removal of oils, fat, grease, and other similar substances (Blackler et al. 1995, Hasenclever 2007a). However, these solvents are not efficient in the removal of particles or water-soluble substances like salt, sugar, most food, and body excrements. Therefore, special detergents are added in the dry-cleaning process. Since many laundry contaminants with odours include water-soluble components, dry-cleaning is less efficient in odour removal than regular water-based laundering (Hasenclever 2007a). Garments that smell of body odour, smoke, mould, or perfume need special attention in addition to the normal dry-cleaning process for complete odour removal. Additionally, solvents used in dry-cleaning can leave a distinct odour in the garments, even after 24 hours of the airing-out process (Rathinamoorthy 2020).

Environmental and health hazards associated with PERC have led to the emergence of various alternative solvents and processes that can reduce the impact of dry-cleaning on the environment and human health (Troynikov et al. 2016). As an alternative to dry-cleaning, professional wet-cleaning is the process of removing soiling and stains from garments and other textile products using fresh water and other agents (Hasenclever 2007b). It uses specialized equipment, programming, detergent, and other additives to minimize adverse effects on textiles, such as mechanical action and temperature. It is followed by drying and restorative finishing procedures, in most cases by steam treatment and/or hot pressing (ISO 3175-4 2018). Professional wet-cleaning is more efficient on the removal of water-based stains than traditional dry-cleaning, and besides, it does not leave chemical smells on the textiles. However, it is less efficient in removal of grease-based stains and may have shrinkage issues with some garments (Troynikov et al. 2016).

10.4.4 Airing and Other Methods Without the Use of Solvents

It is possible to clean textiles without water or other solvents through airing, freezing, or physical removal (brushing, vacuuming, or beating, shaking and hanging outdoors). Physical removal methods used to be more central to keeping clothes clean, whereas today they are mostly used on carpets.

Steam cleaning devises can be used for disinfecting textiles by using the heat of steam vapor to kill germs, microbes, mould, bacteria, and other forms of biocontamination (Kohli 2019).

Moisture content of the steam can range from 5% (dry steam) to 99% (wet steam, a combination of steam and hot water) (Kohli 2019). The moisture, heat, and pressure of steam accesses also tight spaces, and due to its efficient cleaning performance, is also likely efficient in removal of odours.

Airing is a method where textiles are hung in ample space to aerate them, usually outdoors but also indoors to let the odours evaporate. Lately, separate airing programs have been designed for refreshing textiles instead of laundering. These programs have been shown to be better at removal of odour than airing by hanging, but less efficient than laundering (Kalayci et al. 2017). Another study demonstrated that airing reduces odour intensity, and on average, reduced the likelihood of needing to wash the garments by 12–27%, where the lowest rates applied for synthetic garments and the highest for cotton and wool (Klepp et al. 2016).

Freezing can be used for killing dust mites, moths, fur beetles, as well as some microbes. In regions where temperatures can go below zero degrees Celsius, bedding and other items can be pulled in wet snow or hung out for some hours to freeze them. A freezer can also be used. This is common advice given for cleaning raw denim jeans, as well as for second-hand clothing that is potentially infested. However, experiments with freezing odour samples for human ecology research indicated that freezing can be used to preserve many odours without impacting the odour profile (Lenochová, Roberts, and Havlî ek 2009), besides a slight reduction in intensity (Abdul-Bari et al. 2018).

10.4.5 Drying Methods

Laundry can be dried in a tumble drier, drying cabinet, drying room, or by letting the laundry dry 'by itself', i.e., without additional energy consumption. It can be placed flat or on a line either indoors or outdoors. Clothes driers use a lot of energy, but also drying indoors requires energy if done in a heated room (Schmitz and Stamminger 2014).

Microorganisms thrive in moist and warm conditions. If textiles are stored moist for a long time, for example, forgotten in the washing machine, or if the drying process is too slow, a bad smell can develop due to the growth of microorganisms or even mould (Klepp and Tobiasson 2019, Nagoh et al. 2005). To reduce the drying time, it is beneficial if the laundry is spin-dried efficiently at high speed before removed from the washing machine. This will reduce moisture content significantly. Damp textiles should be hung in a way that the surface get lots of air. High temperature, low humidity, and wind contribute to rapid drying. The high temperature of clothes dryers can aid bacterial reduction, but it is not suitable for all materials. High ambient moisture can lead to slow drying conditions, which can lead to the generation of malodour. Nagoh et al. (2005) showed that this can be prevented by using detergent with a high content of protease enzymes as well as bacteriolytic activity. Odour can also be prevented through the construction of the garments, for example, by having looser fit around the underarm area (Rigby 2016). Thin fabrics and open structures provide faster drying and enable the fabrics to stay clean longer.

Drying outdoors has benefits for the odour emitted from laundered clothing, as it can give a fresh pleasant scent to the textiles based on oxidative photochemistry occurring on the textile surface in sunlight (Pugliese et al. 2020). In addition, the direct sun can bleach the colours. This is beneficial for an item such as white cotton bedding, but unwanted on dyed items, and can deteriorate some fibre types like wool, silk, polyamide, or polypropylene. Clothes can also be dried in below zero Celsius degrees. It takes longer time but is not an issue for odour build-up as the microorganisms do not thrive in the cold environment. Freeze drying gives a good smell – and saves energy (Klepp and Tobiasson 2019). Airing and drying outdoors depend on the air quality, including pollution and pollen. In some places, and for some people, drying outside is not always a good option.

10.5 WASHING OF TEXTILES IN DIFFERENT FIBRES

Several studies have looked into the various mechanisms of odour formation and compared the intensity of odours retained in textiles. These have been summarised by two recent review articles (Laing 2019, McQueen and Vaezafshar 2019) that show that there is a general trend in natural fibres laving less intensive odours than synthetics, where wool has the least intensive odour, followed by cotton, viscose, and linen, and synthetic polyester and polyamide garments retain the most intense odour. These properties are discussed in chapter 4. Here, we give a summary of material properties related to how easily they get dirty and how to clean them.

10.5.1 Natural Fibres of Vegetable Origin

Cotton is the most commonly used natural fibre. It is a natural plant seed fibre consisting almost purely of cellulose. Cotton fibres have a porous structure, and it has been estimated that as much as 20–41% of the volume of the fibre is free space (the lumen and spaces between the fibrils in the fibre walls) (Cook 1984b). Due to this and the hydrophilic nature of cotton, it get can get easily dirty, as the soil can enter inside the fibre lumen, as well as in the interfibre spaces (Obendorf, Namasté, and Durnam 1983). It has high moisture adsorbing capacity (Callewaert et al. 2014), and therefore it dries slower than fibres that adsorb less water, such as synthetics. These properties can be altered through various finishes, for example, dirt or water repellents. The fibres are relatively strong and get about 20% stronger when wet; therefore, cotton tolerates well washing at high temperatures and mechanical agitation. Cotton can be washed with alkaline detergents and tolerates bleaching (Blanke 2001; Kjeldsberg, Eilertsen, and Laitala 2011).

Most of the other common natural fibres of vegetable origin are made of the bast or the stem of the plant, including flax (linen), jute, hemp, ramie, bamboo, and nettle. Besides cotton, other seed fibres include coir and kapok, whereas the remainder are leaf fibres, such as sisal, henequen, and abaca (Cook 1984b). In general, leaf fibres are coarser than bast fibres and are less often used in clothing. All of the listed plant fibres are cellulose-based and have similarities with cotton on how they can be cleaned, with some individual variations.

Several studies have looked into the odour properties of cotton, whereas other natural fibres of vegetable origin are less studied. For example, McQueen et al. (2014) investigated whether cotton differed in odour intensity following multiple wear and wash cycles (20 exercise sessions over a 10-week trial period) compared to a polyester fabric. Findings showed that laundering was effective in reducing overall odour intensity and bacterial populations in both materials. However, polyester had the most intense odour following wear and also after washing. The build-up of odour in polyester fabrics may be cumulative as important odorants such as the carboxylicacids are not as effectively removed from polyester compared to cotton (McQueen et al. 2014).

10.5.2 Wool, Hair, and Fur Fibres

Wool comes from sheep, while other hair fibres come mostly from animals of the goat and camelid families, including mohair and cashmere from goats, and alpaca, vicuña, and guanaco from llamas. The most common fur fibre is angora from rabbits. All of these have a similar chemical structure as they are based on keratin (a protein). However, there are great differences in length, fineness, shapes, and internal structures between the wool from different sheep breeds, such as merino and crossbreds, and hairs from other animals (Cook 1984b). Wool is a resilient fibre due to its high elongation at break combined with a high elastic recovery. The thinner fibre types are softer and therefore, in general, weaker.

Wool's outer surface repels water due to thin, waxy lipid coating, whereas the inner cortex is hydrophilic. These properties enable wool to absorb moisture ofabout a third of its weight but still feel reasonably dry. When the outer membrane of wool fibre is damaged, it becomes more easily wetted. The fibres release the moisture slowly, thus reducing the chill caused by temperature drop based on water evaporation.

Wool and other animal fibres tend to have a layer of flat, scale-like cells on the surface. These scales have an ability to lift up like little barbs when wool is wet, and this transformation process is affected by heat and accelerated in alkaline conditions. The scales are aligned with the tip of the fibre. For this reason, when mechanical agitation is applied, the wet fibre slides smoothly in the direction of the hair root, but not back again, and thus get entangled. This property is used in the manufacture of wool felts, but unwanted in laundering, where tumbling or stirring actions of washing machines, or rubbing if washing by hand, can cause felting in wool. Felting shrinkage is irreversible, while relaxation shrinkage can be stretched out after washing.

Felting can be prevented by so-called 'superwash' treatment, where the scales on the fibre surface are either removed, substantially modified, or fixated with a coating (Hassan and Carr 2019; Johansson and Somasundaran 2007).

Due to wool's hydrophobic surface, it is rather resistant towards water-based soils, especially if they are removed before they are absorbed (Kjeldsberg, Eilertsen, and Laitala 2011). Wool also exhibits good properties when it comes to odour resistance based on functional carboxyl groups and ammonia groups that provide sorption sites for odour ammonia and acids (Wang et al. 2019). However, stains on wool are more difficult to get clean than from synthetic materials (higher soil retention) (Kjeldsberg, Eilertsen, and Laitala 2011).

Due to wool's ability to felt, it should be washed with a minimum of mechanical agitation. When washed in a machine, a wool programme should be used, or if not available, a gentle program with long soaking times and reduced mechanical action. If washing wool by hand, soaking and gentle squeezing should be used instead of rubbing.

Wool does tolerate high temperatures if not agitated, but in general, lukewarm water (max 40°C) is recommended when washing it in a machine or by hand.

Wool should not be washed with detergents that include hypochlorite bleaches or protein-degrading enzymes (proteases). Wool is also sensitive to alkaline substances but has good resistance to dry-cleaning and other common solvents.

After washing, wool can be spin-dried at high speed to remove excess water and shorten the drying time (Laitala, Boks, and Klepp 2011). Superwash treated wool can tolerate tumble drying, but in general, line drying or drying flat is recommended. When drying is done outdoors on a line, direct sunshine should be avoided, especially on white woollens to avoid yellowing.

Most research indicates that wool emits the least odour when compared with other textiles. According to Wang et al. (2019), this is related to both sorption and emission properties of wool. Hygroscopic wool fiber may absorb fewer non-polar oily soils on its surface, whereas the presence of carboxyl groups and ammonia groups provide more sorption sites to absorb volatile components.

10.5.3 Silk

Silk is a natural protein fibroid filament spun by silkworms (caterpillars) as they make their cocoons. The majority of silk is based on domesticated Bombyx mori species, but also wild silk (Tussah) is used to some degree (Cook 1984b).

Like wool, silk absorbs moisture readily. It does not tolerate oxidizing agents, such as bleaches, and it loses strength if it is attacked by atmospheric oxygen, especially in sunlight. Therefore, silk has to be stored carefully. Silk has high tenacity, but its wet strength is lower than the dry strength. The filament of silk is fine and easily torn. It can be damaged by chemical action and must be washed with care with a mild detergent. Silk tolerates alkalis slightly better than wool, and weak alkalis such as soap, borax, and ammonia cause little noticeable damage. Silk is insoluble in common dry-cleaning solvents, and some silk garments need to be dry-cleaned. Others can be washed like wool with some exceptions, as one should be more careful with the spin cycle, rinse the textile thoroughly in soft water, and not twist it. After wash, it should be ironed while damp, at moderate heat, preferably with a pressing cloth put between the iron and the silk. Silk stains easily, especially from grease, but also from sweat, acidic liquids, and even water if the water contains impurities or salts (hard water), as silk tends to absorb them, too. Areas exposed to sweat on silk garments should be cleaned regularly to avoid perspiration degrading the fibroins.

10.5.4 Regenerated Cellulose Fibres

Regenerated cellulose fibre is a type of manufactured or man-made fibre that uses cellulose (mainly from wood or plant fibres) as a raw material (Sinclair 2015). These include viscose (known as rayon in North America), lyocell, and cupro, in addition to acetate andtriacetatethat are cellulose-derived fibres and therefore strictly not regenerated cellulose fibres (Sinclair 2015). These fibres are classified by the different production methods, and there are variations of viscose, such as high-wet-modulus (HWM) viscose that has two commonly used brand names, Modal and Polynosic (Sinclair 2015).

Viscose rayon is the most common of these, followed by lyocell that is increasing popularity due to its production methods that are less harmful to the environment. Cupro fibre is much less used than the other regenerated cellulose fibres.

Fabrics made of regenerated cellulose fibre are soft and display high drapability combined with smooth and lustrous appearance and a high water absorption ability (Cook 1984a). However, they also have some disadvantages, including easy stretching with poor elastic recovery, low abrasion resistance, wet shrinkage, and low wrinkle resistance (Sinclair 2015). Lyocell has some beneficial properties in use, such as increased strength (both dry and wet) and lower elongation than regular viscose. This applies also to modal that is as strong wet as dry but is otherwise much like viscose.

Viscose should be washed more gently than cotton. Due to its high water penetration, it loses as much as half its strength when wet and is more easily stretched (Cook 1984a). Viscose does not tolerate chlorine bleaching, should not be tumble dried, and should not be washed in a fully loaded washing machine or spin-dried for a long time (Klepp and Tobiasson 2019). Modal can be tumble dried without damage due to its increased molecular alignment. Acetate requires gentle treatment, not exceeding 40°C, and a gentle washing program, no chlorine, and very gentle tumble drying; it melts at a high temperature and dissolves in acetone.

Odour properties of viscose are less studied than cotton, synthetics, or wool, but the existing studies indicate that it has similar properties to cotton (Laing 2019, Rathinamoorthy et al. 2014). In marketing, viscose made of bamboo is claimed to have anti-microbial properties, but the antibacterial properties are based on lignin content of natural bamboo fibres, which is removed in the viscose process (Afrin et al. 2012).

10.5.5 Synthetic Fibres

Synthetic polymers are mainly produced from primary petrochemicals, generally coming from oil, natural gas, or coal. Polyesters, polyamides (nylons), polyolefin (polyethylene and polypropylene), and polyvinyl derivates (acrylics and modacrylics) constitute 98% of synthetic fibre production and are used in a variety of textile applications (Sinclair 2015). Aramids (aromatic polyamides) are very strong fibres that also have excellent resistance to heat, chemicals, and abrasion, and therefore are mainly used in protective clothing applications. Elastomeric fibres embrace a wide range of constituent polymers: elastane, elastodiene, elastomultiester, and polyolefins. The most common are elastanes, which contain 85% or more by mass of segmented polyurethane (called Lycra or Spandex) (Mather 2015).

In general, synthetic fibres are strong, durable, and resistant to most chemicals, insects, fungi, and rot (Cook 1984a; Sinclair 2015). They have good dimensional stability and usually do not shrink when washed. The care label should be followed for choosing the washing temperature, as laundering above the fibre's glass-transition temperature can cause deformation and permanent set wrinkles (Laurent et al. 2007).

Synthetics have low moisture absorbency and hence are easy to dry. This property also makes them static electric and thus easier to attract dirt/dust, which is especially relevant for polyester that is very hydrophobic (moisture regain only

0.4%) (Abdul-Bari et al. 2018; Cook 1984a). The polyester surface also easily retains oily stains due to the fibres' nonpolar, hydrophobic, and oleophilic nature (Obendorf, Namasté, and Durnam 1983, Pastore and Kiekens 2000). Polyamide is less hydrophobic than polyester but entails many of the same abilities as polyester. Odour studies have shown that synthetic fibres easily take up and develop an odour after exposure to sweat (Abdul-Bari et al. 2018; McQueen et al. 2008, McQueen et al. 2014; Rathinamoorthy et al. 2014). The fatty acids such as triglycerides from the sweat are retained on the surface of the fibre, available to oxidative degradation and form smaller VCs, such as aldehydes (Munk et al. 2000).

Line drying is preferred for polyamide and acrylics, and if tumble-dried, they should be on the cool setting. Acrylic fibres have excellent resistance to the effects of sunlight (Mather 2015), whereas polyamide and polypropylene are sensitive to ultraviolet light, and tend to undergo degradation and yellowing to varying degrees after prolonged exposure to sunlight.

Polyester can be ironed and steamed at hot temperatures, whereas if polyamide is ironed, the temperature should be set to synthetic (below 150°C) to avoid glazing, yellowing, and sticking. If acrylics need to be ironed, it should be done with a cool iron on the reverse side of the fabric (Cook 1984a).

Most synthetic fibres, such as acrylics and polyamide, have good resistance to common organic solvents, including those normally used in dry-clearning (Cook 1984a). Polyethylene is susceptible to attack by oxidizing agents, and it will swell and may ultimately dissolve in some chlorinated hydrocarbons and aromatic solvents, e.g., benzene, toluene, and xylene (Cook 1984a). Acetate, triacetate, and modacrylic fabrics are damaged by acetone or paint thinner. Elastane fibres are quite weak in comparison to other textile fibres, but they possess good resistance to micro-organisms and chemicals, besides chlorine (Mather 2015).

Abdul-Bari et al. (2020) found an accumulation of odorants after multiple soil and wash cycles. Incomplete removal of odorants during washing is the cause for odour build-up, especially on polyester clothing. Strong associations of odorants with polyester contribute to building up a complex odour profile. Therefore, the odour is formed when the clothes are worn again, rewetted, and heated to body temperature. As the microflora are allowed to build up during several washing and wearing cycles, the problem appears to accumulate (Munk et al. 2001).

In a long-term wear trial study by McQueen et al. (2014), the comparison between polyester and cotton fabrics used during exercise showed polyester to be more odorous also after washing. In addition, polyester was found to increase in odour after multiple wash and wear cycles, indicating a cumulative build-up of odour, whereas cotton continued to have a lower odour intensity (McQueen et al. 2014).

10.5.6 Leather, Skin and Fur

Clothing can also be made of non-textile materials such as leather, suede, or fur. In general, these types of garments are cleaned seldom, and often by professionals with dry-cleaning methods. However, some leather garments can also be washed with water: in the washing machine, by hand, or through professional wet-cleaning methods. In these cases, following laundering instructions for wool is beneficial, but also using slower spin-drying speed and special detergents that won't dry out the garment. If a clothes drier is used, it should be set to low temperature, and a dryer ball (or tennis ball) can be added for improved softness. If an odour is a problem, it is advisable first to try to air the clothing out.

10.6 IMPACTS OF LAUNDRY ODOUR ON THE ENVIRONMENT AND CONSUMER BEHAVIOUR

The development of odour in textiles has environmental consequences through various mechanisms. It can impact the following aspects:

1. Increase cleaning frequency
2. Affect the way textiles are cleaned (selection of temperature, washing programme, detergents, etc.)
3. Shorten product lifespans
4. Reduce the use of garments that quickly develop an odour
5. Function as a barrier for sharing and reuse of clothes
6. Increase the use of harmful chemicals in production, for example, the use of various antimicrobial treatments

High washing frequency leads to higher environmental impacts due to increased use of energy, water, and chemicals. The number of days garments are used between washes varies greatly between different types of garments and environments, and is also significant for the total environmental impact from clothing (BSR 2009; Wiedemann et al. 2020). Some clothes are generally washed after each use, whereas others are rarely washed. In hot and humid countries such as India, laundry does not usually accumulate but rather is washed as soon as it is dirty (Honold 2000). This causes more frequent laundering with smaller laundry loads.

Studies have shown that garments worn directly over high-odour body areas, such as underarms, feet, and crotch, tend to be washed most frequently (Laitala, Klepp, and Boks 2012; McQueen et al. 2019). This was evident in a study by McQueen et al. (2019), as items such as underwear and socks were usually worn only once before washing. However, there are some differences in this based on fibre content, as woollen garments are washed less frequently than garments made of other materials (Laitala, Klepp, and Boks 2012, Laitala et al. 2020). Heavily sweated clothing that has been worn during exercise is also likely to be washed more frequently than similar items worn for other activities. In particular, athletic wear is more likely perceived to exhibit 'high' or 'extreme' odour before laundering (McQueen et al. 2019). In addition, a study from Kowton and McQueen (2016) found that participants would increase the wash frequency if any of their clothes appeared odorous. Further, the increased washing frequency is likely to cause additional wear and tear on clothing, which indirectly will also shorten the lifespans of odorous garments. For example, McQueen et al. (2017) found that jeans washed more frequently exhibited a much greater change in colour than jeans with lower washing frequency, and the latter also maintained more of their tensile strength.

Kerr, Rosero, and Doty (2005) found that judgments of cleanliness are related to odour. Thirty-eight percent of the men reported smelling their dirty laundry before washing, compared to 46% of the women. After washing, 79% of the men and 86% of the women smelled their clean laundry (Kerr, Rosero, and Doty 2005). Smelling clothes to decide whether to wash them is more common for some types of items, and less common for those clothes that are washed routinely or ritually according to fixed patterns (Klepp 2005).

McQueen et al. (2019) found that odorous laundry is washed differently from non-odorous laundry. The study showed that respondents were less likely to use cold water and more likely to use hot water when laundering odorous clothing compared to regular laundry. This increases energy consumption and related GHG emissions. However, the respondents were also less likely to use clothes driers for these items. It is not clear if this difference in drying behaviour is due to the type of garments or due to the odour. In another study, focus group participants told about altered practices where they would use additional cleaning or pretreatment techniques for odorous clothing, for example, by adding scented fabric softeners and even fabric sheets in the same laundry (Kowton and McQueen 2016).

Odour can accumulate after multiple soil and wash cycles, as some detergents in combination with low washing temperature are found to not remove all microorganisms with VCs (Abdul-Bari et al. 2020). Such persistent odour has been reported to be one cause for premature clothing disposal. A large share of consumers reported having discarded an item with persistent odour, even though it was otherwise in good condition (Kowton and McQueen 2016, McQueen et al. 2019, McQueen et al. 2020). These items are more likely to be disposed of in the trash instead of being delivered to reuse or recycling, thus additionally contributing to unsustainable clothing consumption (McQueen et al. 2020). Consumers are less likely to use garments that quickly develop malodour. A study on customer experiences of clothing rental revealed that odour was experienced as contamination that was so negative that it had potential to lead to total rejection of the entire rental concept (Clube and Tennant 2020).

Material choice is likely to impact the washing frequencies because synthetic garments, such as those made of polyester or polyamide, are more likely to smell strongly of body odour than clothing composed of natural materials, such as wool or cotton (Klepp et al. 2016; Laing 2019; McQueen and Vaezafshar 2019, McQueen et al. 2007, McQueen et al. 2014). Due to the increasing use of polyester in clothing, the implications for sustainability are significant.

Several odour-control or anti-bacterial treatments have been developed for combatting problems with odour, as discussed in other chapters of this book. Previous studies where athletic wear with odour-control treatment is compared to other materials have indicated that the treatments do reduce odour formation in synthetic materials, but they still have higher odour intensity than wool or cotton garments (Klepp et al. 2016). In addition, Damm (2011) found that respondents who owned antibacterial athletic clothing would wear and wash them as often as their untreated sports clothing. This indicates that the slight reduction in odour is not likely to be enough to reduce consumers' laundering frequencies, especially as athletic clothing worn next to the body is often washed habitually.

10.7 CONCLUSIONS

Odour is an essential reason why we wash clothes. The importance has increased in the last decades and is now one of the main reasons for frequent cleaning of textiles.

The cleaning result is dependent on water, temperature, length of the washing cycle, type and amount of laundry chemicals, and mechanical agitation applied. If one of these elements is reduced, another has to be increased to achieve a similar cleaning result. The optimum mix of these factors will depend on both what the textiles are made of, and thus can withstand, and what kind of dirt needs removal.

Different materials have different abilities to absorb and emit odour, and the ways they can be cleaned varies. Inherent fibre properties affect the soiling characteristics of garments. Comparisons of odours retained in textiles have shown that wool has the least intensive odour, followed by cotton, and synthetic polyester and polyamide garments have the most intense odour. Most textiles can be washed with water and detergents, which are more efficient in the removal of many odorous soils than dry-cleaning. Low-temperature laundering and/or lack of chemical disinfectants can contribute to odour build-up in textiles and in the washing machine.

To avoid odours, choosing materials suitable for the various use areas is essential, such as thin materials in cleaning cloths and towels that enable quick drying. Clothing fit contributes to how quickly clothes get dirty and can be taken into account by designing looser fit around armholes. For environmental improvements, reducing washing frequency, for example, by replacing some of the washing with airing and stain removal is beneficial. In addition, odour accumulation leads to shorter product lifespans due to premature disposal, which contributes to the unsustainable consumption of clothing.

People's relationship to odours has been constantly changing, and it is likely that current norms will change. From a health and environmental perspective, it would be preferable if there was a higher acceptance of natural body odours and reduced preference for artificial perfumes. Washing is important to maintain the boundary between the body's intimate sphere and the public space. This relationship is affected by washing habits and washing technology, our surroundings, nature, and the culture we live in.

REFERENCES

A.I.S.E. 2017. *Low Temperature Washing - A.I.S.E. Factsheet*. International Association for Soaps, Detergents and Maintenance Products.

A.I.S.E. 2020. *A.I.S.E.'s pan-European Habits Survey 2020*. International Association for Soaps, Detergents and Maintenance Products.

Aalto, K. 2003. Kuka pesee Suomen pyykit? Tekstiilienhoito kotitalouksissa ja tekstiilienhoitopalvelut [Who washes the laundry in Finland? Textile care in households and use of textile care services]. In *Publications 11/2003*. National Consumer Research Centre.

Abdul-Bari, M.M., McQueen, R., Paulina de la Mata, A., Batcheller, J.C., and Harynuk, J.J. 2020. Retention and release of odorants in cotton and polyester fabrics following multiple soil/wash procedures. *Textile Research Journal*, 0(0): 0040517520914411. doi: 10.1177/0040517520914411.

Abdul-Bari, M., McQueen, R., Nguyen, H., Wismer, W., Paulina de la Mata, A., and Harynuk, J. 2018. Synthetic clothing and the problem with odor: Comparison of nylon and polyester fabrics. *Clothing and Textiles Research Journal*, 36(4): 251–266.

Afrin, T., Tsuzuki, T., Kanwar, R.K., and Wang, X. 2012. The origin of the antibacterial property of bamboo. *The Journal of The Textile Institute*, 103(8): 844–849. doi: 10.1080/00405000.2011.614742.

Alborzi, F., Schmitz, A., and Stamminger, R. 2016. Long wash cycle duration as a potential for saving energy in laundry washing. *Energy Efficiency*, 10(4): 823–838. doi: 10.1007/s12053-016-9486-z.

Ashenburg, K. 2007. *The Dirt on Clean: An Unsanitized History*. North Point Press.

Bajpai, D., and Tyagi, V.K. 2007. Laundry detergents: An overview. *Journal of Oleo Science*, 56(7): 327–340. doi: 10.5650/jos.56.327.

Blackler, C., Denbow, R., Levine, W., Nemsick, K., and Polk, R. 1995. A Comparative Analysis of Perc Dry Cleaning and an Alternative Wet Cleaning Process. Master's ThesisUniversity of Michigan.

Blanke, H. 2001. *Materialelære: Bind II - Naturfibre, Beklædning og Textil*. Teknologisk Institut.

Bloomfield, S.F., and Scott, E.A. 2003. Developing an effective policy for home hygiene: A risk-based approach. *International Journal of Environmental Health Research*, 13: S57–S66. doi: 10.1080/0960312031000102804.

Bockmühl, D.P. 2017. Laundry hygiene—how to get more than clean. *Journal of Applied Microbiology*, 122(5): 1124–1133. doi: 10.1111/jam.13402.

Boyano, A., Espinosa N., and Villanueva A. 2020. Rescaling the energy label for washing machines: An opportunity to bring technology development and consumer behaviour closer together. *Energy Efficiency*, 13(1): 51–67. doi: 10.1007/s12053-019-09829-4.

Braun, V., and Stamminger, R. 2011. Using fabric softeners, drying and ironing in Germany: A non-representative census of consumer behaviour. *Tenside Surfactants Detergents*, 48(03): 210–220.

Broby-Johansen, R. 1953. *Kropp og klær*. Tiden.

Brown, D.M., Cameron, B.A., Meyer, S.S., and Umber, J.J. 1991. The effects of water hardness level on washing quality using commercial laundry detergents. *Journal of Consumer Studies & Home Economics*, 15(3): 215–222. doi: 10.1111/j.1470-6431.1991.tb00660.x.

BSR. 2009. Apparel industry life cycle carbon mapping business for social responsibility.

Burton, J.L. 1970. The physical properties of sebum in acne vulgaris. *Clinical Science*, 39(6): 757–767. doi: 10.1042/cs0390757.

Callewaert, C., De Maeseneire E., Kerckhof, F-M., Verliefde, A., Van de Wiele, T., and Boon, N. 2014. Microbial odor profile of polyester and cotton clothes after a fitness session. *Applied and Environmental Microbiology*, 80(21): 6611–6619. doi: 10.1128/aem.01422-14.

Chen-Yu, J.H., and Emmel J. 2018. Comparisons of fabric care performances between conventional and high-efficiency washers and dryers. *Fashion and Textiles*, 5(1): 19. doi: 10.1186/s40691-018-0135-3.

Classen, C., Howes D., and Synnott A. 1994. *Aroma: The Cultural History of Smell*. Routledge.

Clube, R.K.M., and Tennant M. 2020. Exploring garment rental as a sustainable business model in the fashion industry: Does contamination impact the consumption experience? *Journal of Consumer Behaviour*, 19(4): 359–370. doi: https://doi.org/10.1002/cb.1817.

Cook, G.J. 1984a. Handbook of Textile Fibres. Vol. II - Man-Made Fibres. Woodhead Publishing Series in Textiles.

Cook, G.J. 1984b. *Handbook of Textile Fibres. Vol. I - Natural Fibres*. Woodhead Publishing Series in Textiles.

Crutzen, A. 2006. Fabric softeners. In Kuo-Yann Lai (ed.), *Liquid detergents* (pp. 487–554). CRC Press, Taylor & Francis Group.

Curran, A.M., Rabin, S.I., Prada, P.A., and Furton, K.G. 2005. Comparison of the volatile organic compounds present in human odor using SPME-GC/MS. *Journal of Chemical Ecology*, 31(7): 1607–1619.

Curran, A.M., Ramirez C.F., Schoon A.A., and Furton K.G. 2007. The frequency of occurrence and discriminatory power of compounds found in human scent across a population determined by SPME-GC/MS. *Journal of Chromatography B*, 846(1–2): 86–97.

Damm, J. 2011. *Silver i "luktfria" kläder – En stinkande lösning: En studie av antibakteriella behandlingar baserade på silversalter utifrån miljö-, hälso- och konsumentperspektiv*. Swedish school of textiles, University of Borås.

Daniel, R.M., Dines, M., and Petach, H.H. 1996. The denaturation and degradation of stable enzymes at high temperatures. *The Biochemical Journal*, 317(Pt 1): 1–11. doi: 10.1042/bj3170001.

Denawaka, C.J., Fowlis I.A., and Dean, J.R. 2016. Source, impact and removal of malodour from soiled clothing. *Journal of Chromatography A*, 1438: 216–225. doi: 10.1016/j.chroma.2016.02.037.

DuPont. 2013. Chinese consumer laundry study, 2013. Accessed July 1, 2017. http://fhc.biosciences.dupont.com/fileadmin/user_upload/live/fhc/DuPont_Infographic_050313.pdf.

Eberle, U., Lange, A., Dewaele, J., and Schowanek, D. 2007. LCA study and environmental benefits for low temperature disinfection process in commercial laundry. *The International Journal of Life Cycle Assessment*, 12(2): 127–138. doi: 10.1065/lca2006.05.245.

Entwistle, J. 2000. *The Fashioned Body: Fashion, Dress, and Modern Social Theory*. Wiley-Blackwell.

Euromonitor International. 2011. Laundry care in Kenya (Sample report).

Gallagher, M., Wysocki C.J., Leyden J.J., Spielman A.I., Sun X., and Preti G. 2008. Analyses of volatile organic compounds from human skin. *British Journal of Dermatology*, 159(4): 780–791.

Garrett, P.R. 2013. *The Science of Defoaming: Theory, Experiment and Applications*. Taylor & Francis Group.

Gattlen, J., Amberg, C., Zinn, M., and Mauclaire, L. 2010. Biofilms isolated from washing machines from three continents and their tolerance to a standard detergent. *Biofouling*, 26(8): 873–882.

Golden, J.S., Subramanian, V., Irizarri, G.M.A.U., White, P., and Meier, F. 2010. Energy and carbon impact from residential laundry in the United States. *Journal of Integrative Environmental Sciences*, 7(1): 53–73. doi: 10.1080/19438150903541873.

Gordon, A.K., Muller, W.J., Gysman, N., Marshall, S.J., Sparham, C.J., O'Connor, S.M., and Whelan, M.J. 2009. Effect of laundry activities on in-stream concentrations of linear alkylbenzene sulfonate in a small rural South African river. *The Science of the Total Environment*, 407(15): 4465–4471. doi: 10.1016/j.scitotenv.2009.04.023.

Gotoh, K., Horibe, K., Mei, Y., and Tsujisaka, T. 2016. Effects of water hardness on textile detergency performance in aqueous cleaning systems. *Journal of Oleo Science*, 65(2): 123–133. doi: 10.5650/jos.ess15168.

Grimsvang, E., and Klonteig, V. 1985. *Husholdningsvask ved 40° og 60°C: Litteraturundersøkelse*. Statens Institutt for Forbruksforskning (SIFO).

Gwozdz, W., Netter, S., Bjartmarz, T., and Reisch, L.A. 2013. *Survey Results on Fashion Consumption and Sustainability among Young Swedes*. Mistra future fashion.

Hartline, N.L., Bruce, N.J., Karba, S.N., Ruff, E.O., Sonar, SU., and Holden, P.A. 2016. Microfiber masses recovered from conventional machine washing of new or aged garments. *Environmental Science & Technology*, 50(21): 11532–11538. doi: 10.1021/acs.est.6b03045.

Hasenclever, K.D. 2007a. D.1 - Dry cleaning of textiles. In I. Johansson, and P. Somasundaran (eds.), *Handbook for Cleaning/Decontamination of Surfaces* (pp. 407–425). Elsevier Science B.V.

Hasenclever, K.D. 2007b. D.2 - Wet cleaning of textiles. In I. Johansson, and P. Somasundaran (eds.), *Handbook for Cleaning/Decontamination of Surfaces* (pp. 427–438). Elsevier Science B.V.

Hassan, M.M., and Christopher M.C. 2019. A review of the sustainable methods in imparting shrink resistance to wool fabrics. *Journal of Advanced Research*, 18: 39–60. doi: 10.1016/j.jare.2019.01.014.

Höfer, D. 2006. Antimicrobial textiles, skin-borne flora and odour. *Current Problems in Dermatology*, 33: 67–77. doi: 10.1159/000093937.

Honisch, M., Brands, B., Weide, M., Speckmann, H-D., Stamminger, R., and Bockmühl, D.P. 2016. Antimicrobial efficacy of laundry detergents with regard to time and temperature in domestic washing machines. *Tenside Surfactants Detergents*, 53(6): 547–552.

Honold, P. 2000. Culture and context: An empirical study for the development of Framework for the elicitation of cultural Influence in product usage. *International Journal of Human–Computer Interaction*, 12(3–4): 327–345. doi: 10.1080/10447318.2000.9669062.

ISO 3175-4. 2018. Textiles—Professional care, drycleaning and wetcleaning of fabrics and garments—Part 4: Procedure for testing performance when cleaning and finishing using simulated wetcleaning. In *International Organization for Standardization.* International Organization for Standardization.

Johansson, I., and Somasundaran, P. 2007. *Handbook for Cleaning/decontamination of Surfaces*. Elsevier Science & Technology.

Kaiser, S.B. 1997. *The Social Psychology of Clothing: Symbolic Appearances in Context.* 2nd edn. Fairchild publications.

Kalayci, B., İlkiz, B.A., Budak, E.H., Altay, P., and Gürsoy, N.Ç. 2017. Design of automatic textile airing program: An approach to sustainable laundering. *Tekstil ve Konfeksiyon*, 27(4): 393–399.

Kao. 2010a. *Consumer Research about Washing in China: Part I (600 Households in Urban Areas) - Estimated Water Amount for Washing*. Kao Corporation.

Kao. 2010b. *Consumer Research about Washing in China: Part II (852 Households in Urban Areas) - Specific Washing Behaviors*. Kao Corporation.

Kerr, K-L., Rosero, S.J., and Doty, R.L. 2005. Odors and the perception of hygiene. *Perceptual and Motor Skills*, 100(1): 135–141.

Kjeldsberg, M., Eilertsen, K., and Laitala, K. 2011. *Shrinkage, Pilling, Stain Removal and Soil Repellence on Wool and Other Fabrics*. Test report 12-2011. National Institute for Consumer Research.

Klepp, I. 2003. *Fra rent til nyvasket : skittent og rent tøy (Clean to new-washed: Dirty clothes - clean habits)*. National Institute for Consumer Research.

Klepp, I.G. 2005. The meaning of cleanliness: Modern demonstrations of female purity. In G. Hagemann and H. Roll-Hansen (eds.), *Twentieth-Century Housewives: Meanings and Implications of Unpaid Work* (pp. 191–216). Unipub.

Klepp, I.G. 2007. Patched, louse-ridden, tattered: Clean and dirty clothes. *Textile: Journal of Cloth and Culture*, 5(3): 254–275. doi: 10.2752/175183507X249459.

Klepp, I.G., Buck, M., Laitala, K., and Kjeldsberg, M. 2016. What's the problem? Odor-control and the smell of sweat in sportswear. *Fashion Practice: The Journal of Design, Creative Process & the Fashion Industry*, 8(2): 296–317. doi: 10.1080/17569370.2016.1215117.

Klepp, I.G., and Tobiasson, T.S. 2019. *Lettstelt: Rene klær med lite arbeid og miljøbelastning*. Solum Bokvennen.

Kohli, R. 2019. Applications of dry vapor steam cleaning technique for removal of surface contaminants. In R. Kohli and K.L. Mittal (eds.), *Developments in Surface Contamination and Cleaning: Applications of Cleaning Techniques* (pp. 681–702). Elsevier.

Koohsaryan, E., Anbia, M., and Maghsoodlu, M. 2020. Application of zeolites as non-phosphate detergent builders: A review. *Journal of Environmental Chemical Engineering*, 8(5): 104287. doi: 10.1016/j.jece.2020.104287.

Kowton, J., and McQueen, R. 2016. The perception of odor in textiles: An exploratory study. *International Textile and Apparel Association (ITAA) Annual Conference Proceedings*. 105.

Kruschwitz, A., Augsburg, A., and Stamminger, R. 2013. How effective are alternative ways of laundry washing? *Tenside Surfactants Detergents*, 50(4): 263–269.

Kruschwitz, A., Karle, A., Schmitz, A., and Stamminger, R. 2014. Consumer laundry practices in Germany. *International Journal of Consumer Studies*, 38(3): 265–277. doi: 10.1111/ijcs.12091.

Laing, R.M. 2019. Natural fibres in next-to-skin textiles: Current perspectives on human body odour. *SN Applied Sciences*, 1(11): 1329. doi: 10.1007/s42452-019-1388-1.

Laitala, K., Boks, C., and Klepp, I.G. 2011. Potential for environmental improvements in laundering. *International Journal of Consumer Studies*, 35(2): 254–264. doi: 10.1111/j.1470-6431.2010.00968.x.

Laitala, K., and Jensen H.M. 2010. Cleaning effect of household laundry detergents at low temperatures. *Tenside Surfactants Detergents*, 47(6): 413–420. doi: 10.3139/113.110096.

Laitala, K., and Kjeldsberg, M. 2012. Cleaning effect of alternative laundry products: A comparison of soap nuts, laundry balls, washing pellets, laundry magnets, water and regular detergent. *Householdand Personal Care today*, 7(4): 53–56.

Laitala, Kirsi, Kjeldsberg, M., and Klepp, I.G. 2012. Troubles with the solution: Fabric softeners and odour properties. *Tenside Surfactants Detergents*, 49(5): 362–368. doi: 10.3139/113.110203

Laitala, K., Klepp I.G., and Boks, C. 2012. Changing laundry habits in Norway. *International Journal of Consumer Studies*, 36(2): 228–237. doi: 10.1111/j.1470-6431.2011.01081.x.

Laitala, K., Klepp, I.G., and Henry, B. 2017a. Global laundering practices: Alternatives to machine washing. *H&PC Today – Household and Personal Care Today*, 12(5): 10–16.

Laitala, K., Klepp, I.G., and Henry, B. 2017b. Use phase of apparel: A literature review for Life Cycle Assessment with focus on wool. In *Professional Report No. 6-2017*. SIFO.

Laitala, K., Klepp I.G., Kettlewell R., and Wiedemann, S. 2020. Laundry care regimes: Do the practices of keeping clothes clean have different environmental impacts based on the fibre content? *Sustainability*, 12(18): 7537. doi: 10.3390/su12187537.

Laitala, K., and Vereide, K. 2010. *Washing Machines' Program Selections and Energy Use*. Project Note 2-2010. National Institute for Consumer Research.

Lamb, J.M., and Kallal, M.J. 1992. A conceptual framework for apparel design. *Clothing and Textiles Research Journal*, 10(2): 42–47. doi: 10.1177/0887302X9201000207.

Larson, E., and Duarte, C.G. 2001. Home hygiene practices and infectious disease symptoms among household members. *Public Health Nursing*, 18(2): 116–127.

Laurent, J.B., Buzzaccarini, F., De Clerck, K., Demeyere, H., Labeque, R., Lodewick, R., and van Langenhove, L. 2007. Laundry cleaning of textiles. In I. Johansson and P. Somasundaran (eds.), *Handbook for Cleaning/decontamination of Surfaces*. Elsevier Science & Technology.

Lenochová, P., Roberts, S.C., and Havlî Ek, J. 2009. Methods of human body odor sampling: The effect of freezing. *Chemical Senses*, 34(2): 127–138.

Lund, H., Kaasgaard, S.G., Skagerlind, P., Jorgensen, L., Jorgensen, C.I., and van de Weert, M. 2012. Correlation between enzyme activity and stability of a protease, an alpha-amylase and a lipase in a simplified liquid laundry detergent system, determined by differential scanning calorimetry. *Journal of Surfactants and Detergents*, 15(1): 9–21. doi: 10.1007/s11743-011-1272-5.

Mather, R.R. 2015. Chapter 6 - Synthetic textile fibres: Polyolefin, elastomeric and acrylic fibres. In R. Sinclair (ed.), *Textiles and Fashion* (pp. 115–138). Woodhead Publishing.

Mauss, M. 1979. *Sociology and Psychology: Essays*. Routledge & Kegan Paul.

McQueen, R., Batcheller, J., Moran, L., Zhang, H., and Hooper, P. 2017. Reducing laundering frequency to prolong the life of denim jeans. *International Journal of Consumer Studies*, 41(1): 36–45. doi: 10.1111/ijcs.12311.

McQueen, R.H., Moran, L.J., Cunningham, C., Hooper, P.M., and Wakefield, K.A-M. 2019. The impact of odour on laundering behaviour: an exploratory study. *International Journal of Fashion Design, Technology and Education*,13(1):20–30. doi:10.1080/17543266.2019.1682687.

McQueen, R.H., Moran, L.J., Cunningham, C., and Hooper, P.M. 2020. Exploring the connection between odour and clothing disposal. *The Journal of The Textile Institute*, 112(11):1859–1866. doi: 10.1080/00405000.2020.1848114.

McQueen, R., Harynuk, J., Wismer, W., Keelan, M., Xu, Y., and Paulina de la Mata, A. 2014. Axillary odour build-up in knit fabrics following multiple use cycles. *International Journal of Clothing Science and Technology*, 26(4): 274–290. doi: 10.1108/IJCST-05-2013-0064.

McQueen, R., Laing, R.M., Brooks, H.J.L., and Niven, B.E. 2007. Odor intensity in apparel fabrics and the link with bacterial populations. *Textile Research Journal*, 77(7): 449–456. doi: 10.1177/0040517507074816.

McQueen, R., Laing, R.M., Delahunty, C.M., Brooks, H.J.L., and Niven, B.E. 2008. Retention of axillary odour on apparel fabrics. *Journal of the Textile Institute*, 99(6): 515–523. doi: 10.1080/00405000701659774.

McQueen, R., and Vaezafshar, S. 2019. Odor in textiles: A review of evaluation methods, fabric characteristics, and odor control technologies. *Textile Research Journal*, 90(9–10): 1157-1173. doi: 10.1177/0040517519883952.

Miracle, G.S., Randall, S.L., Liu Z., Brogden, D.W., Ketcha, M.M., Good, D.A., Johnson, M.B., Stenger, P.C., Hertz, P.R., and Meli, F. 2020. Copper chelants and antioxidants in laundry detergent formulations reduce formation of malodor molecules on fabrics. *Journal of Surfactants and Detergents*, 23(6): 1125–1134. doi: 10.1002/jsde.12467.

Morales-Garcia, A.L., Hayward, A.S., Malekpour, A.K., Korzycka, K.A., Compson, R., Gori, K., and Lant, N.J. 2020. The application of a nuclease enzyme to clean stubborn soils and odors in laundry. *Journal of Surfactants and Detergents*, 23(4): 797–807. doi: 10.1002/jsde.12398.

Motwani, M., Rhein, L.D., and Zatz, J.L. 2001. Differential scanning calorimetry studies of sebum models. *Journal of Cosmetic Science*, 52(4): 211–224.

Munk, S., Johansen, C., Stahnke, L., and Adler-Nissen, J. 2001. Microbial survival and odor in laundry. *Journal of Surfactants and Detergents*, 4(4): 385–394. doi: 10.1007/s11743-001-0192-2.

Munk, S., Munch P., Stahnke, L., Adler-Nissen, J., and Schieberle, P. 2000. Primary odorants of laundry soiled with sweat/sebum: Influence of lipase on the odor profile. *Journal of Surfactants and Detergents*, 3(4): 505–515.

Nagoh, Y., Tobe, S., Watanabe, T., and Mukaiyama, T. 2005. Analysis of odorants produced from indoor drying laundries and effects of enzyme for preventing malodor generation. *Tenside Surfactants Detergents*, 42(1): 7–12.

Namligoz, E.S., Bahtiyari, M.I., Hosaf, E., and Coban, S. 2009. Performance comparison of new (dendrimer, nanoproduct) and conventional water, oil and stain repellents. *Fibres & Textiles in Eastern Europe*, 17(5): 76.

Nayak, R., and Ratnapandian, S. 2018. *Care and Maintenance of Textile Products Including Apparel and Protective Clothing*: CRC Press.

Nix, I.D., Frontzek, A., and Bockmühl, D.P. 2015. Characterization of microbial communities in household washing machines. *Tenside Surfactants Detergents*, 52(6): 432–440. doi: 10.3139/113.110394.

Obendorf, S.K., Namasté, Y.M.N., and Durnam, D.J. 1983. A microscopical study of residual oily soil distribution on fabrics of varying fiber content. *Textile Research Journal*, 53(6): 375–383. doi: 10.1177/004051758305300609.

Okada, N., Fujii, T., and Minagawa, M. 1984. Removal of oily soils (Part 4) effect of fatty acid soaps on removal of oily soils and yellowing of the fabrics washed in hard water. *Journal of Home Economics of Japan*, 35(5): 322–330.

Okada, N., Fujii, T., and Okuyama, H. 1976. Removal of oily soils. Part 2. The relation of yellowing and residue with hard ions on fabrics. *Kaseigaku Zasshi J Home Econ*, 27: 215–220.

Pakula, C., and Stamminger, R. 2009. Comparison of resources used for laundry washing worldwide. WFK 44th Int. Detergency Conf., Düsseldorf, May 12th to 14th, 2009.

Pakula, C., and Stamminger, R. 2010. Electricity and water consumption for laundry washing by washing machine worldwide. *Energy Efficiency*, 3(4): 365–382. doi: 10.1007/s12053-009-9072-8.

Pastore, C.M., and Kiekens, P. 2000. *Surface Characteristics of Fibers and Textiles*. Vol. 94. Taylor & Francis Group.

Patra, J.K., and Gouda, S. 2013. Application of nanotechnology in textile engineering: An overview. *The Journal of Engineering and Technology Research*, 5(5): 104–111. doi: 10.5897/JETR2013.0309.

Pugliese, S., Jespersen, M.F., Pernov, J.B., Shenolikar, J., Nygaard, J., Nielsen, O.J., and Johnson, M.S. 2020. Chemical analysis and origin of the smell of line-dried laundry. *Environmental Chemistry*, 17(5): 355-363. doi: 10.1071/EN19206.

Rathinamoorthy, R., Thilagavathi, G., Brindha, S., Gayathri, P., Poornakala, N.S., and Pradeep, B. 2014. Odour control studies on apparel fabrics finished with methanol extract of Terminalia chebula. *Fibers and Polymers*, 15(8): 1669–1676.

Rathinamoorthy, R. 2020. Effect of repeated dry-cleaning process on physical properties of cotton, silk and wool fabrics. *Indian Journal of Fibre & Textile Research*, 45(1): 116–122.

Rigby, E.D. 2016. *Fashion Design and Laundry Practices: Practice-Orientated Approaches to Design for Sustainability*. University of the Arts London.

Schmitz, A., and Stamminger, R. 2014. Usage behaviour and related energy consumption of European consumers for washing and drying. *Energy Efficiency*, 7(6): 937–954. doi: 10.1007/s12053-014-9268-4.

Shove, E. 2003. *Comfort, Cleanliness and Convenience: The Social Organization of Normality*. Berg.

Sinclair, R. 2015. *Textiles and Fashion: Materials, Design and Technology*. Woodhead Publishing Limited.

Sinner, H. 1960. *Über das Waschen mit Haushaltwaschmachinen*. 2nd edn. Haus+Heim Verlag.

Smith, J.A., Neil, K.R., Davidson, C.G., and Davidson, R.W. 1987. Effect of water temperature on bacterial killing in laundry. *Infection Control & Hospital Epidemiology*, 8(5): 204–209.

Stamminger, R. 2010. Reinigen. In: *Lebensmittelverarbeitung Im Haushalt, 28*. aid infodienst.

Stapleton, K., Hill, K., Day, K., Perry, J.D., and Dean, J.R. 2013. The potential impact of washing machines on laundry malodour generation. *Letters in Applied Microbiology*, 56(4): 299–306.

Takeuchi, K., Hasegawa, Y., Ishida, H., and Kashiwagi, M. 2012. Identification of novel malodour compounds in laundry. *Flavour and Fragrance Journal*, 27(1): 89–94. doi: 10.1002/ffj.2088.

Takeuchi, K., Yabuki, M., and Hasegawa, Y. 2013. Review of odorants in human axillary odour and laundry malodour: The importance of branched C7 chain analogues in malodours perceived by humans. *Flavour and Fragrance Journal*, 28(4): 223–230.

Terpstra, P. 2001. The impact of changing sustainable technology and changing consumer habits on infectious disease transmission in the domestic setting. *Journal of Infection*, 43(1): 99–102. doi: 10.1053/jinf.2001.0860.

Terpstra, P.M.J. 1998. Domestic and institutional hygiene in relation to sustainability. Historical, social and environmental implications. *International Biodeterioration & Biodegradation*, 41(3-4): 169–175.

Toedt, J., Koza, D., and Cleef-Toedt, K.V. 2005. *Chemical Composition of Everyday Products*. Greenwood Press.

Troynikov, O., Watson, C., Jadhav, A., Nawaz, N., and Kettlewell, R. 2016. Towards sustainable and safe apparel cleaning methods: A review. *Journal of Environmental Management*, 182: 252–264. doi: 10.1016/j.jenvman.2016.07.078.

Wang, J., Lu, X., Wang, J., and Wang, X. 2019. Quantitative and sensory evaluation of odor retention on polyester/wool blends. *Textile Research Journal*, 89(13): 2729–2738. doi: 10.1177/0040517518801183.

Wang, L., Ding, X., Huang, R., and Wu, X. 2014. Choices and using of washing machines in Chinese households. *International Journal of Consumer Studies*, 38(1): 104–109. doi: 10.1111/ijcs.12070.

Wiedemann, S.G., Biggs, L., Nebel, B., Bauch, K., Laitala, K., Grimstad Klepp, I., Swan, P.G., and Watson, K. 2020. Environmental impacts associated with the production, use, and end-of-life of a woollen garment. *The International Journal of Life Cycle Assessment*, 25(8): 1486–1499. doi: 10.1007/s11367-020-01766-0.

11 Odour Evaluation Techniques in Textiles Area: Introduction of E-nose as a Potential Alternative Tool

Sima Shakoorjavan
Division of Chemistry and Physical Chemistry of Polymers, Institute of Materials Science of Textiles and Polymers Composites, Lodz University of Technology, Lodz, Poland

Textile Engineering Department, School of Materials and Advanced Processes Engineering, Amirkabir University of Technology (Tehran Polytechnic), Tehran, Iran

Somaye Akbari
Textile Engineering Department, School of Materials and Advanced Processes Engineering, Amirkabir University of Technology (Tehran Polytechnic), Tehran, Iran

Dawid Stawski
Division of Chemistry and Physical Chemistry of Polymers, Institute of Materials Science of Textiles and Polymers Composites, Lodz University of Technology, Lodz, Poland

CONTENTS

DOI: 10.1201/9781003141426-11

11.1 INTRODUCTION

11.1.1 Odour Definition

The word "odour" in its literal sense means "the property of a substance that activates the sense of smell" which can be referred to both pleasant and unpleasant smell (Odour, Definition n.d.). In scientific sense, odours are non-ionic and hydrophobic volatile inorganic or organic chemical compound with very low concentration and molecular weight generally 20 ppmv or less than 300 Daltons (Da), respectively; which generally have an unpleasant smell. When these volatile chemical compounds are pleasing, energizing, refreshing, or calming in nature, they are named aroma or fragrance (Aroma Compounds and Their Odors n.d.; Govind n.d.; Schiffman and Pearce 2003). Odour perception is caused by stimulation of the olfactory system by odourant substances. In other words, odour sensation takes place when specific receptors in the olfactory system interact with odourants that can enter and reach to those receptors in olfactory epithelium. Therefore, the fundamental physical requirement of odour perception is to have low molecular weight and very low concentrations; there is a report that only 2% of volatile compound is enough to be perceived by human nose. After odourant entrance to nostril, molecules are required to be transferred form gaseous phase into aqueous media of the nasal mucus in olfactory epithelium in the top of the nasal cavity. In the following, each olfactory epithelium transfers its signals directly to olfactory bulb. In the molecular level, the chemical signal transforming to the electrophysiological message, and then processed in our brain as an odour (Schiffman and Pearce 2003; Triller et al. 2008; Zeki 1993). Until now, the largest discovered and known odourant molecules refer to labdane with molecular weight of 296 Daltons (Ohloff 1994). As a molecular weight of odourant increase, specific anosmia become more frequent since they do not meet the biological size requirement to reach nose. In addition, volatility (vapor pressure) of a molecule decreases as molecular size increase. Therefore, molecule with large molecular size are more likely to be odourless (Turin and Yoshii 2003).

Sense of smell, unlike other senses such as taste and sight, has not been understood well and does not have identical and distinctive dimension of category. Although huge progress and development was achieved in the knowledge of olfactory biochemistry and physiology, still there is obscure understanding of odour quality and its relationship with molecular properties (Schiffman and Pearce 2003). The next section is dedicated to describe aroma compound chemical structures, to point out aroma classification, and possible relationship between odour chemical structure and the molecules odour based on finding until now.

11.1.2 Chemical Structures of Aroma Compounds

Odourants molecules can encompass very wide range of chemical structures and classes including alcohols, organic acid, aldehydes, aromatics, amines, amides, ethers, esters, hydrocarbons, halogenated hydrocarbons, ketones, phenols, nitriles, and compounds containing nitrogen or sulfur, which all are uncharged molecules (Schiffman and Pearce 2003). Despite of substantial advancement in the knowledge of olfactory biochemistry and physiology, still there is a big gap between odour quality and its molecular properties; however, many attempts were made to build a bridge between them (Turin and Yoshii 2003). According to literature, there are two main theories describing the relationship between the properties of a chemical and its odour that were supported by quantitative measurements with statistical analysis. The first theory, introduced by Amoore (Amoore et al. 1967), is the so-called theory of odotypes, linking the presence of specific functional groups and shape of the molecule with its smell. The second theory is the vibration theory, which tries to relate the vibrational structure of a molecule, defined by the FTIR spectrum, with smell (Wright and Robson 1969). Both theories, published in Nature in the 1960s, aim to link a selected element of a molecule's structure with the smell it emits.

Odotype theory is based on a similar mechanism to that used in the description of ligand-receptor interactions. According to this theory, receptors cannot detect the specific molecules but the functional groups such as hydroxyl, aldehyde, carbonyl, sulfone, etc. It assumes that the binding of the ligand to the receptor is caused by electrostatic and dipole interactions, hydrogen bonds, hydrophobic interactions, and the shape of the molecule; and the strength of a molecule's binding to the receptor is related to its shape and functional groups on the surface, which subsequently affect the strength of the smell. Therefore, odotype theory allows to some extent to link the presence of functional groups in chemical compounds with the smell they emit (Turin 2002). For example, compounds containing the -SH group will smell sulfur, the presence of the hydroxyl group (-OH) will be detected as a characteristic sharp spirit smell, and compounds with the -CN group have a metallic character similar to the smell of almonds, and nitrites (-NO2) are used as flavor and smell enhancers in food products. An interesting case are aldehydes, which have a distinct, strong smell. Depending on the number of carbon atoms in the structure, the smell is different: (C8 - sweet orange pulp; C9 - rose; C10 - bitter orange peel; C11 - coriander; C12 - violet and lilac; C13 - aroma of hot wax with a hint of grapefruit; C14 - fresh peach). Despite the connection between a specific functional group and the emission of a specific odour (Turin 2002) sometimes it happens that compounds with different chemical structure have a similar odour such as ethyl citronellyl oxalate and cyclopentadecanolide. Both compounds have a similar scent of musk while having different chemical structure. According to the odotype theory, these compounds assume a similar conformation under room conditions and therefore have a similar aroma (Greenberg 1979). Another fact that contradicts this theory is the concentration dependence of the detected odour on small gas molecules. Compounds such as sulfur dioxide, ozone or sulfur hexafluoride smell completely different in the whole concentration range (Turin and Yoshii 2003). Moreover, enantiomers are a contradiction of the odotype theory. It turns out that some pairs have identical smells, while others have completely different aromas. This may mean that recipients are equipped with two types of receptors (they are chiral), so they will perceive the smells of optically active compounds differently. This assumption is correct, when a pair of enantiomers has the same odour, they stimulate both receptors in the same way. This assumption is contradicted by the situation in which they have different smells (Sugawara et al. 2000). As can be seen, the odotype theory is not able to predict the aroma of new compounds with a satisfactory accuracy. It requires refinement and further research.

The vibrational theory assumes that the olfactory receptors selectively resonate with odourous substances, and thus, knowing the infrared spectrum of a given compound, its smell can be predicted (Wright and Robson 1969). There were several disadvantages to this theory from the beginning: it could not explain the different smells of optical isomers (which actually have identical infrared spectra), and it did not explain the presence of odourless compounds. Additionally, the mechanism on which the perception of chemical compounds through their vibrational spectrum was based was unknown. In 1996 Luca Turin (Turin 1996) proposed updating the vibration theory. Using inelastic electron tunneling spectroscopy, he showed correlation between the obtained tunneling spectrum and odour in structurally unrelated molecules. As predicted, molecules with very similar structure but different vibrations smell different. The great advantage of the vibration theory in relation to the previous theory is the ability to predict the smell of molecules. Turin (Turin 1996) divided the frequency spectrum of particle vibrations into four areas where the identification of the spectral bands is sufficient to determine the aroma with high probability. The method developed by Turin has never easily miss-defined the type of smell. The author explained possible mistakes with errors in the literature determining the aroma of the substance, errors in the calculation of vibration spectra or in the algorithm. It turned out that the presented method is perfect for determining the aroma of new compounds before synthesis, by comparing the calculated spectrum of the new substance with that of the standard odour.

A number of experiments were carried out to determine which theory is the closest to the truth. Most have shown that the vibration theory is far more correct. However, it has one significant drawback compared to the odotype theory. It cannot in any way predict the strength of the smell and why some substances do not have it at all. If we determine the truth of a theory by its ability to predict the smell of new compounds, it will turn out that neither of the two theories can predict with certainty the flavor of the new substance. However, the vibrational theory is much more successful in this field than odotype theory. On the other hand, the odotype theory is able to some extent to explain the scent power of new compounds. Therefore, Luca Turin and Fumiko Yoshii (Turin and Yoshii 2003) adopted a provisional theory assuming that the type of smell is felt on the basis of the vibration theory, and its power is felt on the basis of the odotype theory.

11.1.3 Odour Types in Textile Materials

We can categorize the odour of textiles in Three classes that is summarized in Figure 11.1. First class refer to inherent odour of textiles such as inherent odour of

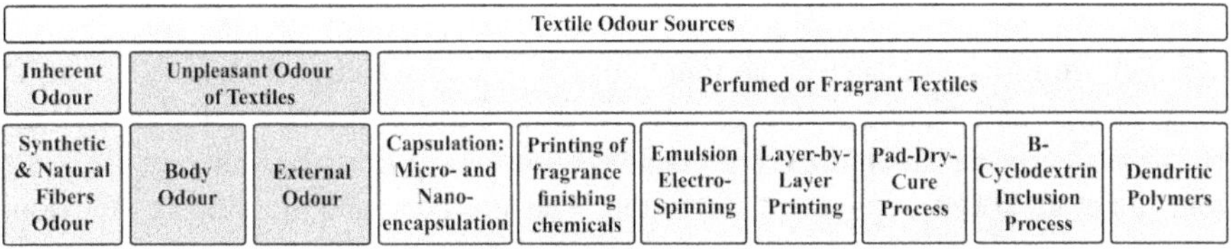

FIGURE 11.1 Classification of textiles odour types.

synthetic fibre using in carpet production or natural smell of wool specially when they are wet (Haeringer and Goschnick 2008). second classes and types of textiles odour would be associated with the unpleasant smell of textiles that can stem from two sources. The first source is related to the emission of body odours; secretions of the human body such as emissions from the sweat glands, urine, feces, genitals and skin. Bacteria was shown to be the main factor responsible for generating most body odours. Human skin is physiologically inhabited by microorganisms constantly present on the skin, and through decomposition of sweat by microorganism malodour created which can be picked up by clothing (Laing 2019; Muszyński 2010; Preti, Spielman, and Leyden 1998). Human behavior can further increase the emission of odours. This type of behavior includes applying of hygienic procedures (or not), eating specific foods (garlic, onions), drinking alcohol, and taking certain medications (Henkin 1995; Labows and Preti 1992). The second source of unpleasant textile odours is external, clothing can pick up the odour not only from body, but also from environment; for example, tobacco smoke on textiles (Ochiai et al. 2020). The amount of scent particles that can be retained on textiles depends on the type of material (natural, artificial or synthetic fibres), the type of product (woven, knitted fabric, non-woven fabric), its structure (type of weave, surface weight), and the duration of exposure to external factors (Hammond 2013). Even washing, which is the primary way to eliminate negative odours, may, with improper operation and maintenance of the washing machine, transfer old fragrances, contaminants or even bacteria from the appliance to the textile material (Denawaka, Fowlis, and Dean 2016). Fragrant textiles or perfumed textiles can be considered as the third class of odour types in textile. Fragrances can influence our well-being, awareness and social behavior; therefore, novel and special finishing, which can import fragrance to textile with long-lasting and control released behavior, has attracted most of researcher's attention (Martel et al. 2002; Tulshyan and Dedhia 2021). Despite of the importance of textile odour (original or artificial odour), measuring and assessment of textiles odour in qualitative or quantitative sense is important issue as well which is discussed in the following sections.

11.2 MEASURING METHODS

In textile industry, for many years, odour has been considered as an great indication to recognize and evaluated the efficiency of a process; for instant, analysing laundering effectiveness in malodour removal and minimize using of washing agents can be facilitated through monitoring the odour of contamination (Haeringer and Goschnick 2008). Also, in forensic science, odour, as an example the odour of human hand on a textile, is a solid evidence that can be admitted to court (Prada,

Curran, and Furton 2011). Moreover, aroma finishing and producing fragrant textiles with long-term release as an innovative finishing which can directly effect on the customer satisfaction and experience is a favourable commercial goal and has been considered as an engineering challenge for a long time (Saifullah et al. 2019; Yan Liu et al. 2008). As a result, detection, measurement, and monitoring the odour release from textiles have to be strongly considered.

Measurement and assessment of odour release can be a challenge since an odour does not consist of a single compound but a mixture of hundreds of odourants compound, and odour sensation is not understood as well as others sense like sense of taste or sense of sight (Schiffman and Pearce 2003). Generally, we can divide odour measurement methods into two categories. First is analytical method which is consist of human assessment as a qualitative method known as "sensory panel", and the second is analysing method as quantitative method which is related to identification and detection of chemical compounds of an odour by one by one. The later method is not exactly the method of monitoring or assessment of odour release but the detection method which associate with odourant compound of an odour and they are not always applicable for textiles. Therefore, scientists introduce new technique of monitoring the odour release from textile which recognizes a whole pattern of all odourant components not the components of samples. The following sections are dedicated to describe above-mentioned method of odour measurements.

11.2.1 Analytical Method

11.2.1.1 Sensory Panel Analysis as a Qualitative Method (Human Nose)

Generally, sensory evaluation refers to measurement and analysing technique of features which can be perceived and understood by human senses like sight, touch, hearing, taste and smell (Stone, Bleibaum, and Thomas 2012). Sensory evaluation regarding odourant which deal with sense of smell is mostly utilized in food (Domenico et al. 2020), beverage (Voss et al. 2021) and cosmetic (Carranza et al. 2020) industries. In textile industry, sensory analyses as primary method to measure the odour release from textiles were utilized by Signe Munk et al. (Munk et al. 2000) on 2000. They tried to study the influence of lipase added in detergent on human sweat removal from cotton and polyester swatches after mild washing process through sensory evaluation performed by nine trained judges (assessors) using ranking test and statistically Friedmans test. Assessors who are selected to perform sensory panel analysis should be trained and taught to know how to rank samples and differentiate intensity and quality of odourants (McQueen and Vaezafshar 2020).

Generally, there is three discriminations and scaling methods for odour evaluation of textiles. The first and simples method is paired comparison (applied on non-woven textile) (Willard and Beaverson n.d.) in which samples can be differentiated directly from one another when there is small difference between samples. This method strongly depends on assessor memory to remember the sensation of the previous stimuli. Therefore, this method is not applicable for multiple samples and comparisons (McQueen et al. 2007). Miller (Miller 2002) introduced an efficient method of sensory evaluation named "Quad analysis" that is more time-efficient than traditional paired comparison method and is appropriate for multiple samples comparisons. The principle is based on a series of forced-choice paired comparison between four samples (quads) selecting randomly from a larger group. As it is shown in Figure 11.2, the quads are force-ranked from the best or highest intensity (4) to the worst or lower intensity (1) through paired comparison (pair 1 and pair 2). From pair 1 and pair 2 the most intense samples paired in pair 3 and the 2 remained samples paired in pair 4. In the last step, the least intense sample from pair 3 and the most intense sample from pair 4 couple in pair 5, and 2 remained samples retain in their position (McQueen et al. 2007; McQueen and Vaezafshar 2020). The major defect of quad analysis method is that the magnitude of difference is not considered and cannot be measured. To address this problem the "line scale" is introduce for sensory evaluation (Stefanowicz 2013). MacQueen et al. (McQueen et al. 2007) evaluated axillary odour on cotton, wool, and polyester through nine assessors according to quad analysing (QA) and line scaling (LS). The results showed that QA is time-efficient in comparison traditional paired-comparison method, but still it takes a lot of time and effort in comparison to scaling methods. LS method showed to be more appropriate technique for odour measurement of samples that are likely to change over time, and it is easy and does not required repeat smelling.

Odour sensory panel evaluation as a subjective method required serious considerations to prevent potential biases that may occur during measurement

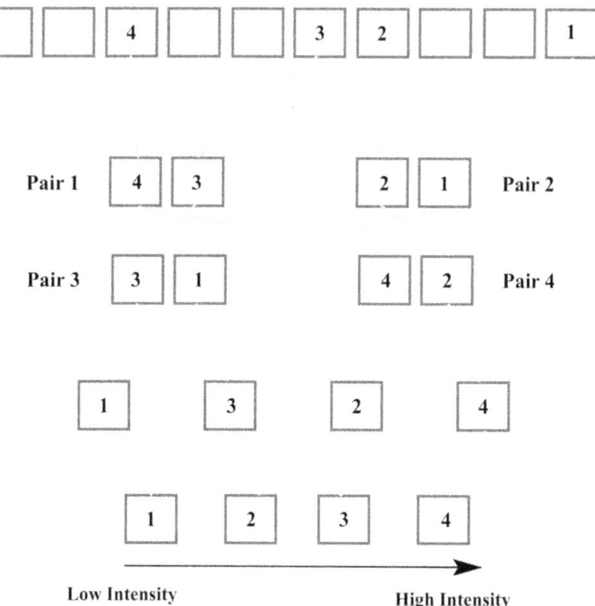

FIGURE 11.2 Quad Analysis scheme as a quantitative odour assessment of textiles.

process. The selection of a proper test method and trained sensory panels along with preparing appropriate environment for performing the measurement are all crucial concerns (McQueen and Vaezafshar 2020). ISO 8586 is a reference for selecting and training assessor for odour sensory evaluation. The minimum proposed number of trained assessors are 10 but based on details of the sensory panel measurements and the level of sensitivity number of panellist can be larger (ISO 8586:2012, Sensory Analysis—General Guidelines for the Selection, Training and Monitoring of Selected Assessors and Expert Sensory Assessors 2012). However, odour sensory evaluation is affected by various factors such as sensory panel fatigue, physical and mental health, and is strongly depended on intrinsic differences in human sensitivity to odourant and adaptation to odour during measurement process. To address the problems associated with human assessment, instrumental analyses and electronic nose are alternative tools to control odour release from textiles (McQueen and Vaezafshar 2020; Nagle, Schiffman, and Gutierrez-Osuna 1998).

11.2.2 Analysing Tool for Odour Identification/Release Assessment of Aroma and Volatile Compounds

To circumvent problems associated with human odour assessment instrument analyses are the best solution. However, most of the instrumental analysis as a quantitative and objective methods try to identify and detect the chemical compounds of an odour including analysing concentration and types of odourants dealing with an odour (McQueen and Vaezafshar 2020) rather than measuring the intensity or the quality of the odour release form textile which are concerned in qualitative and subjective human assessment. In the following section the most typical instrumental analyse used in odour detection from textiles are described.

11.2.2.1 Gas Chromatography Coupled with Different Detectors

One of the most typical and common analytical technique to separate and detect organic molecules or gases in a mixture along with determination of their percentage is gases chromatography (GC) (Dewulf, Van Langenhove, and Wittmann 2002). However, limited studies regarding odour evaluation in textiles industry using this technique was investigated since odour collection from textile being detectible by GC is not simple, and additional device such as headspaces are required or it should be coupled with other detectors such as mass spectrum and flam ionization which are described in the following.

There is a record that GC was employed to identify the amount of aroma content in protective cotton fabric treated with N-doped TiO_2 embedded citral microcapsule using pad-roll method as well as monitoring sustained-aroma release properties of the treated fabric. The aroma value was measured using portable gas meter according to Triangel odour bag method (GB/T 14675-93). The results showed the amount of citral content is 2.90 mg/g (the weight of essence to the weight of fabric) and the aroma release after 12 weeks was about 50% that provide promising application of the functional cotton fabric in air purification and personal protection (Wang et al. 2020). By using dynamic headspace analysis, there is a chance to quantify and measure individual volatiles in a mixture as they separated by gas chromatography. In order to have non-volatile materials, such as polymers in headspace analysis, a minimum amount of samples is required to be prepared. In a study, evaporation or release of fragrance from the synthesised amphiphilic multiarm star-block copolymers as hyperbrached polymers (HBPs) was measured using dynamic headspace analysis in which evaporation rate of fragrance form copolymers in the presence of and non-ionic, anionic, cationic surfactant as well as ethanol was measured. Briefly, the polymer was dissolved in ethanol solution (85%), then small amount of sample was located in the sampling cell of headspace. Through pumping air across samples, the evaporated volatiles were trapped on cartridges at certain time intervals, and the adsorbed volatile in cartridge were thermally desorbed and measured by gas chromatography. The results confirmed controlled release of fragrance in the presence of surfactant and the amphiphilic polymer could provide controlled released of bioactive materials in under typical application condition (Ternat et al. 2008).

Nikolova et al. (Nikolova and Bayryamov 2019) reported three different characterise techniques including sensory panel analysis, gas chromatography with headspace or coupled with mass spectroscopy, and electronic nose to examine the efficiency of encapsulation, capability of immobilization, and properties of microcapsule coated on textiles. In the case that microcapsules can be vaporized without decomposition, GC analysis is a great technique to separate microcapsules component and determining their relative amount with high accuracy. In order to analysis the internal phase of microcapsules without direct sampling, headspace-CG would be an appropriate method. To find the relationship between chromatographic profiles with odour characteristics, GC-MS is the best solution. A synoptic of technical procedure, finished textiles with microcapsules need to be cut into small pieces and put to enclosed vials that are immersed water bath at certain temperature for certain time. Then, the samples take to heated chromatograph injector of GC-MS, and the peak area of the GC-MS would be considered as initial results.

The most common odour evaluation using GC in textiles area associated with human sweat and axillary odour (McQueen et al. 2014) and laundry effectiveness in malodour removal (Denawaka, Fowlis, and Dean 2016) when coupled with mass spectrum. Rathinamoorthy et al. (Rathinamoorthy and Thilagavathi 2016) tried to characterize the odour-forming compounds on worn cotton knitted fabric using gas chromatography and mass spectrum (GC-MS). In this study, GC-MS was utilized to identify and measure the amount of the most well-known sweat components that are responsible for odour forming in the axillary including volatile short chain fatty acids and alcohols such as a saturated fatty acid of myristic acid and derivative of myristyl alcohol, respectively. For GC-MS analysis, the sweat components were extracted from the worn cotton swatches in which one was plain fabric and the other was treated with methanolic extract of Terminalia chebula fruit. The obtained results confirmed their previous finding that bacterial population and their secretion play pivotal roles in creating and formation of new odour compounds in fabric surface. They could identify the major odour forming source from axilla including steroidal fractions of 5α-androst-16-ene-3-one and cholesterol on worn textile as well as other prominent odour forming compounds such as fatty acids and alcohols. Also, it was revealed that finished cotton had minor amount of components which were responsible for odour formation.

Gas chromatography-flame ionization is another method that has been investigated to consider odour retention and release from textile. Abdul-Bari et al. (Mukhtar Abdul-Bari et al. 2020) analysed odour release and retention of knitted polyester and cotton fabrics soiled with three different odourant with different lipophilicites and polarities (octanoic acid, 2-nanenal, dodecane) using gas chromatography coupled with flame ionization detector during several soil/wash cycles. Also, they evaluated the effectiveness of two commercial detergents on odour removal. The results showed that polyester fabrics release more soil odour above headspace than cotton fabric before laundering since polyester showed higher peak areas of volatiles release per mass. However, the same laundering procedure was more effective on soil odourant removal from cotton fabric than polyester. It was also determined that laundering agents are sufficient in removing polar octanoic acid.

11.2.2.2 Proton Transfer Reaction-Mass Spectrometry

Another fast quantitative tool to detect and identify known volatile organic compounds is proton-transfer reaction mass spectrometry (PTR-MS) which works based on ionising reaction of H_3O^+ with detected volatile organic compounds by non-dissociative proton transfer. This technique has a great potential in food, medical and environmental researches (Ellis and Mayhew 2013; Yuan et al. 2017). In textiles area, PTR-MS were utilized to analysis the volatile profile of body and axilla odour as well but less commonly than other above-mention techniques. For long time, it was accepted that unpleasant odour of axilla result from certain gram-positive bacteria activities which are present on skin surface since axillary sweat is odourless by itself, and textile are great substrate for bacterial growth and to adsorb and emanate body odour (Mcqueen and Laing 2007; Senol and Fireman 1999). As a results, many researches were carried out to produce antibacterial textiles to avoid unpleasant odour emission from textiles (Chen, Wang, and Yeh 2010; Moses and Venkataraman 2017; Velmurugan et al. 2014). However, it was proved that intensity of axilla odour in fabrics depended on not only on the numbers of the bacteria population but also the chemical and physical structure of the fabrics (Mcqueen and Laing 2007; Rathinamoorthy and Thilagavathi 2014). In this regard, to investigate the effect of fibre chemical and physical structure on axilla odour retention on textile, McQueen et al. (McQueen et al. 2008) tried to measure and analysis human axilla retention on and emanation from three interlock fabrics (cotton, wool, and polyester) using an online monitoring instrument PTR-MS. To analyse samples (after 1 and 7 days of wearing) with the PTR-MS system, samples were placed in 250 mL bottles, and then connected to the PTR-MS inlet flow. After 1 hour and reaching headspace equilibrium, sampling was performed in which headspace was drawn at certain flow rate through a heated capillary in to the proton-transfer reaction mass spectrometry. Mass spectrometric data was recorded for five times, and the mean of data was reported and analysis through principle component analysis (PCA). The results indicated that odour intensity of axilla in fabric inversely depended on fibre hydroscopicity. In another study, adsorption profile of body odour compounds on three different yarn/fibre types, including cotton, wool, and polyester fabrics, was investigated by Yao et al. (Yao et al. 2015). In this study, six compounds that are responsible for body odour, including decanal, cyclohexanone, ethylbenzen, butanoic acid, methyl ester, and phenol, were selected, and the profile of each compounds was, first, checked by PRT-MS to confirm their consistency with published references. To measure odour compounds on textile the experiment was repeated 15 times, and the mean of the data was reported. The obtained results revealed which type of fibre adsorbed which odourant compounds. For example, cotton adsorbed all odourant compounds less than other yarns (wool and polyester), and phenol, in comparison to other odourant compounds, was highly adsorbed by all three fibre types. Somewhere else, adsorption and release behaviour of other body odour compounds, including dimethyl disulfide, 2-propanethiol, benzaldehyde, nonanal, butanoic acid, and 3-methyl-2-hexenoic acid on three types of fibre, the same as the previously mentioned study (cotton, wool, and polyester), was investigated by Richter et al. (Richter et al. 2018). The results indicated that although wool fibres have high adsorption in comparison to other fibres, the odour release is almost lower. Polyester showed relatively high adsorption and release behaviour, and cotton has low adsorption and consequently low overall release.

11.2.2.3 Solid-Phase Micro-Extraction

In a study, headspace solid-phase micro-extraction (SPME) and comprehensive two-dimensional gas chromatography time-of-flight mass spectrometry (GC × GC-TOFMS) was employed to detect human axillary sweat chemical compounds and odourants release from worn cotton and polyester fabric (de la Mata et al. 2017). Prada et al. (Prada, Curran, and Furton 2011) also investigated human hand odours on cotton, polyester, rayon, and wool fabric using SPME-GC-MS.

Third-hand smoke as a consequence of cigarette smoking has great potential to attach to textile products and fabrics resulting in unpleasant odour release from the textile for months. Ochiai et al. (Ochiai et al. 2020) successfully developed non-woven fabric embedded three different photo-catalyst which are responsive to ultra violet (UV) and visible light and capable of adsorbing and decomposing odour and tar components of tobacco smoke after specific visible or UV irradiation periods. The evaluations were performed in the terms of color and odour changes before and after UV exposure. Odour assessments and total volatile organic compounds (TVOC) concentration associated with smoke were calculated using gas chromatography and mass spectrum of solid-phase extraction. The results showed that the nonwoven fabric treated with Fe/TiO_2 photocatalyst had the best results in contamination removal after 24 h of irradiation, with removal ratio of TVOC of 55%.

11.2.2.4 Other Instrumental Techniques

Inverse gas chromatography is a powerful and sensitive fast technique to investigate physicochemical properties and parameters at any required concentration specially at infinite dilution region of any surfaces including powders, films, and fibres

(Mohammadi-Jam and Waters 2014). In a study, scientist tried to characterize cotton fabric and fragrance molecule interaction, including surface energy and specific free energy of fragrance molecules on cotton fabrics using inverse gas chromatography. However, there was no record that the odour evaluation was investigated in the study (Reutenauer and Thielmann 2003).

There is a record that infrared-based (IR) spectrometer and sensory analysis were employed to investigate the influence of wool in polyester/wool blends in body odour retention and emission. The principle of the technique is based on differences in wavelength absorption. In other words, according to the number and the types of functional groups the light wavelength absorption would be different. Therefore, IR-based technique could be a good tool to quantitative study of odour compounds. In this study, the utilized system was consisting of three essential parts: a gas pump, a sample chamber, and infrared-based spectrometer. Briefly, 2 gram of fabric was located into the sample chamber, and certain amount of odour compound was injected into the chamber using gas pump and circulated in the sealed system. Real-time odour sorption behaviour of each fabric was screened through IR-based spectrometer. In order to measure odour release from fabric, 2 gram of fabric was placed into the sealed container, which was filled with a certain volume of odour for 8 hours at 20°C, and the sample was removed and transfered to sealed chamber. The odour emission was controled by the IR-based spectrometer. The results showed that the fabric blend containing 20% wool had considerably lower body odour release in comparison to 100% polyester (Wang et al. 2019).

Based on research done by Fieber et al. (Fieber et al. 2007), hyperbrached polymers (HBPs) with water-soluble core-shell structures are great materials to entrap fragrance molecules in order to boost their longevity and durability. In the next step, they fully focused on investigation of fragrance release from the synthesised amphiphilic multiarm star-block copolymers as HBPs using thermogravimetry (TGA) and dynamic headspace analysis in which evaporation rate of fragrance form copolymers in the presence of and non-ionic, anionic, cationic surfactant, as well as ethanol was measured. Weight loss as a result of fragrance evaporation was measured by TGA in the presence and absence of amphiphilic multiarm star-block copolymers. The results confirmed that the presence of the amphiphilic core-shell structure decreased the fragrance evaporation rate, especially at the high ethanol content in the solution (Ternat et al. 2008).

11.3 INNOVATIONS IN ODOUR ASSESSMENT IN TEXTILES

Odour evaluation of textiles (intrinsic odour, sweat odour, fragrance) through gas chromatography and mass spectroscopy are expensive and time-consuming methods, and they do not exactly refer to odour measurement of textiles, but to detect and characterize odourant components. In addition, odour of fragrant textile consists of a mixture of volatile gases in which interpretation of results may be very complicated. In this regard, electronic nose as a simple and inexpensive alternative would be an appropriate solution to monitor odour release from textiles. This apparatus has the capability to analyse a whole pattern of components of an odour, not to recognize one by one odourant components (Nagle, Schiffman, and Gutierrez-Osuna 1998; Staerz et al. 2020).

This new invention, e-nose, has a great application in a variety of commercial industries, including biomedical and pharmaceutical industries, cosmetics, agricultural, manufacturing, and food industries, as well as military application, environmental monitoring, and quality control research (Patel 2014; Sharmitha et al. 2014). In the case of the textile industry, there is a record that the commercial electronic nose, Bloodhound® ST214, which is based on 24 organic semiconducting sensors, has been utilized to measure anti-odour and anti-microbial capabilities of fabrics (Gibson et al. 2009). However, current and available commercial electronic noses are too expensive to be used in textile field; therefore, a self-designed e-nose would be a more efficient and cost-effective system to be employed in the textiles industry according to specific and certain aim and application (Zhang et al. 2020), such as evaluation of fragrance finished textiles (Sharmitha et al. 2014) and other applications, which are reported in the following section.

11.3.1 Electronic Nose as a Quantitative Odour Assessment

Electronic nose, briefly E-nose, is considered to be an artificial nose that mimics the mammalian olfactory system (Scheme 11.1). It is a system comprised of three fundamental components, including a sample handler, an array of chemical gas sensors, and a signal processing system. The basic idea is that each chemical sensor in the array has different sensitivity to an odourant resulting in different response and pattern for each odour referring to an odour fingerprint (Patel 2014). The fundamental procedures that should be happened to cause E-nose to detect and identify an odour are summarized below (Patel 2014; Wilson and Baietto 2009):

1. Exposure of gases to the sensors
2. Reversible chemical reaction of gas molecules with sensing materials of sensors, resulting in reversible changes in electrical resistance response
3. Signal creation and transmitting to neural network or other pattern recognition (PARC) mechanism
4. Matching the signal pattern with references or library database of known materials in E-nose system
5. Cleaning sensors for the next process and sampling

Two main parts of the electronic nose are the sensing system, serving as human olfactory nerves, and the pattern recognition system, serving as human brain (Patel 2014). The most common gas sensors utilizing in E-nose sensing system are chemo-resistive gas sensors, which are believed to be adaptable to a wide range of gasses (oxidant and reducing), and have low fabrication cost and high thermal and temporal stability. Chemo-resistive gas sensors are comprised of conductive polymers (CPs) and metal oxide, which operate based on conductance variation of sensing element that occurred by reversible oxidant or reduction reaction of sensor surface when exposed to a gas (Arshak et al. 2004; Korotcenkov 2013). Conductance variation as sensors response is used to be analysed through pattern-recognition

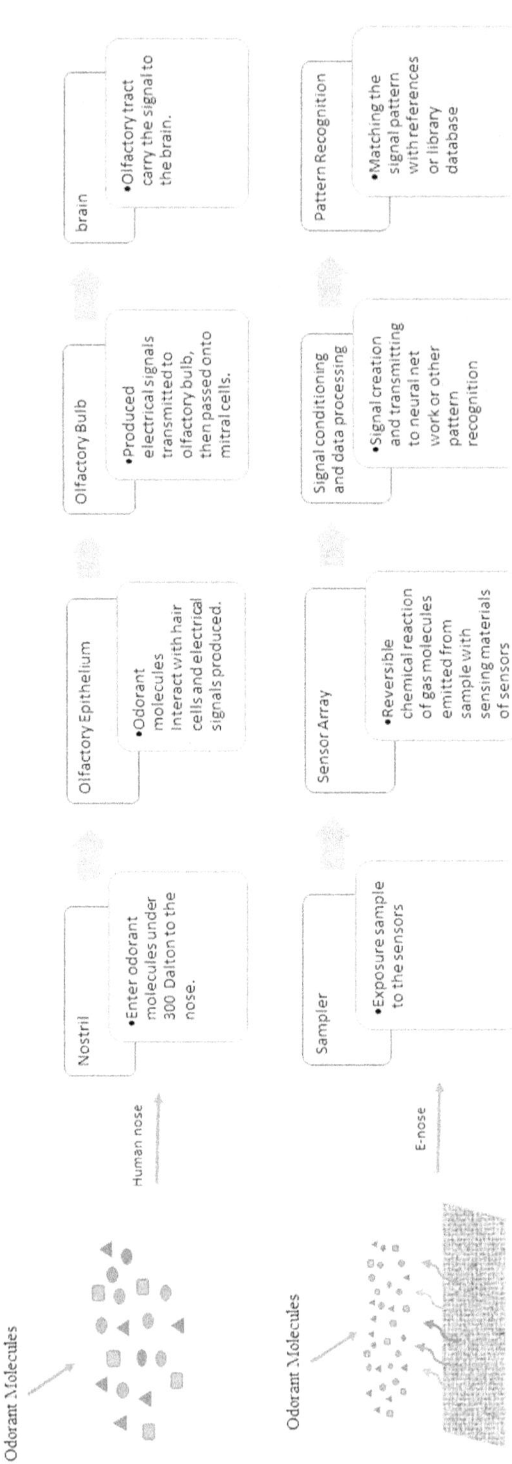

SCHEME 11.1 The mechanism of how Electronic nose and mammalian olfactory system detect an odour.

method. Identification and classification of an unknown analyte is based on recognition of the specific odour signature and pattern of collective sensor responses according to available database. Indeed, first, a reference library and database of known samples should be created (training process), and then sampling (test process) of unknown samples can be operated. The most common recognition-pattern method for data analysing is principle component analysis (PCA), which is a linear features extraction technique to shorten the dimension of data with the least information loss (Patel 2014; Wilson and Baietto 2009).

Daniel Haeringer et al. (Haeringer and Goschnick 2008) employed an electronic nose to monitor laundry odour to minimize the amount of washing agent while the cleanliness is at on optimum. It is claimed that E-nose was integrated in the host washing machine to monitor online gas information, which is a very promising tool for obtaining details of laundry characterization, which are prerequisite for the intelligent washing process. The unique gradient of microarray of 38 metal oxide sensors as Karlsruhe Micronose KAMINA was developed for fabric discrimination and odour determination. Gas sensor microarray (GSMA) is not compromised of separate sensors, but a monolithic metal oxide layer with parallel electrode strips into sensor segments leading to economical fabrication, great gas performance and high robustness. The principle of odour discrimination is based on changes in electrical conductivities response according to ambient gas stem from gradients of the surface temperature and the thickness of permeable membrane that differentiated sensor segments. In this study, cotton, wool, and polyester were selected as fabric model, and three different odours, including cigarette smoke, rotten milk, and human sweat, were chosen as representative of real odour. In all cases, 10 cm of fabric stripes were placed in heatable glass cylinder (20–75°C), and the released odour from fabric was transfered to the KAMINA chip board through a constant flow of synthetic air with 50% humidity. Then, data processing and evaluation were performed through linear discriminant analysis (LDA) model.

Textiles and clothing can pick up and adsorb body odour as well as environment odour, leading to creation of unpleasant odour in textiles. In this regard, scientists were impelled to develop textiles with anti-odour and anti-bacterial properties. In a study, the researchers successfully introduced anti-odour properties to polyester and cotton fabric with activated carbon as the most common adsorption materials through pigment printing and coating technique. To evaluated anti-odour properties of the treated fabrics, they were exposing to onion odour for 24 hours, and the odour release was measured and monitored using portable electronic nose 3 (PEN 3) devices. The logic behind selecting onion odour is that changes in body odour by any physiological changes such as stress are described as oniony. All samples were placed in a screw-top glass and the analyses were conducted, and the E-nose data was analysed with LDA technique. Onion odour was recorded first as a reference odour, and the distance of samples odour to onion odour as a reference indicated how and which samples absorb more and less onion odour (Eza, Ahmad, and Ahmad 2012).

Producing value-added functional textile and impart functional properties such as aroma and fragrance with long-lasting properties is one of the innovative finishes in the textile field of study that causes a pleasant sensational effect on the consumer.

In this regard, measuring aroma released from textiles as well as evaluation of the effectiveness of aroma finishing technique, and its durability need special attention (Akbari et al. 2016; Sharmitha et al. 2014; Singh and Sheikh 2021). There are some records that E-nose is a great alternative tool to evaluate odour release of fragrant textiles, which is summarized here. Maria M. Miro Specos et al. (Miró Specos et al. 2010) attempted to enhance the durability of fragrances in cotton using two different types of microcapsules containing lemon oil. The electronic nose as well as sensory panels (judges) were employed to measure oil content in a treated fabric; in other words, to measure aroma release from textiles before and after laundering. The electronic nose system consists of 12 gas sensors with different sensitivity, and a whole set of responses is considered as a fingerprint referring to the sample odour. The data analysis was performed by principle component analysis in which the radar area was proportional to average intensity as you can see in Figure 11.3.

P. Sharmitha et al. (Sharmitha et al. 2014) successfully fabricated an electronic nose device as a simple, convenient, and reliable objective measurement method to evaluate odour release of jasmine finished cotton. The aim was to monitor aroma emitted from fragrant cotton fabric over time (immediately, after 5 and 10 days) and the influence of the washing process (after 1 and 2 washing cycle) on odour durability on finished textile using both electronic nose and sensory panel analysis. In this study, the electronic nose system was comprised of a developed test box with electrical connection for sensor characterization, a small fan to homogenously circulate fragrance, and a metal oxide (SnO_2) sensor model MQ 135 containing 6 pin, 4 of them for fetching signals and the other 2 for providing heating current. Before, the aroma measurement learning process was conducted by three different jasmine fragrance concentration and 36 sets of samples to create a reference database library. Then, the device was utilized to recognize and quantify measurement of odour release. Briefly, all samples in the size of 5*5 cm^2 were placed in specific location design in a developed text box and exposed to the sensor. Reaction of an odour with sensing material of the sensor cause changes in physical parameters of the sensor, resulting in producing multiple curves in relation to change in voltage vs. exposure time. The maximum value of each curve was filtered and transformed into multidimensional voltage response pattern, and then concentration according to the multidimensional pattern is obtained as electronic nose results. The results showed that retention intensity measured by E-nose is strongly correlated with

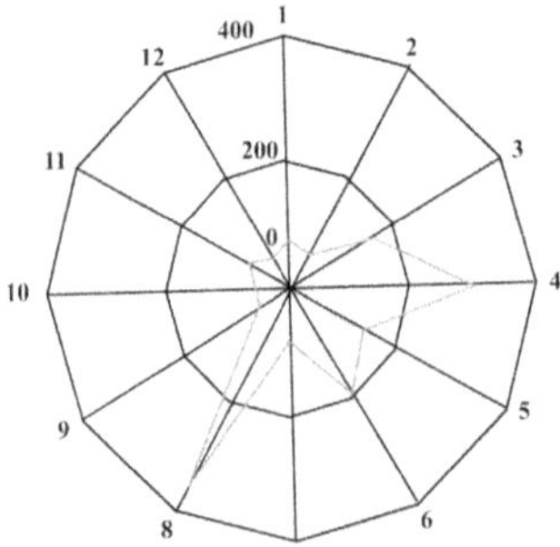

FIGURE 11.3 Radar analysis; radar area was proportional to average intensity.

human rating scores, and E-nose showed greater ability for odour evaluation of fragrant samples in comparison to human panels.

Despite all of the advantages and positives features, metal oxide-based gas sensors have low selectivity, leading to use of an array of gas sensors in the E-nose system (Barsan, Koziej, and Weimar 2007). In addition, discrimination capability of a sensor array is affected by sensors ageing, leading to slight changes in sensor frequency, known as drift, which is temporal sensor response changes (Arshak et al. 2004; Yuwono and Lammers 2004). To address these problems, various techniques were used; the most common one was temperature modulation involving controlling semiconductor surface temperature through applying different voltage to the sensor heater, leading to difference in heater temperature (Hossein-Babaei and Amini 2012). Shakoorjavan et al. (Shakoorjavan et al. 2016) successfully fabricated a single-sensor gas diagnosis device made by temperature modulation technique (4 heating steps) acting as a virtual sensor array in order to quantitatively measure odour emission from fragrant textiles. In this study, cotton, wool, polyester, and worsted fabrics were fragranced with jasmine and thyme essence using second-generation polypropylene imine dendrimer (PPI-G2), and accuracy and capability of fabricated virtual sensor array acting as E-nose to discriminate fragrant fabric based on fabric types, odour types, and intensity was studied. Before each sampling, the system was run several times in clean air to make the virtual sensor array free from any contamination. Repeatability and appropriate operation of the device was examined by conducting 14 experiments and confirmed by determination of correlation coefficient in four heating steps. Odour discrimination capability of the device was studied by conduction training and test experiments of thyme and jasmine fragrant wool fabrics, and the E-nose response was analysis by principle component analyse (PCA) and linear discriminant analysis (LDA) pattern recognition technique, as you can see in Figure 11.4. The results showed that the virtual sensors array is capable of distinguishing thyme and jasmine essence on wool fabric, and test experiments were located in a restricted area of the training experiment, which indicates the appropriate operation of the device and its recognition capability. In addition, the device was able to discriminate not only a certain essence (thyme) on 4 different fabrics based on the fabric types (Figure 11.5) but also different thyme essence concentration (10, 40, and 80 g/l) on wool fabric. Another achievement of this study was that although human recognition (performance-based quad analysis) of odour intensity was different from one person to another, there was a strong correlation with 0.96 coefficients between weighted human results and the virtual sensor array results. At the end, a single-sensor gas device as a virtual sensor array acting an E-nose was proposed to be a great alternative tool to sensory panel analysis for fragrant fabrics. The same fabricated electronic nose (Shakoorjavan et al. 2016) was utilized in other studies as an quantitative odour measurements. Asafi Fard et al. (Fard, Shakoorjavan, and Akbari 2018) tried to investigate controlled release of thyme essence from fragrant cotton fabric using PPI dendrimer, and the relationship between odour intensity and antibacterial properties. The E-nose results confirmed that PPI dendrimer can provide controlled release of fragrant cotton fabric, and odour intensity can be considered as an antibacterial indication since there was a strong correlation of $r = 0.988$ between

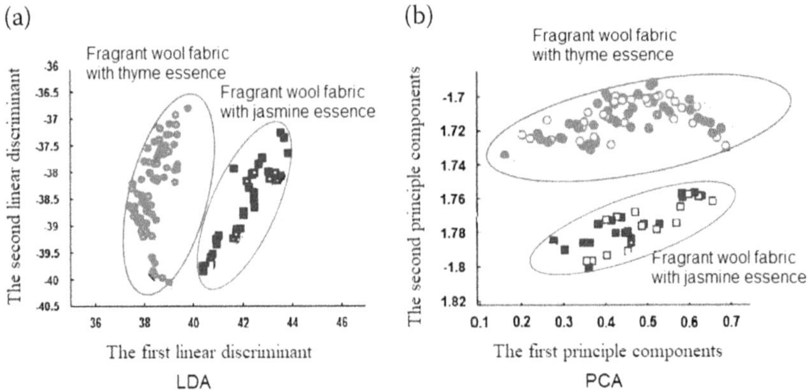

FIGURE 11.4 Test and train data analysis of wool fragrant textile with jasmine and thyme based on (a) LDA and (b) PCA model; Blank circles and squares show trained results, and filled ones show test results.

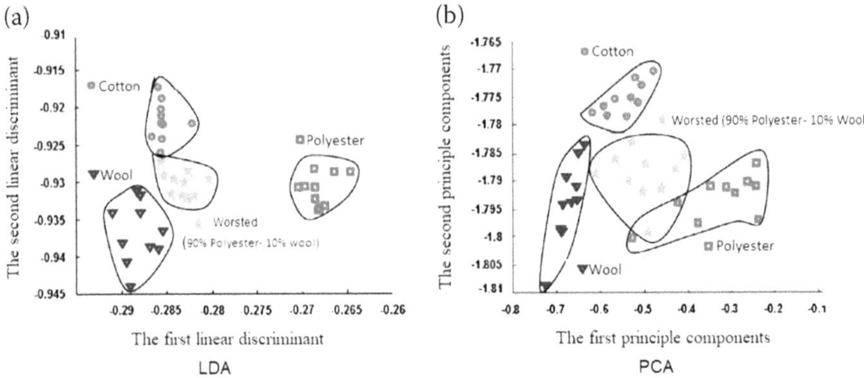

FIGURE 11.5 Data analysis of thyme-fragrant textiles based on (a) LDA and (b) PCA model to study fabric type on odour emission.

antibacterial activity and odour intensity of thyme-fragrant cotton fabric. Mounesan et al. (Mounesan, Akbari, and Brycki 2021) successfully produced electrospun fibre-embedded tea tree and eucalyptus essential oils using polyamidoamine (PAMAM) dendrimer, and the E-nose results indicated that PAMAM can extend essential oil release from 5 to 27 days. Kamrudi et al. (Kamrudi, Akbari, and Haghighat Kish 2020) reported the kinetic trend release of thyme essential oil-embedded in nylon 6 electrospun fibre using a single-sensor gas diagnosis device.

11.4 CONCLUSION

Odour evaluation of textiles includes qualitative sensory panel analysis and quantitative instrumental analysis. Although sensory panel analysis performed by humans – judges, panellists, and assessors – is a simple and applicable method, it is subjective and highly depends on human physical and emotional status. There is always a risk of biased assessment. In order to circumvent sensory analysis defects, objective and quantitative measurement were utilized. According to present research, there is a different instrumental method to identify and characterize odourant molecules trapped in textiles. The most common technique is gas GC with headspace or coupled with different directors such as mass spectrometry or flame ionization. In the case that concentration and types of odourant is a concern, headspace analysis is an appropriate technique. If the identification of an individual compound in a mixture is required to have analysis, GC-MS is a solution, and in the case of analysing a known compound, a flame-ionization detector would help. Besides GC, solid-phase microextraction (SPME)-GC time-of-flight mass spectrometry is also reported to be utilized as a technique to monitor human hand or body odour released from textiles. Proton-transfer reaction mass spectrometry is the next analysis and instrumental tool that was reported as sensitive and a fast technique to detect and identify known volatile compounds, such as body and sweat odorant compounds transferred to a garment next to the body. Unlike the GC technique that is expensive and time-consuming, this technique is in real-time and is quite a fast analysis without the need of sample pre-treatment. In addition to the above-mentioned technique, inverse GC, infrared-based spectrometer, and thermogravimetry are also reported in textile odour emission analysis. All of the above techniques are associated with identification and characterization of volatile components that are responsible for an odour, and they would be useless when the whole pattern of an odour is a concern. An odourant and fragrance are made of a complex mixture of up to 40 or more ingredients; therefore, odour assessment of fragrant textiles as an added-value product with long-lasting properties with the above technique would be complicated. To address this problem, electronic nose as a quantitative and simple technique comprised of a sensor array and pattern recognition algorithm was introduced. The principle of the E-nose is based on chemoresistive gas sensors with different sensitivity that respond differently to an odour. As a result, E-nose is capable of monitoring and recognition of an odour as a whole component, or fingerprint. In other words, it analyses and produces a whole pattern of components of an odour, rather than recognizing individual odourant components one by one. E-nose is an artificial nose that mimics mammalian olfactory obviate the all defects of sensory analysis performed by humans, and since there was a good correlation between sensory panel analysis results and E-nose response, E-nose was introduced as an appropriate alternative tool to sensory analysis as a highly applicable technique in odour assessment of textiles.

REFERENCES

Akbari, S., Akbari, M., Kish, M.H., and Mazaheri, F.M. 2016. Preparation of long-lasting fragrant worsted fabrics using Polypropylene-Imine (PPI) dendrimer. *International Journal of Clothing Science and Technology*, 28(5): 699–711. 10.1108/IJCST-07-2015-0081.

Amoore, J.E., Palmieri, G., and Wanke, F. 1967. Molecular shape and odour: Pattern analysis by PAPA. *Nature*, 216: 1084–1087.

Aroma Compounds and Their Odors. n.d. Accessed November 21, 2020. https://www.thoughtco.com/aroma-compounds-4142268.

Arshak, K., Moore, E., Lyons, G.M., Harris, J., and Clifford, S. 2004. A review of gas sensors employed in electronic nose applications. *Sensor Review*. 10.1108/02602280410525977.

Barsan, N., Koziej, D., and Weimar, U. 2007. Metal oxide-based gas sensor research: How to? *Sensors and Actuators, B: Chemical*, 121(1): 18–35. 10.1016/j.snb.2006.09.047.

Carranza, K., Rodriguez, C., Esenarro, D., Veliz, M., and Arteaga, J. 2020. Sensory evaluation of a perfume made of orange essential oil. *International Journal of Chemical Engineering and Applications*, 11(4): 89–92.

Chen, C.-C., Wang, C.-C., and Yeh, J.-T. 2010. Improvement of odor elimination and antibacterial activity of polyester fabrics finished with composite emulsions of nanometer titanium dioxide-silver particles-water-borne polyurethane. *Textile Research Journal*, 80(4): 291–300. 10.1177/0040517508100626.

Denawaka, C.J., Fowlis, I.A., and Dean, J.R. 2016. Source, impact and removal of malodour from soiled clothing. *Journal of Chromatography A*, 1438(March): 216–225. 10.1016/j.chroma.2016.02.037.

Dewulf, J., Van Langenhove, H., and Wittmann, G. 2002. Analysis of volatile organic compounds using gas chromatography. *TrAC - Trends in Analytical Chemistry*, 21(9–10): 637–646. 10.1016/S0165-9936(02)00804-X.

Domenico, J.D., Machado-Lunkes, A., Prado, N.V., Weber, C.I., and Lucchetta, L. 2020. Reduction of sodium content in Pork Coppa: Physicochemical, microbiological and sensory evaluation. *Scientia Agricola*, 78(6): 1–9. 10.1590/1678-992x-2020-0153.

Ellis, A.M., and Mayhew, C.A. 2013. *Proton Transfer Reaction Mass Spectrometry: Principles and Applications*. Wiley.

Eza, T.S.M., Ahmad, W.Y.W., and Ahmad, M.N. 2012. The activated carbon as anti-odour coated and pigment printed fabric. In *BEIAC 2012 - 2012 IEEE Business, Engineering and Industrial Applications Colloquium* (pp. 210–215). 10.1109/BEIAC.2012.6226054

Fard, P.A., Shakoorjavan, S., and Akbari, S. 2018. The relationship between odour intensity and antibacterial durability of encapsulated thyme essential oil by PPI dendrimer on cotton fabrics. *Journal of the Textile Institute*, 109(6): 832–841. 10.1080/00405000.2017.1376820.

Fieber, W., Herrmann, A., Ouali, L., Velazco, M.I., Kreutzer, G., Klok, H.A., Ternat, C., Plummer, C.J.G., Manson, J.A.E., and Sommer, H. 2007. NMR diffusion and relaxation studies of the encapsulation of fragrances by amphiphilic multiarm star block copolymers. *Macromolecules*, 40(15): 5372–5378. 10.1021/ma070222i.

Gibson, T., Chandler, R., Hallam, V., Simpson, C., and Bentham, M. 2009. Industrial applications of electronic nose technology in the textiles industry. In *AIP Conference Proceedings*, 1137:235–238. 10.1063/1.3156515.

Govind, R., and Narayan, S. 2005. Biotreatment of organic and inorganic odors. Selection of bioreactor media for odor control. In*Biotechnology for Odor and Air Pollution Control*(pp. 65–100). Springer.

Greenberg, M.J. 1979. Dependence of odor intensity on the hydrophobic properties of molecules. A quantitative structure odor intensity relationship. *Journal of Agricultural and Food Chemistry*, 27(2): 347–352. 10.1021/jf60222a054.

Haeringer, D., and Goschnick, J. 2008. Characterization of smelling contaminations on textiles using a gradient microarray as an electronic nose. *Sensors and Actuators B: Chemical*, 132(2): 644–649.

Hammond, C.J. 2013. Chemical composition of household malodours-an overview †. *Wiley Online Library*, 28(4): 251–261. 10.1002/ffj.3163.

Henkin, R.I. 1995. Body odor. *Journal of the American Medical Association*, 273(15): 1171–1172. 10.1001/jama.1995.03520390025014.

Hossein-Babaei, F., and Amini, A. 2012. A breakthrough in gas diagnosis with a temperature-modulated generic metal oxide gas sensor. *Sensors and Actuators, B: Chemical*, 166–167(May): 419–425. 10.1016/j.snb.2012.02.082.

ISO 8586:2012, Sensory Analysis—General Guidelines for the Selection, Training and Monitoring of Selected Assessors and Expert Sensory Assessors. 2012.

Kamrudi, N., Akbari, S., and Haghighat Kish, M. 2020. Enhanced control release of thyme essential oils from electrospun nanofiber/polyamidoamine dendritic polymer for anti-bacterial platforms. *Polymers for Advanced Technologies*, 31(8): 1719–1731. 10.1002/pat.4899.

Korotcenkov, G. 2013. *Handbook of Gas Sensor Materials: Properties, Advantages and Shortcomings for Applications Volume 1: Conventional Approaches*. Vol. 1. Springer Science & Business Media. https://scholar.google.com/scholar?hl=en&as_sdt=0%2C5&q=+Handbook+of+gas+sensor+materials%3A+properties%2C+advantages+and+shortcomings+for+applications&btnG=.

Labows, J.N., and Preti, G. 1992. Human semiochemicals. In S. Van Toller and G.H. Dodd (eds.), *Fragrance: The Psychology and Biology of Perfume* (pp. 69–90). Elsevier Applied Science Publishers.

la Mata, A.P.de, McQueen, R.H., Nam, S.L., and Harynuk, J.J. 2017. Comprehensive two-dimensional gas chromatographic profiling and chemometric interpretation of the volatile profiles of sweat in knit fabrics. *Analytical and Bioanalytical Chemistry*, 409(7): 1905–1913. 10.1007/s00216-016-0137-1.

Laing, R.M. 2019. Natural fibres in next-to-skin textiles: Current perspectives on human body odour. *SN Applied Sciences*, 1(11). 10.1007/s42452-019-1388-1.

Martel, B., Morcellet, M., Ruffin, D., Vinet, F., and Weltrowski, M. 2002. Capture and controlled release of fragrances by CD finished textiles. *Journal of Inclusion Phenomena*, 44: 439–442. 10.1023/A:1023028105012.

McQueen, R.H., Harynuk, J.J., Wismer, W.V., Keelan, M., Xu, Y., and Paulina De La Mata, A. 2014. Axillary odour build-up in knit fabrics following multiple use cycles. *International Journal of Clothing Science and Technology*, 26(4): 274–290. 10.1108/IJCST-05-2013-0064.

Mcqueen, R.H., and Laing, R.M. 2007. Odor intensity in apparel fabrics and the link with bacterial populations. *Article in Textile Research Journal*, 77(7): 449–456. 10.1177/0040517507074816.

McQueen, R.H., Laing, R.M., Delahunty, C.M., Brooks, H.J.L., and Niven, B.E. 2008. Retention of axillary odour on apparel fabrics. *Journal of the Textile Institute*, 99(6): 515–523. 10.1080/00405000701659774.

McQueen, R.H., Laing, R.M., Wilson, C.A., Niven, B.E., and Delahunty, C.M. 2007. Odor retention on apparel fabrics: Development of test methods for sensory detection. *Textile Research Journal*, 77(9): 645–652. 10.1177/0040517507078792.

McQueen, R.H., and Vaezafshar, S. 2020. Odor in textiles: A review of evaluation methods, fabric characteristics, and odor control technologies. *Textile Research Journal*, 90(9–10): 1157–1173. 10.1177/0040517519883952.

Miller, R.W. 2002. Subjective property characterization by 'quad' analysis: An efficient method for conducting paired comparisons. *Textile Research Journal*, 72(12): 1041–1051. 10.1177/004051750207201202.

Miró Specos, M.M., Escobar, G., Marino, P., Puggia, C., Victoria Defain Tesoriero, M., and Hermida, L. 2010. Aroma finishing of cotton fabrics by means of microencapsulation techniques. *Journal of Industrial Textiles*, 40(1): 13–32. 10.1177/1528083709350184.

Mohammadi-Jam, S., and Waters, K.E. 2014. Inverse gas chromatography applications: A review. *Advances in Colloid and Interface Science*. Elsevier. 10.1016/j.cis.2014.07.002.

Moses, J.J., and Venkataraman, V.K. 2017. Study of K/S, anti-bacterial, UV-protection, anti-odor and SEM on chemical treated cotton fabrics. *Journal of Engineered Fibers and Fabrics*, 12(2): 155892501701200. 10.1177/155892501701200205.

Mounesan, M., Akbari, S., and Brycki, B.E. 2021. Extended-release essential oils from Poly (Acrylonitrile) electrospun mats with dendritic materials. *Industrial Crops and Products*, 160(February): 113094. 10.1016/j.indcrop.2020.113094.

Mukhtar Abdul-Bari, M., McQueen, R.H., de la Mata, A.P., Batcheller, J.C., and Harynuk, J.J. 2020. Retention and release of odorants in cotton and polyester fabrics following multiple soil/wash procedures. *Textile Research Journal*, 90(19–20): 2212–2222. 10.1177/0040517520914411.

Munk, S., Münch, P., Stahnke, L., Adler-Nissen, J., and Schieberle, P. 2000. Primary odorants of laundry soiled with sweat/sebum: influence of lipase on the odor profile. *Journal of Surfactants and Detergents*, 3(4): 505–515. 10.1007/s11743-000-0150-z.

Muszyński, Z. 2010. Drobnoustroje Skóry Człowieka–Wskazówki Dla Kosmetologów. *Homines Hominibus*, 6: 55–64.

Nagle, H.T., Schiffman, S.S., and Gutierrez-Osuna, R. 1998. How and why of electronic noses. *IEEE Spectrum*, 35(9): 22–34. 10.1109/6.715180.

Nikolova, M., and Bayryamov, A.P.S. 2019. Characterization techniques for microcapsules immobilized on textiles. *Proceedings of University of Ruse*, 58: 96–102.

Ochiai, T., Aoki, D., Saito, H., Akutsu, Y., and Nagata, M. 2020. Analysis of adsorption and decomposition of odour and tar components in tobacco smoke on non-woven fabric-Supported photocatalysts. *Catalysts*, 10(3): 304. 10.3390/catal10030304.

Odour, Definition. n.d. Https://Www.Dictionary.Com/Browse/Odor.

Ohloff, G. 1994. *Scent and Fragrances. The Fascination of Odors and Their Chemical Perspectives*. Springer-Verlag.

Patel, H.K. 2014. *The Electronic Nose: Artificial Olfaction Technology*. Biological and Medical Physics, Biomedical Engineering. Springer India. 10.1007/978-81-322-1548-6.

Prada, P.A., Curran, A.M., and Furton, K.G. 2011. The evaluation of human hand odor volatiles on various textiles: A comparison between contact and noncontact sampling methods. *Journal of Forensic Sciences*, 56(4): 866–881. 10.1111/j.1556-4029.2011.01762.x.

Preti, G., Spielman, A.I., and Leyden, J.J. 1998. The structure, origin and function of human axillary odours. In *Fragrances* (pp. 21–27). Springer. 10.1007/978-3-642-80340-6_3.

Rathinamoorthy, R., and Thilagavathi G. 2014. Effect of antimicrobial finish on odor control properties of apparel fabric. *Journal of Textile & Apparel Technology & Management (JTATM)*, 9(1): 1–15.

Rathinamoorthy, R., and Thilagavathi, G. 2016. GC-MS analysis of worn textile for odour formation. *Fibers and Polymers*, 17(6): 917–924. 10.1007/s12221-016-5891-3.

Reutenauer, S., and Thielmann, F. 2003. The characterisation of cotton fabrics and the interaction with perfume molecules by inverse gas chromatography (IGC). *Journal of Materials Science*, 38: 2205–2208. 10.1023/A:1023788400431.

Richter, T.M., Bremer, P.J., Silcock, P., and Laing, R.M. 2018. Textile binding and release of body odor compounds measured by proton transfer reaction–mass spectrometry. *Textile Research Journal*, 88(22): 2559–2567. 10.1177/0040517517725126.

Saifullah, Md, Shishir, M.R.I., Ferdowsi, R., Rahman, Md R.T., and Van Vuong, Q. 2019. Micro and nano encapsulation, retention and controlled release of flavor and aroma compounds: A critical review. *Trends in Food Science and Technology*. 10.1016/j.tifs.2019.02.030.

Schiffman, S.S., and Pearce, T.C. 2003. Introduction to olfaction: Perception, anatomy, physiology, and molecular biology. In *Handbook of Machine Olfaction: Electronic Nose Technology* (pp. 1–32). 10.1002/3527601597.ch1.

Senol, M., and Fireman, P. 1999. Body odor in dermatologic diagnosis. *Europe PMC*, 63(2): 107–111.

Shakoorjavan, S., Akbari, S., Kish, M.H., and Akbari, M. 2016. Correlation of sensory analysis with a virtual sensor array data for odour diagnosis of fragrant fabrics. *Measurement: Journal of the International Measurement Confederation*, 90: 396–403. 10.1016/j.measurement.2016.04.074.

Sharmitha, P., Kavitha, S., Srinivasan, D.J., and Scholar, M.T. 2014. Fabrication of electronic nose for evaluation of performance of fragrance finished textiles. *International Journal of Advanced Information Science and Technology (IJAIST)*, 3(3). 10.15693/ijaist/2014.v3i3.57-60.

Singh, N., and Sheikh, J. 2021. Microencapsulation and its application in production of functional textiles. *Indian Journal of Fiber and Textile Research*, 45(4): 495–509.

Staerz, A., Roeck, F., Weimar, U., and Nicolae, B. 2020. Electronic nose. In *Surface and Interface Science* (pp. 335–379). Wiley. 10.1002/9783527822492.ch67.

Stefanowicz, P. 2013. Sensory evaluation of food principles and practices. *Journal of Wine Research*, 24(1): 80– 80. 10.1080/09571264.2013.764662.

Stone, H., Bleibaum, R.N., and Thomas, H.A. 2012. *Sensory Evaluation Practices*. Science. https://books.google.com/books?hl=en&lr=&id=ZqNcZYNUXWIC&oi=fnd&pg=PP2&dq=Sensory+Evaluation+Practices&ots=IX2uAD0-bR&sig=Q38X3NKR4IpP1NbQWxVuRceyj2c.

Sugawara, Y., Hara, C., Aoki, T., and Sugimoto, N. 2000. Odor distinctiveness between enantiomers of linalool: difference in perception and responses elicited by sensory test and forehead surface potential wave. *Chemical Senses*, 25(1): 77–84.

Ternat, C., Ouali, L., Sommer, H., Fieber, W., Velazco, M.I., Plummer, C.J.G., Kreutzer, G., Klok, H.A., Månson, J.A.E., and Herrmann, A. 2008. Investigation of the release of bioactive volatiles from amphiphilic multiarm star-block copolymers by thermogravimetry and dynamic headspace analysis. *Macromolecules*, 41(19): 7079–7089. 10.1021/ma801366m.

Triller, A., Boulden, E.A., Churchill, A., Hatt, H., England, J., Spehr, M., and Sell, C.S. 2008. Odorant - receptor interactions and odor percept: A chemical perspective. *Chemistry and Biodiversity*, 5(6): 862–886. 10.1002/cbdv.200890101.

Tulshyan, A., and Dedhia, E. 2021. An overview of microencapsulation technology in the application of aroma and antibacterial finishes. *International Journal of Home Science*, 7(1): 34–39. 10.22271/23957476.2021.v7.i1a.1105.

Turin, L. 1996. A spectroscopic mechanism for primary olfactory reception. *Chemical Senses*, 21(6): 773–791.

Turin, L. 2002. A method for the calculation of odor character from molecular structure. *Journal of Theoretical Biology*, 216(3): 367–385. 10.1006/jtbi.2001.2504.

Turin, L., and Yoshii, F. 2003. Structure-odor relations: A modern perspective. In *Handbook of Olfaction and Gustation* (pp. 275–294).

Velmurugan, P., Cho, M., Lee, S.-M., Park, J.-H., Seo, S.-K., Myung, H., Bang, K.-S., and Oh, B.-T. 2014. Antibacterial activity of silver nanoparticle-coated fabric and leather against odor and skin infection causing bacteria. *Article in Applied Microbiology and Biotechnology*, 98(19): 8179–8189. 10.1007/s00253-014-5945-7.

Voss, G.B., Monteiro, M.J.P., Jauregi, P., Valente, L.M.P., and Pintado, M.E. 2021. Functional characterisation and sensory evaluation of a novel synbiotic okara beverage. *Food Chemistry*, 340(March): 127793. 10.1016/j.foodchem.2020.127793.

Wang, J., Lu, X., Wang, J., and Wang, X. 2019. Quantitative and sensory evaluation of odor retention on polyester/wool blends. *Textile Research Journal*, 89(13): 2729–2738. 10.1177/0040517518801183.

Wang, S., Ding, H., Zhao, Y., Li, Y., and Wang, W. 2020. Fabrication of protective textile with n-doped tio2 embedded citral microcapsule coating and its air purification properties. *Fibers and Polymers*, 21(2): 334–342. 10.1007/s12221-020-9352-7.

Willard, E.W., and Beaverson, N.J. n.d. US8241747B2 - Malodor absorbent polymer and fiber. Accessed January 11, 2021. https://patents.google.com/patent/US8241747B2/en.

Wilson, A., and Baietto M. 2009. Applications and advances in electronic-nose technologies. *Sensors*, 9(7): 5099–5148. 10.3390/s90705099.

Wright, R.H., and Robson, A. 1969. Basis of odour specificity: Homologues of benzaldehyde and nitrobenzene. *Nature*, 222(5190): 290–292.

Yan Liu, F.T., Balasubramian, K., Pierce, J.D., and Dugan, J. 2008. Scent infused textiles to enhance consumer experiences. *Journal of Industrial Textiles*, 37(3): 263–274. 10.1177/1528083707083791.

Yao, L., Laing, R.M., Bremer, P.J., Silcock, P.J., and Leus, M.J. 2015. Measuring textile adsorption of body odor compounds using proton-transfer-reaction mass spectrometry. *Textile Research Journal*, 85(17): 1817–1826. 10.1177/0040517515576325.

Yuan, B., Koss, A.R., Warneke, C., Coggon, M., Sekimoto, K., and De Gouw, J.A. 2017. Proton-transfer-reaction mass spectrometry: Applications in atmospheric sciences. *Chemical Reviews*, 117(21): 13187–13229. 10.1021/acs.chemrev.7b00325.

Yuwono, A.S., and Lammers, P.S. 2004. Performance test of a sensor array-based odor detection instrument. *Agricultural Engineering International: CIGR Journal*, 1–16.

Zeki, S. 1993. *A Vision of the Brain*. Blackwell Scientific Publications.

Zhang, W., Liu, T., Ueland, M., Forbes, S.L., Wang, R.X., and Su, S.W. 2020. Design of an efficient electronic nose system for odour analysis and assessment. *Measurement: Journal of the International Measurement Confederation*, 165(December): 108089. 10.1016/j.measurement.2020.108089.

12 Odour Testing Methods and Regulatory Norms

Anton P. van Harreveld
Sensenet Holding BV, The Netherlands

Rajal Shinkre
Odournet Holding India Pvt. Ltd., Goa, India

Carmen Villatoro
Odournet SL, Barcelona, Spain

Charlotte Tournier
Odournet France, France

Saisha Naik
Odournet Holding India Pvt. Ltd., Goa, India

CONTENTS

DOI: 10.1201/9781003141426-12

12.1 INTRODUCTION: THE PURPOSE OF ODOUR TESTING IN TEXTILES

Improving the olfactive attributes of any consumer product always involves the questions, "What unwanted olfactive attribute should I suppress?" or "Which desirable olfactive attribute should I enhance or even add?" This challenge can apply to the product or material itself, or it can be applicable within the context of its intended usage, as is the case in "active textiles," which aim to remove unwanted odours in the user context, e.g., garments that reduce body odours caused by perspiration or textiles used in upholstery or curtains that aim to remove unwanted odours from indoor air, such as cooking odours or tobacco smoke odour. To determine a course of action in improving the olfactive attributes of a textile product, it is helpful to understand the interaction of the odour as human perception and the odorants that are involved on a molecular level. Ultimately, all odour perceptions are caused by volatile odorants, which can reach the olfactive neurons at the top of our nasal cavity to trigger olfactory receptor neurons, causing these to interact with the odorant to fire a neurological signal that is interpreted in the brain as an odour perception.

Both the molecular and the sensory perspectives need to be considered when designing a testing approach to meet an olfactive attribute challenge or an olfactory performance test for a textile product. In the end, for product formulation, identifying the molecules associated with the intended sensory modification is essential to decide how to reach the intended improvement.

12.2 SENSORY TESTING PARAMETERS AND TESTING METHODS

Psychophysics is the scientific discipline that is directed at expressing the attributes of the perception of an odorous gas in quantitative or qualitative parameters. The perception of odour can be expressed in terms of four attributes or dimensions:

- detectability
- intensity
- hedonic tone
- odour quality

Odour annoyance potential has been proposed as the fifth attribute (van Harreveld 2001) to express the propensity of an odour to cause odour annoyance in an environmental exposure context. No standardized accepted method of this parameter is available, although research pursuing the method for this attribute is ongoing:

- annoyance potential

These attributes and the existing testing methods for measurement are described in more detail in the sections below.

12.2.1 Testing Method for Detectability

Detectability (or odour threshold) expresses the mass concentration of an odorant (or mixture of odorants) that is able to provide the stimulus that can just be detected by olfaction by a certain percentage of the test population. To determine the odour threshold experimentally, an odorous gas sample is diluted in an olfactometer and presented to a panel of human test subjects (Figure 12.1). The detection threshold is determined as the dilution factor where 50% of the panel members can just discern the difference between the diluted odorous gas and odourless neutral gas. This dilution factor also defines the magnitude of the odour concentration C_{od}, which is a multiple of the mass concentration of odorant(s) at the detection threshold. The odour concentration at the detection threshold is 1 odour unit per meter cubed ($ou_E \cdot m^{-3}$) by convention. Odour concentration is used to express the olfactive stimulus quantity contained in an odorous gas sample.

The odour concentration can also be expressed using the signal detection approach proposed by Alexander Graham Bell, which is commonly used to express noise levels in decibels. The decibel is 10 times the logarithm of the ratio between the stimulus and the detection threshold for that stimulus. It is expressed mathematically as:

$$L = 10 \cdot \log(I/I_0)$$

FIGURE 12.1 Laboratory setup for measurement of odour concentration using dynamic olfactometry with human panel members, according to EN13725. (Photo reprinted with permission of Sensenet Holding BV, Odournet Group.)

Where I is the strength of the signal, and I_0 is the signal at the detection threshold.

Whereas in noise measurement the detection threshold sound pressure is a conventional quantity value, in an odorous gas sample this value is specific for each composition of odorants. This implies that the detection threshold must be measured experimentally. The ratio between the odour concentration in $ou_E \cdot m^{-3}$ and the odour concentration at the detection threshold, which is, by definition, the quantity 1, provides the basis for expressing odour stimulus levels or odour concentration in odour decibels (dB_{od}):

$$L_{od} = 10 \cdot \log_{10} C_{od} / 1$$

Where C_{od} is the odour concentration in $ou_E \cdot m^{-3}$, measured by dynamic olfactometry.

A European standard method EN13725 "Air quality – Determination of odour concentration by dynamic olfactometry" for measuring odour concentration is widely used for testing odour concentration. The European odour unit (ou_E) is traceable to the mass of a primary or secondary odorant reference material. The traceability is achieved by carefully selecting panel members with an olfactory acuity that falls within a narrow band of values for the odorant reference material. The primary reference material defined in the EN13725 standard is n-butanol. The primary reference odorant material has allowed linking the unit of odour concentration to the unit of mass, based on the definition of the conventional quantity value (Varela Bruce and Antileo 2021; Oetti, Moshammer, and Mandl 2018; Schulz and van Harreveld 1996):

$$1 \ \mathrm{ou_E}.\ \mathrm{m}^{-3} \equiv 40 \ \ \mathrm{ppb/v} \ \ \mathrm{n{-}butanol}$$

In the revised standard EN13725:2021, a standard procedure was introduced to determine the secondary reference odour mass (SROM) value for any odorant or defined odorant gas (mixture), by comparison of the response of a panel of qualified panelists to the primary reference odorant and any secondary reference odorant or defined odorant gas. This opens the opportunity to establish odour detection threshold values of high quality, traceable to the unit of mass for the primary reference odorant. The standardisation of olfactometric measurement procedures, started in the 1990s, in combination with proficiency testing using reference odorant materials, has produced significant improvements to the reproducibility of odour detection thresholds and odour concentration (van Harreveld and Heeres 1995) (Figure 12.1).

12.2.1.1 Measurement of Odour Concentration Using Olfactometry

According to the EN13725 standard method, the odour concentration of an odorous gas sample is determined by presenting a panel of selected and screened human subjects with that sample, in varying dilutions with neutral gas, to determine the dilution factor at the 50% detection threshold (D_{50}). The odour concentration of the examined sample is then expressed as multiples of 1 European Odour Unit per cubic meter [$\mathrm{ou_E}\cdot\mathrm{m}^{-3}$] at standard conditions for olfactometry (temperature 293 K, atmospheric pressure 101.3 kPa on a wet basis). The measurement of odour concentration is a biometric method, with significant inherent variance, caused by differences in olfactory acuity for one panel member, over time, and between panel members. The geometric precision under repeatability conditions of the sensory calibration with n-butanol is required to be below the requirement of EN13725, which is $r \leq 3$. The corresponding limits of the confidence interval for a single measurement result x, with a cover factor $k = 2$, is:

$$x\cdot 2.09^{-1} \leq x \leq x\cdot 2.09$$

In other words, if the real concentration is 1000, the result of analysis will be that 95% of cases lie in the interval between 478 and 2090 $\mathrm{ou_E}\cdot\mathrm{m}^{-3}$. Analysing more than one replicate of a sample can reduce the uncertainty. Figure 12.2 shows the 95% confidence interval for replicated measurements, for the repeatability that is required in the EN13725 standard.

For the assessment of the efficiency of odour mitigation, repeatability is an important consideration. Again assuming the repeatability required in the CEN standard EN13725, Figure 12.3 gives confidence intervals for the filter efficiency concerning the number of samples taken both before and after the abatement unit n, when the actual efficiency is 90%.

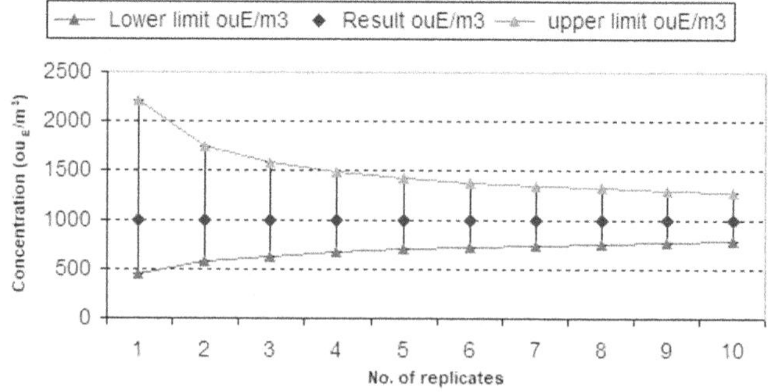

FIGURE 12.2 Confidence intervals for replicated measurements using dynamic olfactometry according to EN13725. The true value is assumed to be 1000 $ou_E{\cdot}m^{-3}$. (Graph by A. van Harreveld, used with permission by Sensenet Holding BV.)

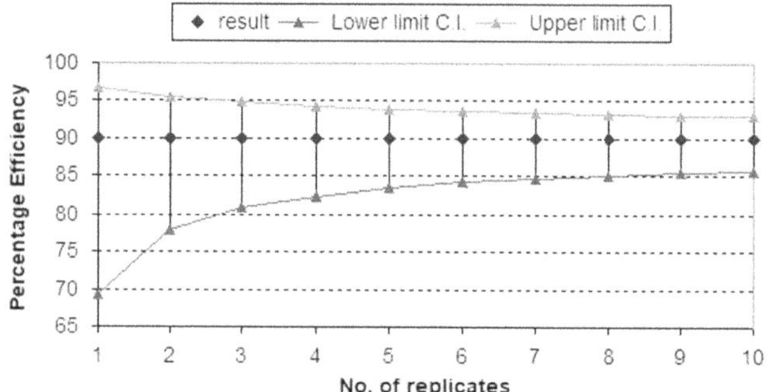

FIGURE 12.3 Confidence intervals for determining the abatement efficiency of an odour abatement system, as a function of the number of samples collected both before and after treatment. (Graph by A. van Harreveld, used with permission by Sensenet Holding.)

12.2.2 Test Methods for Intensity

Intensity expresses the perceived strength or magnitude of the odour sensation. Intensity increases with increasing odorant concentration. The relationship between perceived intensity and the logarithm of odour concentration is linear. The relationship between perceived intensity I and stimulus can be described as a theoretically derived logarithmic function according to Weber-Fechner Law:

$$S = k_w \cdot \log I/I_o$$

where

S is the perceived intensity of sensation (theoretically determined)
I is the physical intensity (odour concentration)
I_o is the threshold concentration

k_w is the Weber-Fechner coefficient
or as a power function according to Steven's law:

$$S = k \cdot I^n$$

where
S is the perceived intensity of sensation (empirically determined)
I is the physical intensity (odour concentration)
n is Stevens' exponent
k is a constant

A commonly used guideline for testing the odour intensity of an odorous sample is:

- VDI 3882:1992, part 1, *Determination of Odour Intensity,* Düsseldorf, Germany

The odorous sample is presented to human assessors in an odour panel, at varying dilution factors, hence varying the odour concentration and the associated perceived odour intensity.

The assessors are asked to rank perceived intensity at each presentation as a value of the perceived intensity I on a seven-point intensity scale:

0	no odour
1	very faint odour
2	faint odour
3	distinct odour
4	strong odour
5	very strong odour
6	overwhelming odour

The values for I are then plotted against the logarithm of the odour concentration or the dilution factor. The regression line is characteristic for the relationship between perceived intensity and odour concentration for that odorous gas composition. The point where the regression line intersects with the horizontal axis is equivalent to the detection threshold. The rate of increase of intensity with increasing odour concentration is different for different odorous gas samples. The slope of the regression line for different odorous gas samples provides the intensity characteristics. Some odours cause rapid increase in perceived intensity (e.g., H_2S). Other odours cause only a slow rise of perceived intensity, such as commercial toilet air fresheners that are designed to be perceived at a similar intensity, regardless of dilution. An example of the relationship between intensity and concentration is presented in Figure 12.4.

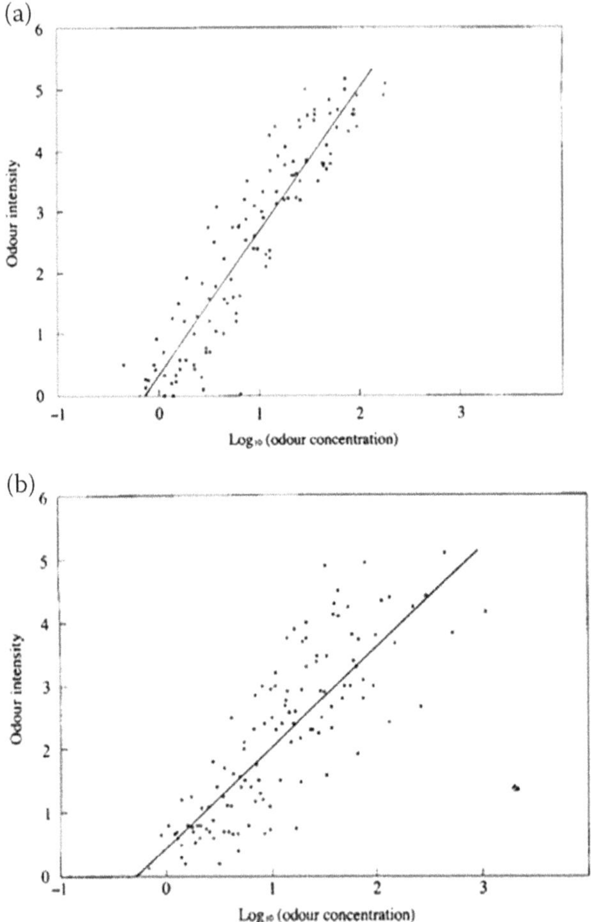

FIGURE 12.4 Relationship between odour concentration and perceived intensity, for broiler house odour (a) and the odour of pig slurry after application (b). (From Misselbrook, Clarkson, and Pain 1993.)

12.2.3 Test Methods for Odour Quality (Descriptive)

Odour quality or odour type is expressed in semantic descriptors, i.e., words that describe the odour of the odorous gas sample. Several semantic descriptors may apply, for example *fruity*, *woody*, *animalistic*, etc. There is, to date, no uniform, standardized set of semantic descriptors to describe all odours. Different professions use different sets of semantic descriptors, more or less uniformly understood by the sensory testing community in a certain domain, e.g., perfume creation, wine, olive oil, coffee, etc. Such specific sets of semantic descriptors for a specific purpose can be neatly expressed graphically in an odour wheel. An example of an odour wheel is provided in Figure 12.5.

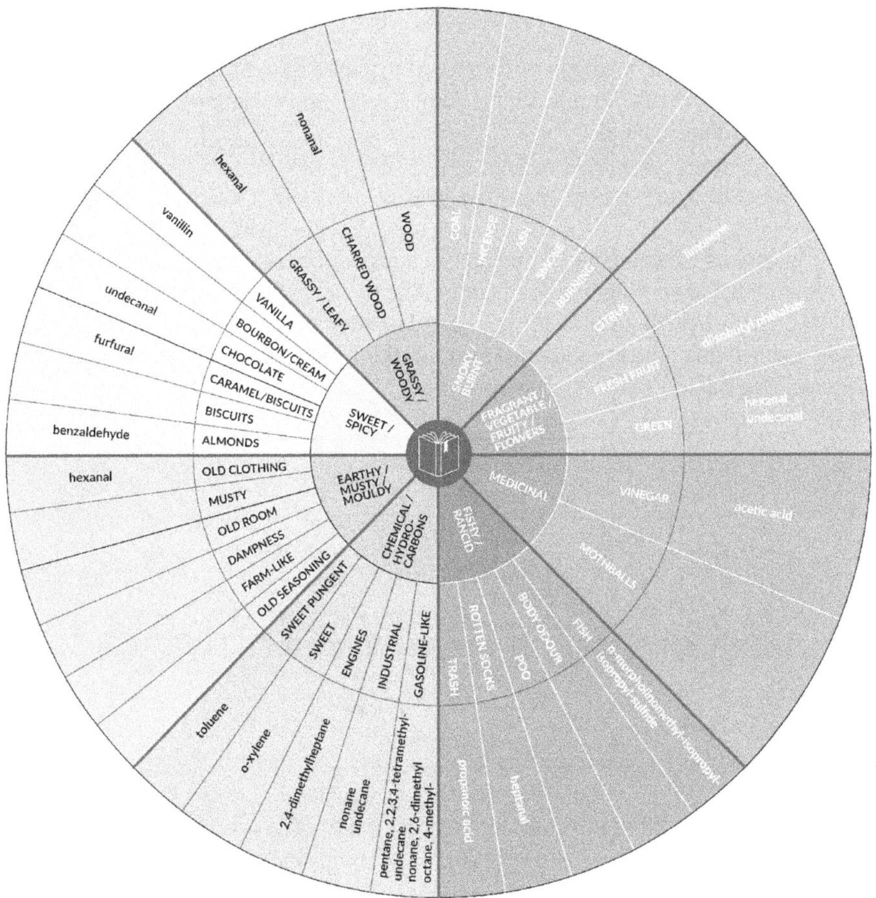

FIGURE 12.5 Odour wheel, developed for characterising the smell of historic books, developed by Bembibre and Strlic (2017). (Reprinted with permission by the authors.)

The American Society for Testing and Materials (ASTM) has developed a standardised set of 146 descriptors to characterise an odour (Dravnieks 1985). In the evaluation procedure, human panel members are asked to indicate which descriptors apply to the odour in question. Their responses are collated into an index for each descriptor. An odour atlas has been prepared using this method, establishing descriptive odour profiles for over 100 odorous compounds, using a panel of 120 individuals. Dravnieks et al. (1984) proposed a method to derive a hedonic tone score from the odour profile of descriptors and demonstrated that a reasonably strong correlation exists between the hedonic tone derived from descriptors and that derived from actual hedonic tone measurements, as described in the following section.

12.2.4 Test Methods for Hedonic Tone

Hedonic tone refers to a category judgement by human assessors of the relative like (pleasantness) or dislike (unpleasantness) of the odour. A commonly used guideline for testing hedonic tone is:

- VDI 3882, part 2; *Determination of Hedonic Tone,* Düsseldorf, Germany

The principle of measurement is the presentation of the odour to human assessors in an odour panel, at varying dilution factors. The perceived intensity and hedonic tone will vary with the dilution factor applied. The human assessors are asked to rank perceived hedonic tone at each presented dilution factor as a value on the nine-point hedonic tone scale:

+4	very pleasant
+3	pleasant
+2	moderately pleasant
+1	mildly pleasant
0	neutral odour/no odour
−1	mildly unpleasant
−2	moderately unpleasant
−3	unpleasant
−4	offensive

This test is typically performed by a larger panel of assessors, e.g., n ≥ 20, and typically performed using a non-expert panel. The mean of the values for *H* of all assessors at each dilution factor is plotted against the odour concentration in $ou_E{\cdot}m^{-3}$, as shown in the fictitious example in Figure 12.6.

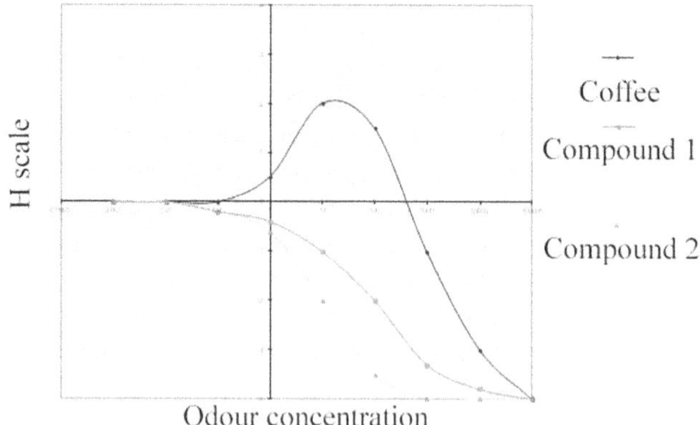

FIGURE 12.6 Hedonic tone as a function of odour concentration. (Graph by A. van Harreveld, published with permission by Sensenet Holding BV, Amsterdam, the Netherlands.)

12.2.5 Test Methods for Annoyance Potential

Annoyance potential is an attribute proposed to express the propensity of an odour to cause annoyance within a population exposed to this odour intermittently, over a long period of time. Annoyance is a concept that differs from like or dislike in that it includes the context of perception. The proposed definition of *annoyance* (Van Harreveld 2001) is:

- The complex of human reactions that occurs as a result of an immediate exposure to an ambient stressor (odour) that, once perceived, causes negative cognitive appraisal that requires a degree of coping.

The associated definition of *annoyance potential* (Van Harreveld 2001) is:

- Annoyance potential is the attribute of a specific odour (or mixture of odorants) to cause a negative appraisal in humans that requires coping behaviour when perceived as an ambient odour in the living environment.

The annoyance potential parameter is principally relevant to the assessment of the risk of causing annoyance as an air quality issue in the environment. Annoyance potential is likely to be a function of both odour quality and hedonic tone in addition to perceived intensity in a particular context. As explained above, hedonic tone and perceived intensity both are a function of the odour concentration for a particular odour quality, expressed as dB_{od} or $ou_E{\cdot}m^{-3}$. The exact nature of the interaction between the dimensions of odour, which combine in the parameter odour annoyance potential, is yet to be clarified. Currently, the parameter odour annoyance potential is of limited practical use for odour testing in textiles.

12.3 SAMPLE PREPARATION

Sample preparation for textile testing can be a complex undertaking. The first stage consists of creating the standardised conditions of use of the textile. The following test setups can be used in the test design, as required:

- In-wear tests, where human subjects wear the textile
- Laundry tests
- Exposure room tests
- Exposure chamber tests

The second stage consists of collecting the textile sample. This may involve collecting specific areas of the garment, or from garments with different textiles and characteristics for comparisons. The third stage generally consists of sample containment, to create a headspace atmosphere under standard conditions, for subsequent odour parameter testing. For molecular analysis of the volatile odorants in the headspace, the sampling techniques commonly used for gas chromatography–mass spectrometry (GC-MS) analysis can be applied: adsorption tubes (Tenax, activated carbon) or solid-phase microextraction (SPME). This sorption step is typically followed by thermal desorption (TD) and GC-MS analysis.

12.3.1 Odorant Sample Collection on a Textile

When defining a sampling protocol, the objective is to create conditions that resemble the real use conditions as closely as possible. To do so, different sampling tools are available, depending on the test objective.

12.3.1.1 Evaluation of the Odour Reduction Efficiency of a Textile

To test the effectiveness of odour reduction using active textiles, the malodour under usage conditions must be generated, to compare the resulting odour after interaction with the textile. Typically, these tests are conducted as a comparison between active textiles and neutral textiles. Specific technical standards for testing odours and odorants relevant for textiles exist, including:

- ISO 17299-1:2014 Textiles—Determination of deodorant property. ISO 17299 consists of the following parts:
  - Part 1: General principle
  - Part 2: Detector tube method
  - Part 3: Gas chromatography method
  - Part 4: Condensation sampling analysis
  - Part 5: Metal-oxide semiconductor sensor method

Part 1: General principle establishes definitions of the major component odorant chemicals in body odours. The ISO17299 standards were developed by Technical Committee ISO/TC 38, *Textiles*. Other relevant international standards cover relevant methodologies for sensory testing, many of which were developed for testing food products by ISO/TC 34, *Food products*, Subcommittee SC 12, *Sensory analysis* or the European Committee for Standardization (CEN) Technical Committee CEN/SS C01, *Food Products*. However, the general principles established in these standards may be relevant for sensory testing of textiles as well. Examples of such standards are:

- ISO 8586:2012 Sensory analysis—General guidelines for the selection, training, and monitoring of selected assessors and expert sensory assessors
- ISO 13300-1:2006 Sensory analysis—General guidance for the staff of a sensory evaluation laboratory
- ISO 6658:2017 Sensory analysis—Methodology—General guidance
- ISO 5492:2008 Sensory analysis—Vocabulary
- ISO 11136:2014 Sensory analysis—Methodology—General guidance for conducting hedonic tests with consumers in a controlled area
- ISO 4120:2021(en)
- Sensory analysis—Methodology—Triangle test

- ISO 8589, Sensory analysis—General guidance for the design of test rooms
- ISO 5496:2006 Sensory analysis—Methodology—Initiation and training of assessors in the detection and recognition of odours

Finally, the ISO16000 standards prepared by Technical Committee ISO/TC 146, *Air quality*, Subcommittee SC 6, *Indoor air* include relevant test methods, in particular concerning the testing of emissions from materials to air:

- ISO 16000–5:2007 Indoor air—Part 5: Sampling strategy for volatile organic compounds
- ISO 16000–6:2011 Indoor air—Part 6: Determination of volatile organic compounds in indoor and test chamber air by active sampling on Tenax TA® sorbent, TD, and GC using MS/FID
- ISO 16000–9:2006—Part 9: Determination of the emission of volatile organic compounds from building products and furnishing—Emission test chamber method
- ISO 16000–10:2006—Part 10: Determination of the emission of volatile organic compounds from building products and furnishing—Emission test cell method
- ISO 16000–11:2006—Part 11: Determination of the emission of volatile organic compounds from building products and furnishing—Sampling, storage of samples, and preparation of test specimens
- ISO 16000–28:2020 Indoor air—Part 28: Determination of odour emissions from building products using test chambers

12.3.1.2 In-vivo or In-wear Tests

The most representative method for collecting body odours in fabrics is through human wear trials, as this is how odour would be transferred to fabric in real life. Human body odours like sweat can be generated in-vivo by an in-wear panel, where human subjects wear garments for specified periods, under specific conditions, to simulate the intended usage. Every human being has a specific "volatilome," which also depends on external factors, such as diet, health, personal hygiene, exercise levels, skin microbiome, and stress. Therefore, to obtain a representative source of relevant odorants, a sufficiently large panel of human subjects is required to take into account inter- and intra-individual variations. The number of human in-wear subjects and their selection and the protocol for preparation and behaviour instruction needs to be carefully designed.

The protocol details behavior: 7 to 10 days prior to the trials, the test subjects cannot use any type of fragrance, deodorant, cosmetics, body care product, etc. on their skin and apparel and are allowed to wash their armpits with an unscented neutral gel. Simple methods for collecting odour include sewing or pinning fabric swatches into T-shirts, taping textile pads to the axillary vault, and wiping the axillary vault and/or other parts of the body with a textile swatch following excessive sweating (Domingues et al. 2016). Bisymmetrical T-shirts with a different fabric on each side of the body have been used when multiple wear and wash cycles were needed for odour to develop over time. Human wear trials can be limited by high inter- and intra-individual variability of odour emitted from human subjects. Therefore, human wear trials, although realistic and representative of how odour is generated and transferred to fabrics, are time-consuming and can lack repeatability. To counter many of these issues, in-vitro, or lab-based, methods may be carried out, setting suitable controls in place.

12.3.1.3 In-vitro Malodour Exposure Tests

When testing the capacity to reduce environmental odours, such as tobacco smoke odour, garbage, or cooking odours, the test textile can be exposed by placing it in a test chamber in contact with the odorant gas. The odorant gas can be generated by creating actual test conditions or by generating a simulated malodour gas mixture of known composition. Real-world sources of malodours can vary significantly, both in odorant concentration and in composition. The nature of the odorant compounds generated from a "garbage" source can vary according to the waste present in the dustbin just as the "kitchen" type odour would vary according to the food cooked. Such variability can limit the repeatability of the experiments, leading to noncomparable results for studies conducted at different times. This variability can be controlled by using synthetic mixtures of certain odorant molecules typical of the odour source to be studied. Annex B of standard ISO 17299-1:2014 Textiles—Determination of deodorant property—Part 1: General principle describes certain mixtures applicable to different sources, such as sweat, cooking, or tobacco smoke odours.

For the application of these odour mixtures, several options are also possible. A direct application of the malodorous liquid can be done, or textiles can be placed in an exposure chamber with the odorant gas. The malodorous mixture is then dispersed in a standardised manner into a chamber of defined volume into which the textiles to be tested are then introduced, for a defined duration. In order to assess effectiveness, a control should be prepared under the same conditions as the test sample, which may be the fabric to be tested before application of the odour reduction treatment or a conventional nonfunctional fabric on the market.

12.3.1.4 Standardised Laundry Tests

When designing a test protocol that includes the effects of a domestic laundry and/or drying cycle, the following international standard can provide a starting point:

- ISO/DIS 6330 Textiles—Domestic washing and drying procedures for textile testing

12.3.2 Collecting the Textile Sample

The methodology for collecting the textile sample for analysis needs to be carefully considered and is an integral part of the testing protocol. For determining the intrinsic olfactory property of a textile material, the general guidelines for obtaining a sample from a bulk product should be considered. Guidance can be found in the following international standard:

- ISO 5089:1977 Textiles—Preparation of laboratory test samples and test specimens for chemical testing

12.3.3 Obtaining a Sample of the Inherent Odorant Gas Emitted by a Textile

12.3.3.1 Evaluation by Direct Smelling

The most straightforward approach for assessing the inherent odour of a textile product is to ask trained panelists to directly smell the textile material (McGinley and McGinley 2017).

12.3.3.2 Obtaining a Standardised Static Headspace Sample of Odorant Gas

A more elaborate approach is required if we want to achieve more standardised sample conditions. To do so, a sample of textile material can be placed into a container for a determined time and at a certain temperature, to generate a static headspace of odorant gas, which is in equilibrium with the odorants attached to the textile sample. The headspace can be contained in polymer film odorant gas sample bags, for example, using PET film (Nalofan®) or polyfluoride film bags. Alternatively, glass containers can be used. The material of the container shall be odourless and inert to odorant molecules. This headspace that is created in this manner can be captured on sorbent tubes for subsequent molecular analyses or presented to panelists for sensory evaluation. Figure 12.7 illustrates how a volume of the odorant gas headspace, with the volatile organic compounds emitted by the textile sample, is transferred to a thermodesorption tube through a manual pump.

12.3.3.3 Dynamic Headspace Extraction of Odorant Gas

Dynamic headspace extraction (DHE) is a widely used method for the extraction of volatiles, especially for chemical analysis using GC. The advantage of this technique is that more concentrated extracts can be obtained as the volatiles are continuously swept onto a trap containing an adsorbent resin. Traps can be desorbed using a TD system. Micro-Chamber/Thermal Extractor™ (μ-CTE™) is a standalone sampling accessory for dynamic headspace sampling of organic vapours from a wide variety of materials. Operation is simple, with short sampling times (typically <60 minutes) and the capability to analyse up to four samples at once. The μ-CTE, see Figure 12.8, allows rapid collection of the entire vapour profile from products, making it ideal for assessment of screening, quality-control, and monitoring. The sensitivity enhancement of this dynamic headspace technique is raised

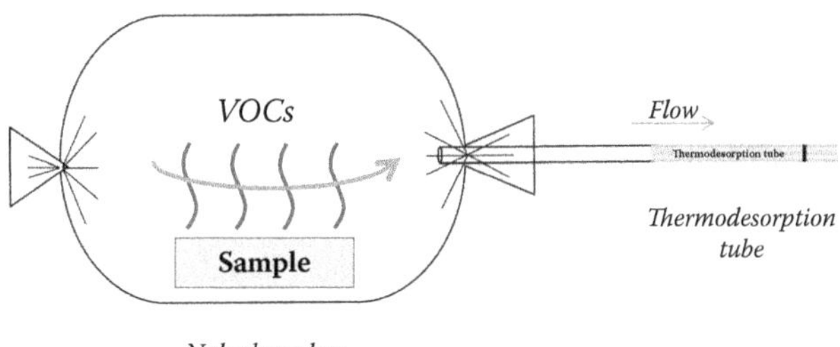

FIGURE 12.7 Schematic representation of headspace sampling. (Reprinted with permission, © Sensenet holding BV, Amsterdam.)

FIGURE 12.8 Micro-chambers with samples of cheese for headspace sampling. (Reprinted with permission. © Markes International.)

further by combining it with TD, making it a powerful technique when trace-level aroma-active compounds are under scrutiny.

12.4 SAMPLE PRESENTATION

Once samples of headspace are collected, the sample is ready for presentation to test panel members. This can be done in a very straightforward way (direct sniffing), using devices for standardized headspace presentation, e.g., the PureSniff (Olfasense GmbH) or using dynamic dilution olfactometers.

12.4.1 Direct Sniffing

The simplest method for odour appraisal consists of directly sniffing the material. Typically, such a test procedure is led by a test leader ensuring the proper identification of the samples and also that these samples are presented to all assessors in the same manner. Potential disadvantages of the direct sniffing approach are that the airflow conditions may differ from assessor to assessor, due to differences in the distance between test item and nose, differences in inhalation velocity. Another confounding issue may involve observational bias caused by visual cues (McGinley and McGinley 2017).

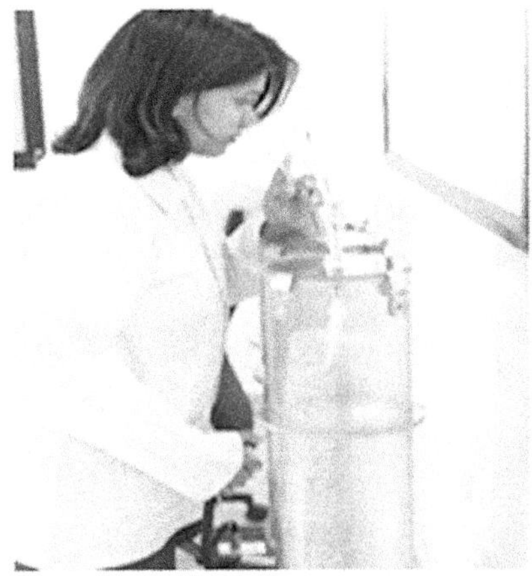

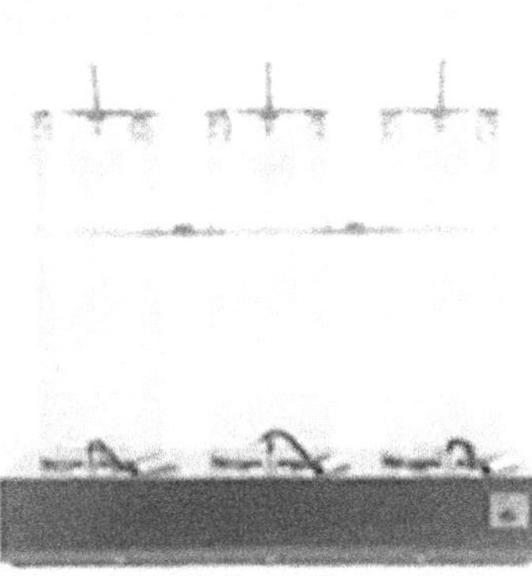

FIGURE 12.9 Direct sniff device for presentation of headspace sample without dilution. (Reprinted with permission. © Olfasense GmbH, Kiel, Germany.)

12.4.2 Direct Sniffing Under Standardised Conditions

Sample presentation apparatus providing a standardized flow from a headspace with controlled conditions can unify the appraisal conditions for the assessors. In this way the assessor smells a flow of odorous air sample instead of the test object as such. To achieve such conditions, test items can be placed in gas sampling bags which are subsequently filled with odourless neutral gas and kept under controlled conditions of a defined time interval. The sample bag can then be placed in an apparatus such as the PureSniff III, which will present the headspace sample of odorous gas through a presentation port to the assessor for inhalation, and appraisal allows the presentation of a gas sample at a defined volume flow rate and exposure time to ensure that all the assessors are presented with samples under comparable presentation conditions (McGinley and McGinley 2017). This approach provides for a more standardized presentation condition and hence more comparable conditions for odour appraisal.

The PureSniff apparatus is shown as an example in Figure 12.9.

12.4.3 Dynamic Olfactometer

A dynamic olfactometer, according to the European Standard EN 13725 for odour concentration measurements, can also be used as a presentation device for supra threshold appraisal, including intensity, hedonic tone and odour quality. The apparatus allows conducting the presentations at a controlled dilution, which in turn allows comparisons at similar odour stimulus levels (in dB_{od} or $ou_E{\cdot}m^{-3}$), which is useful for avoiding measurement bias by differences in stimulus strength. An example of a dynamic olfactometer apparatus is shown in Figure 12.10.

The TO8 olfactometer software is able to conduct tests for the following parameters:

- Odour concentration (see 12.2.1)

FIGURE 12.10 Dynamic olfactometer for measurement of odour concentration according to EN13725. (Reprinted with permission. © Olfasense GmbH, Kiel, Germany.)

- Odour intensity (see 12.2.2)
- Hedonic tone (see 12.2.4)

12.5 MOLECULAR ODORANT ANALYSIS USING GC-MS AND GC-O

Odorant compounds can be identified and their odor contribution assessed by using classical GC-MS and GC-olfactometry (GC-O) techniques. GC-MS is a powerful technique to unravel the chemical composition of odourous gas samples (Domingues et al. 2016). When combined with a TD unit, it allows concentration of air samples on adsorption tubes for subsequent analysis. This approach allows analysis of gas samples with very low concentrations. The volatile compounds present in the air sample are separated and flow into a detector at distinct retention times, after passing through the GC column, driven by an inert gas. The signal is recorded in a chromatogram, which in combination with a library of retention times and signal fingerprints for a large range of molecules allows identification of all chemical species (as far as these are detectable) included in the sample.

Samples of real odours can contain hundreds of different molecular species. This raises the challenge to determine which of these molecular species are relevant as odorants. To determine which peaks represent odorants at relevant concentrations for olfactive detection and which are irrelevant, the GC-sniffing technique is used. GC-Olfactometry (GC-O) or GC-sniffing involves a trained human assessor who sniffs the effluent of the GC column, and describes the odor perceived in function of the retention time. The intensity of the peak and also a semantic description is recorded, with the retention time. The graphical result is named olfactogram (or aromagram), which mirrors the gas chromatogram, as shown in Figure 12.11 and 12.12. Using appropriate sample preparation methods such as headspace and TD, the composition of the malodorous odorant mix released by fabric samples can be unraveled based on the molecular and sensory determination.

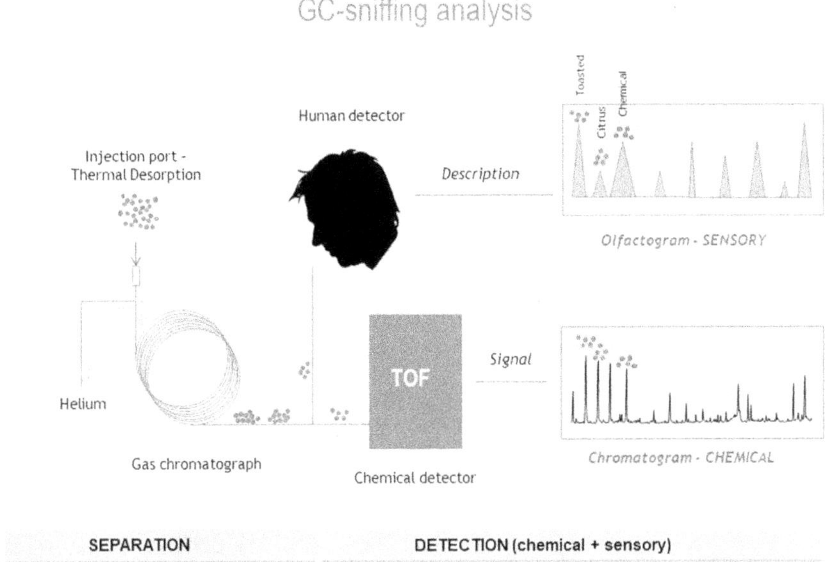

FIGURE 12.11 Schematic representation of the GC-MS analysis combined with GC-sniffing, using a human panel member to create an olfactogram. (Image reprinted with permission. © Sensenet Holding BV, Amsterdam, the Netherlands.)

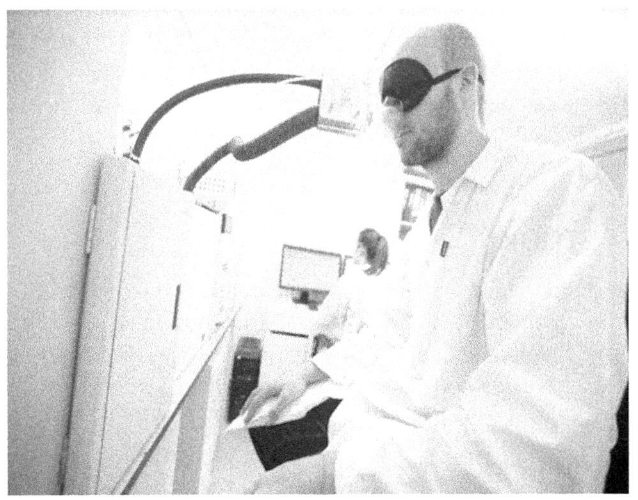

FIGURE 12.12 Human panel member sniffing the GC-O port of a GC column, for GC-sniffing. (Image reprinted with permission. © Sensenet Holding BV, Amsterdam, the Netherlands.)

12.5.1 Expressing Odour Stimulus Relevance of Odorants as Odour Activity Value (OAV)

The perception of odour is triggered when odour-active chemical compounds (odorants) in the textile product interact with olfactory receptor cells located in our nasal cavity. Some odorants can be perceived while others are not, because their concentration is below the olfactory detection threshold. The minimal concentration necessary to allow detection of the odour of a chemical compound is expressed as the odour threshold value (OTV). The odour detection threshold is the lowest concentration of a certain odour compound that is detectable by the human sense of smell. In practical terms, this is a very variable value, because human olfactive sensitivity varies over a wide range. That is why the determination of a reproducible OTV requires a large panel of observers (e.g., $n > 10$) in order to obtain a stable mean value. This uncertainty due to inter-individual variation has been reduced by making OTVs traceable to an agreed reference stimulus, in the EN13725 European olfactometry standard. The revised 2021 edition of this standard includes a standard procedure to determine OTVs, as Secondary Odour Reference Mass (SROM), traceable to the European Reference Odour Mass (EROM), defined for the primary reference odorant n-butanol. The olfactory relevance of a specific odorant in an odorous gas can be expressed by dividing its mass concentration by its odour detection threshold value (OTV). This calculation results in the Odour Activity Value, which is an estimate of the specific odour concentration for that particular odorant in an aroma composition. This concept was first introduced by Patton and Josephson (Patton and Josephson 1957) and led to the definitions of aroma value (Rothe and Thomas 1963) and odor unit (ou) (Guadagni et al. 1966). To know the 'olfactory stimulus weight' of perception of an odorant, the Odour Activity Value (OAV) is often used. This is calculated from the abundance of that odorant molecule (or mass concentration) divided by the odour detection threshold, or Odour Threshold Value (OTV).

12.6 CASE STUDIES

The four case studies summarised below were conducted by Sensenet's Laboratory in Barcelona for commercial clients. The results are presented anonymously and are intended to give practical examples of experimental approaches to olfactive optimization challenges, using both sensory and molecular approaches as outlined in the previous sections in this chapter.

12.6.1 Case Study 1: Odour Reduction Efficacy of Textile Treated with Odour Control Technology

The aim of the study was to assess the odour reduction efficacy of a textile treated with odour control technology. The odour reduction efficacy of the treated textile was compared against an untreated reference textile. The malodour source used to imbue the textile samples was a commercial artificial sweat mix, composed of acetic acid, lactic acid, butyric acid, ammonium chloride, and pyruvic acid. The testing procedure consisted of a combination of sensory odour assessment and chemical profiling of volatile organic compounds (VOCs). The sample preparation protocol was:

- Fabric swatches of the treated and untreated cotton samples were cut to size (15 cm × 10 cm) for triplicate sampling and analysis. A measured quantity of the artificial sweat mix was applied on each fabric sample and the samples were placed in 10L Nalophan bags, filled with neutral gas and stored at room temperature for 4 hr.
- After 4 hours, the headspace of both treated and untreated samples was transferred to new odourless Nalophan sampling bags (10L) for odour evaluation.
- The odour concentration analyses (ou_E/m^3) were conducted by dynamic olfactometry, according to the European Standard EN13725, using four qualified panel members. The objective was to determine the odour level emitted into the headspace by each textile sample. The odour concentration values

TABLE 12.1
Odour reduction efficacy results (table reprinted with permission, © Sensenet Holding BV, Amsterdam, the Netherlands)

Parameters	*Untreated cotton*	*Treated cotton*
Odour concentration Geomean [ou_E/m^3]	3010	40
Confidence interval ($p < 0.05$) [ou_E/m^3]	$1905 \le C_{od} \le 4757$	$25 \le C_{od} \le 63$
Odour reduction efficacy [%]	98.67	

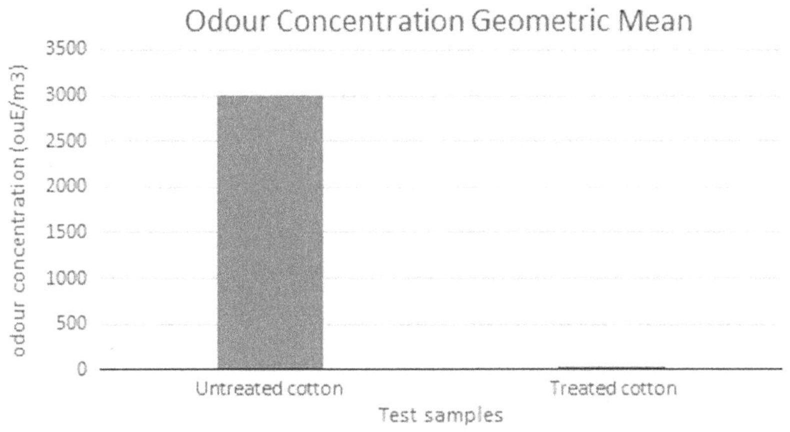

FIGURE 12.13 Graphic representation of odour concentration values of untreated and treated cotton fabric samples. (Graph reprinted with permission. © Sensenet Holding BV, Amsterdam, the Netherlands.)

TABLE 12.2
Results of compounds targeted by GC-MS (Table reprinted with permission, © Sensenet Holding BV, Amsterdam, the Netherlands)

Compounds	CAS no.	Concentration (ug/m^3)		OTV
		Untreated cotton	*Treated cotton*	
L-Lactic acid	79–33–4	1.4	1.0	–
Acetic acid	64–19–7	147.6*	1.8*	15
Butanoic acid	107-92-6	799.2*	26.1*	1
Pyruvic acid	127–17–3	–	–	–

Note
* The concentrations exceed the odour threshold value (OTV). The results correspond to the average of two replicates.

(ou$_E$/m^3) for the treated and untreated samples were used to calculate the odour reduction efficacy of the test sample as a percentage.

- For the analysis of VOCs, the air samples from 10L Nalophan bags were transferred onto sorbent tubes using the Easy VOC manual sampling pump (Markes) and analysed by Gas Chromatography, Time of Flight Mass Spectrometry (GC-TOF-MS) targeted for the VOCs in artificial sweat.

The odour concentration geometric mean value of the untreated cotton sample was 3010 ou$_E$/m^3 whereas that of the treated cotton fabric was 40 ou$_E$/m^3 as seen in Table 12.1 and Figure 12.13. The difference was statistically significant ($p < 0{,}05$). The odour reduction efficacy of the treated cotton sample was 98.67% when compared with the untreated cotton sample.

The VOC analysis results (Table 12.2) displayed a trend similar to the trend shown by sensory analysis, with a 98.8% reduction in acetic acid, followed by that of butanoic acid (96.7%). A slight, though not significant, reduction was observed in lactic acid, whereas pyruvic acid was not detected in either of the samples. The treated cotton sample was found to be effective in reducing the odorants in the headspace of the simulated artificial sweat mix.

12.6.2 Case Study 2: Molecular Determination of Compounds Responsible for Malodours in Fabric Samples by TD-GC-Sniffing-ToFMS

For this study, sweat tainted laundry was used. The sweat taint was achieved by an in-wear procedure involving human subjects wearing shirts during high physical activity. The volatile organic VOCs emitted by sweat-tainted laundry were then analysed using combined gas chromatography and time-of-flight mass spectrometry (GC-ToFMS). Determining the chemical and sensory composition of molecular

TABLE 12.3
Total VOCs (μg/m³) in each fabric sample classified by chemical groups

	Concentration μg/m³		
Chemical groups	**S1**	**S2**	**S3**
Alcohols	142.7	211.5	749.8
Aldehydes	581.5	457.5	290.9
Aliphatic hydrocarbons	56.7	33.8	94.9
Aromatic alcohol	35.7	14.5	1.8
Aromatic compounds	24.8	16.3	92.2
Esters	389.5	265.3	561.3
Ether	186.6	1.6	1.6
Ketones	109.8	110.2	54.0
Nitrogen-containing compounds	93.5	0.4	2.6
Organic acids	845.4	47.3	4.2
Oxygen-containing compounds	4 795.9	17.4	0.0
Terpenes	30.8	3.9	125.7
Heterogroups	255.8	236.7	23.0
Siloxanes-containing compounds	174.7	119.6	1 817.0
Total VOCs	7 723.6	1 536.1	3 818.9

species generated in fabric samples through GC-MS and GC-sniffing can provide an important source of information to detect the quality of the product in terms of odour aspects. The study was designed to study fabric-textile products using a combined technology, GC-Sniffing and GC-ToFMS to determine the sensory information of the products based on its chemical composition. Table 12.3 shows the summarized information of VOCs concentration found in the samples based on the total concentration per chemical groups of compounds. The predominant families in the samples are oxygen compounds (S1), aldehydes (S2), and siloxanes (S3).

In total, 21 distinct odorants were chemically detected and identified, as listed in Table 12.3. According to the GC-O results, the odours perceived with greater impact during sniffing were: 24.9 minutes (butanoic acid), 27.1 minutes (butanoic acid, 3-methyl), 32.7 minutes (hexanoic acid) and 36.7 min (p-cresol), with odour intensities of 4 (strong). These compounds are responsible for cheesy, rancid, sour, sweaty, phenolic, and animal odour notes.

The individual odours perceived during the analyses were grouped according to a more general description, resulting in 10 new categories. A general value of intensity was obtained by adding up the values of intensity of all odours belonging to a new category, see Table 12.4. The sensory profile in the samples is shown in Figure 12.14. The most prominent odour categories are fatty/waxy, cheesy, phenolic and earthy. Sample S1 showed higher intensities for these odour notes (Table 12.5).

TABLE 12.4
Odorants perceived during the sniffing process and chemically identified by GC-ToFMS

Kovats Index (*)	RT	Descriptor	Odor Intensity (**)			MS Identification	CAS
			S1	S2	S3		
420	6,3	ethereal pungent	2	2	2	Acetaldehyde	75-07-0
440	17,0	sour vinegar	2			Acetic acid	64-19-7
526	17,3	nutty cacao fatty	2	3	2	Butanal, 3-methyl-	590-86-3
530	21,2	acid cheesy	2			Propanoic acid	79-09-4
540	22,5	chemical solvent	2			1,3-Dioxolane, 2-ethyl-4-methyl-	4359-46-0
593	22,8	sulfur fatty	3			Thiophene, 3-methyl-	616-44-4
631	24,1	green leafy fresh	3	3		Hexanal	66-25-1
694	24,9	cheesy sour rancid	4			Butanoic acid	107-92-6
702	27,1	cheesy sour rancid	4	3	3	Butanoic acid, 3-methyl-	503-74-2
728	28,2	fatty green	2	2	2	Heptanal	111-71-7
737	31,1	mushroom earthy	3	3	3	1-Octen-3-one	4312-99-6
819	31,2	earthy mouldy	2	2	3	1-Octen-3-ol	3391-86-4
840	31,5	rubber green		2		5-Hepten-2-one, 6-methyl-	110-93-0
844	32,0	citrus fresh	3	3	3	Octanal/Limonene	124-13-0/138-86-3
858	32,7	sweaty cheesy fatty	4	2		Hexanoic acid	142-62-1
900	35,4	waxy aldehydic	3	3	3	Nonanal	124-19-6
908	36,7	phenolic animal	4	4	3	p-Cresol	106-44-5
916	37,8	fatty green	2	3	2	2-Nonenal, (E)-	18829-56-6
943	38,2	acid cheesy	3	2	2	Octanoic acid	124-07-2
978	38,6	waxy aldehydic	2		2	Decanal	112-31-2
1026	45,1	phenolic tobacco	3	3	3	p-Cresyl isovalerate	55066-56-3

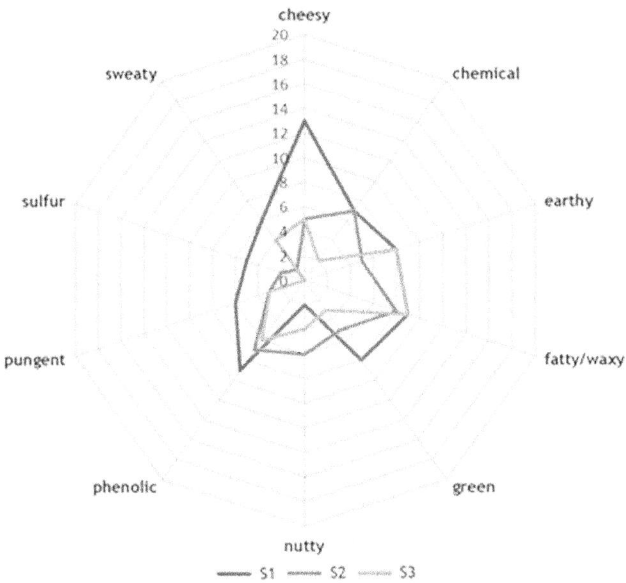

FIGURE 12.14 Spider chart representing a general profile and a sensory comparison of the samples tested. (Graph reprinted with permission. © Sensenent Holding BV, Amsterdam.)

TABLE 12.5
Odorants classified by general categories. Intensity value represents the total sum of individual odours

	Odor Intensity		
Category	**S1**	**S2**	**S3**
cheesy	13	5	5
chemical	7	7	2
earthy	8	5	8
fatty/waxy	9	8	9
green	8	5	3
nutty	2	6	4
phenolic	9	7	6
pungent	6	3	3
sulfur	5	2	0
sweaty	6	1	4

12.6.3 Case Study 3: Evaluation of the Effectiveness of a Steamer to Reduce Odours on Textile

The objective of the study was to assess whether a steamer could help to reduce unpleasant odours on fabrics and clothes. The protocol involved tests using 3

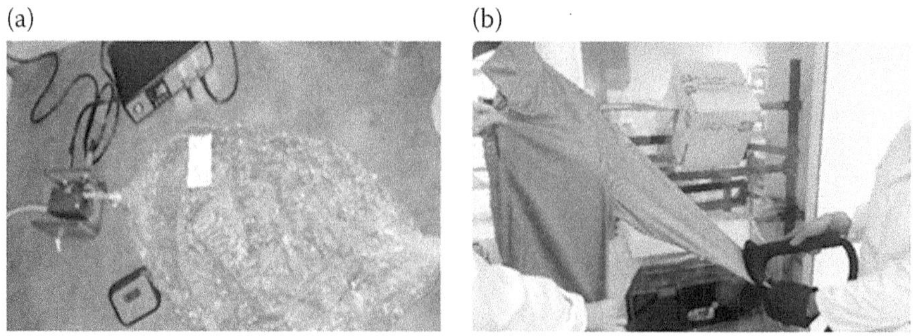

FIGURE 12.15 Textile sample preparation. (a) Contamination of shirts (b) Steam treatment of a half-shirt. (Photos reprinted with permission. © Sensenent Holding BV, Amsterdam.)

common real-life malodours: cooking, cigarette and sweat. The textile material used was 100% cotton shirts cut in half, to compare a control (no steam treatment) to a sample treated with the steamer. The application protocol was representative of normal usage conditions. The sample preparation protocol was: (Figure 12.15)

- Synthetic malodour mixtures were prepared according to the standard ISO 12999-1 and internal experience.
- The two half-shirts were tainted with the synthetic malodour using a pollution chamber.

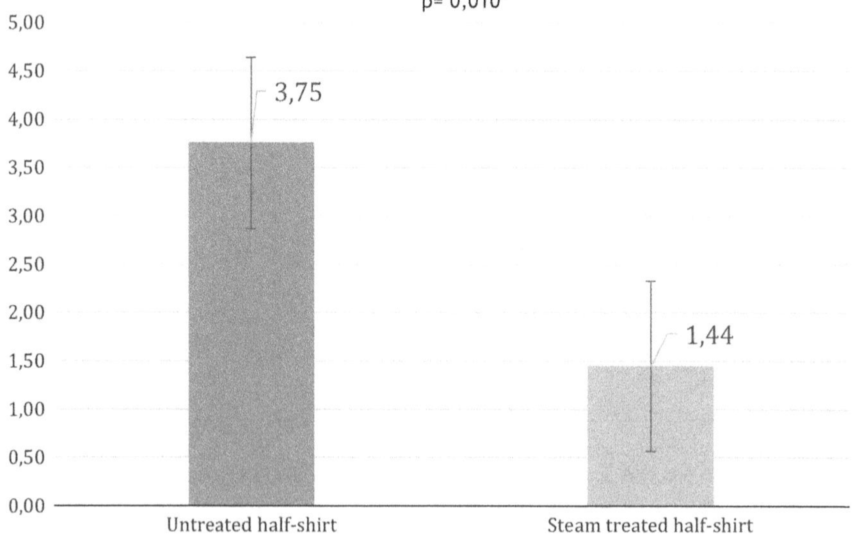

FIGURE 12.16 Reduction of malodour intensity after steam treatment. (Graph reprinted with permission. © Sensenent Holding BV, Amsterdam.)

TABLE 12.6
Results of targeted GC-ToFMS molecular analysis before and after steam treatment of a malodorous textile sample (by Sensenet Holding BV, Amsterdam, reprinted with permission)

Compound	CAS N°	Concentration (μg/m³)		Percentage change
		Untreated half shirt	Steam treated half shirt	
Acetic acid	64–19–7	**65,0**	6,8	–89,5%
Valeric acid	109-52-4	**136,2**	**30,7**	–77,5%
Butyric acid	107–92–6	**13,9**	**<0.01**	–99,9%
Propionic acid	79–09–4	**49,3**	8,6	–82,5%

- The steamer treatment was applied to one of the half shirts.
- The shirts were then placed in glass jars for odour profiling by trained panel members.
- A headspace sample of both treated and untreated textile samples was also generated in Nalophan gas sample bags for molecular analyses, using GC-ToFMS targeted on the most olfactory relevant odorant molecules in the synthetic mixtures.

12.6.3.1 Results

The results presented below are for the odour "sweat".

Sensory analysis demonstrated a statistically significant decrease in malodour intensity after steam treatment with the steamer. (Figure 12.16)

The targeted GC-TofMS molecular analyses, carried out to assess the impact of steam treatment on the main odorant molecules, showed the results listed in Table 12.6.

The concentrations in bold exceed the odour threshold value (OTV).

The data confirmed the claim of efficacy of the steamer to remove daily life malodours impregnated on clothes and home fabrics. New efficacy claims and additional consumer benefits can be assigned to the steamer treatment in addition to its main function, which is the removal of fabric wrinkles. Sensory analysis results were supported by molecular analysis results, showing a decrease of 68% to 99% for the most relevant malodorous compounds concentrations.

12.6.4 Case Study 4: Odour Test for Textile Floorcoverings

In a worst-case scenario, the test method applied for textile floorcoverings simulates the release of odorants at increased temperatures (caused by direct exposure to sunlight or floor heating). The test was conducted in compliance with the Swiss standard SNV 195651 for the determination of olfactive nuisance caused by textiles. After the sample conditioning, a team of odour panel members assessed the resulting odour of the samples after treatment. The panel members were tasked with defining the type of odour as well as the hedonic tone. The primary objective of the odour test was to ensure a clear distinction between acceptable new-product odour (= grade 3) and unacceptable odour (= grade 4). For a period of 15 hours, a textile sample of 144 cm^2 is stored in an air-sealed desiccator (volume about 2 liters) at 37°C and 50% relative humidity. The humidity is adjusted by means of a saturated magnesium nitrate solution (approx. 100 ml). Under these conditions, at least seven odour panel members will subsequently open the desiccator briefly and assess the hedonic tone of the odour perceived by direct sniffing of the headspace. The hedonic tone of the odour is graded using the following scale:

1 = no odour
2 = not unpleasant
3 = slightly unpleasant
4 = unpleasant
5 = very unpleasant
6 = extremely unpleasant

The mean value of the grades for the hedonic tone of the odour by the panel members must be a grade < 4.

12.7 SUMMARY

The introduction of this chapter explains the interaction of odorants in the volatile organic compound mixture generated in the skin-textile interaction and how this is turned into a perception of odour. The difference between sensory and molecular odour testing is reviewed, indicating that the study of odorants, using chemical and combined chemical and sensory methods, such as GC-MS and GC-O, can provide very useful data on the identity and concentration of relevant odorants and their sensory weight in the composition of odorant mixtures. However, to establish the extent of the differences in perceived odour, sensory methods are essential. The different parameters that can be used to characterise an odour are explained:

- *Detectability*, expressed as odour concentration, in odour units per volume, as a measure of the quantity of olfactory stimulus present in air (standard method: EN13725)
- *Odour intensity*, or the perceived strength of the odour (standard method: VDI3882 Part 1)
- *Hedonic tone*, or the perceived pleasantness or unpleasantness of the odour under study (method: VDI3882 Part 2)
- *Odour type*, providing a semantic description of the odour under study (odour wheel, champ d'odeur)

Sample preparation for odour testing in textiles, critical to the test outcomes, explains the need for defining sampling protocols to create conditions as close to the real use conditions of the textile as possible. In-vivo and in-vitro test methods can be used to evaluate the odour performance of treatments used on textiles. Molecular odorant analysis using GC-MS and GC-O, valuable for the identification of odorant compounds, are described. Specific standards for testing odours and odorants relevant for textiles are reviewed, and case studies relevant to odour and textiles are presented.

REFERENCES

Bembibre, C., & Strlič, M. 2017. Smell of heritage: a framework for the identification, analysis and archival of historic odours. Heritage Science, 5: 2. doi: 10.1186/s40494-016-0114-1.

Domingues, R., Villatoro, C., Companioni, E., Meacham, A., and Vera, L. 2016. Chemical and sensory study of human body odours released by two types of textiles. *Household and Personal Care Today*, 11(5): 74–77.

Dravnieks, A. 1985. *Atlas of Oodor Character Pprofiles.* ASTM publication. PCN 05-061000-36, ISBN 0-8031-0456-1.

Dravnieks, A., Masurat, T., and Lamm, R.A. 1984. Hedonics of odors and odor descriptors. *Journal of the Air Pollution Control Association*, 34(7): 752–755. doi: 10.1080/00022470.1984.10465810

Guadagni, D.G., Buttery, R.G., and Harris, J. 1966. Odour intensities of hop oil components. *Agriculture Publishes Original Research and Critical Reviews*, 17(3): 142–144. doi: 10.1002/jsfa.2740170311

McGinley, M.A., and McGinley, C.M. 2017. Methods for odor evaluation of textiles and other materials. *AATCC International Conference Proceedings*, 250–264.

Misselbrook, T.H., Clarkson, C.R., and Pain, B.F. 1993. Relationship between concentration and intensity of odors for pig slurry and broiler houses. *Journal of Agricultural Engineering Research*, 55: 163–169.

Oettl, D., Moshammer, H., and Mandl, M. 2018. Towards a harmonized odour guideline for some austrian provinces. *Chemical Engineering Transactions*, 68: 25–30. doi: 10.3303/CET1868005

Patton, S., and Josephson, D.V. 1957. A method for determining significance of volatile flavour compounds in foods. *Journal of Food Science*, paper 2118. doi: 10.1111/1365-2621.1957.tb17017.

Rothe, M., and Thomas, B. 1963. Aroma of bread: Evaluation of chemical taste analysis with the aid of threshold value. *Z. Lebensm-Unters Forsch*, 119: 302–310.

Schulz, T.J., and van Harreveld, A.P. 1996. International moves towards standardisation of odour measurement using olfactometry. *Water Science and Technology*, 34(3-4): 541–547. doi: 10.2166/wst.1996.0474

Steven's Power law.

Van Harreveld, A., and Heeres, P. 1995. Quality control and optimisation of dynamic olfactometry using n-butanol as a standard reference odorant. *STAUB -Reinhaltung der Luft*, 55: 45–50. Spinger Verlag.

Van Harreveld, A.P. 2001. From odorant formation to odour nuisance: New definitions for discussing a complex process, *Water Science & Technology* 44(9): 9–15. ISSN: 0273-1223

Varela Bruce, C., and Antileo, C. 2021. Assessment of odour emissions by the use of a dispersion model in the context of the proposed new law in Chile. *Journal of Environmental Management*, 295: 113–208.

Weber Fechners law.

Index

A

D

E

F

I

M

N

O

P

T

U

V

W

X

Y

Z

CPSIA information can be obtained
at www.ICGtesting.com
Printed in the USA
BVHW032034290622
640978BV00001B/1